Envisioning the Survey
Interview of the Future

BICENTENNIAL
1807
⊛WILEY
2007
BICENTENNIAL

THE WILEY BICENTENNIAL—KNOWLEDGE FOR GENERATIONS

*E*ach generation has its unique needs and aspirations. When Charles Wiley first opened his small printing shop in lower Manhattan in 1807, it was a generation of boundless potential searching for an identity. And we were there, helping to define a new American literary tradition. Over half a century later, in the midst of the Second Industrial Revolution, it was a generation focused on building the future. Once again, we were there, supplying the critical scientific, technical, and engineering knowledge that helped frame the world. Throughout the 20th Century, and into the new millennium, nations began to reach out beyond their own borders and a new international community was born. Wiley was there, expanding its operations around the world to enable a global exchange of ideas, opinions, and know-how.

For 200 years, Wiley has been an integral part of each generation's journey, enabling the flow of information and understanding necessary to meet their needs and fulfill their aspirations. Today, bold new technologies are changing the way we live and learn. Wiley will be there, providing you the must-have knowledge you need to imagine new worlds, new possibilities, and new opportunities.

Generations come and go, but you can always count on Wiley to provide you the knowledge you need, when and where you need it!

WILLIAM J. PESCE
PRESIDENT AND CHIEF EXECUTIVE OFFICER

PETER BOOTH WILEY
CHAIRMAN OF THE BOARD

Envisioning the Survey Interview of the Future

Edited by

FREDERICK G. CONRAD

Institute for Social Research
University of Michigan
Ann Arbor, Michigan

MICHAEL F. SCHOBER

New School for Social Research
New York, New York

WILEY-
INTERSCIENCE

A JOHN WILEY & SONS, INC., PUBLICATION

Published by John Wiley & Sons, Inc., Hoboken, New Jersey.
Published simultaneously in Canada.

For general information on our other products and services or for technical support, please contact our
Customer Care Department within the United States at (800) 762-2974, outside the United States at
(317) 572-3993 or fax (317) 572-4002.

Wiley also publishes its books in a variety of electronic formats. Some content that appears in print may
not be available in electronic formats. For more information about Wiley products, visit our web site at
www.wiley.com.

Wiley Bicentennial Logo: Richard J. Pacifico

Library of Congress Cataloging-in-Publication Data:

Envisioning the Survey Interview of the Future / [edited by] Frederick G. Conrad,
    Michael F. Schober.
        p.   cm.—(Wiley series in survey methodology)
    Includes bibliographical references.
    ISBN 978-0-471-78627-6 (cloth : alk. paper)
        1. Social surveys—Technological innovations.   2. Social surveys—Data processing.
3. Questionnaires.   4. Interviewing.   I. Conrad, Frederick G. (Frederick George), 1955–
II. Schober, Michael F., 1964–
    HM538.E68   2007
    303.48′330723—dc   2007013723

Printed in the United States of America

10 9 8 7 6 5 4 3 2 1

Contents

Preface

The ways in which we communicate are changing almost daily, largely because communication technologies that were unimaginable even a decade ago are now commonplace. For example, in order to edit this book, we conducted regular meetings by desktop video conference, a technology that has been promised commercially since the "picture phone" was unveiled at the New York World's Fair in 1964 but is only now becoming widely available. Mobile phones that also allow web browsing and video downloads are now a reality. Help desks staffed by avatars—animated, conversational agents—are here. And the way we use new communication technologies is changing how we engage with one another. Prior to widespread text messaging, it was unthinkable to be simultaneously involved in several multiway conversations, but for those who have grown up with this technology this is quite normal.

All this change creates vast possibilities for social researchers who obtain data about the world by communicating with survey respondents. But a moment's thought about how to exploit these new media for collecting survey data reveals just how little we know about the way people (i.e., respondents) will use them and how these media will affect the data we obtain through them. For example, if an animated interviewing agent (i.e., a piece of software) presents spoken questions about a sensitive topic, will respondents shade the truth as they sometimes do with human interviewers or will they answer with greater candor as they seem to under computerized self-administration? Will the effect of these new survey modes differ across cultures; for example, do some cultures give greater weight to the facial expressions of conversational partners than do others, leading to different patterns of interaction in text-chat as opposed to video-mediated interviews? And so on. The relevant studies simply have not been done. Part of the explanation is probably that the technologies are just too new to have been adapted to domains beyond the ones for which they were designed. But the relevant studies have also not been carried out because the two relevant communities—survey methods and communication technology researchers—are relatively unaware of each other.

As a first step to bring together members of these highly productive groups, we organized a workshop in November 2005 at the Institute for Social Research, University of Michigan (Ann Arbor). Because technologies become obsolete as quickly as new ones are invented, we tried to focus the discussion at the level of the issues to consider when adopting a new technology for survey data collection, irrespective of the particular technology. The workshop goals included exposing the participants to the problems that their counterparts work on and to promote the kind of conversation that might lead to future collaborations. In the longer term, we hope that the workshop and the current book might accelerate the evaluation and adoption of new technologies by survey methodologists, and convince communication technologists that survey research is an intellectually engaging domain in which to apply their ideas.

The current book is a collection of chapters *based* on workshop presentations, but which have evolved substantially since their original delivery. The chapters in the current volume are written mostly by researchers from one community or the other, but all the authors have stretched their thinking by grappling with issues raised by their counterparts in the other community. At least this group of authors can now claim membership to some degree in both of the communities represented at the workshop and in the book. We hope the joint membership grows as a result of the book.

This book consists of fourteen chapters that reflect the workshop goals and are written much more in the spirit of promoting dialogue between survey methodologists and communication technologists than in providing the final word on any one medium or any one survey problem. Schober and Conrad first lay out a series of questions that one must ask when thinking about adopting emerging technologies for interviews. Then Schaeffer and Maynard present conversational examples of respondent–interviewer interaction that could signal communication breakdowns—the kinds of dialogues that future interviewing systems will need to be able to address or resolve or prevent. In his chapter, Couper reviews the evolution of survey data collection technologies and discusses the likely impact of future technologies on survey participation and response quality. Fuchs considers web surveys on mobile phones and how this might overcome problems of current telephone and web surveys; he explores the possibility of using video invitations. Anderson considers how current knowledge about video-mediated communication (e.g., video conferences and experimental, collaborative tasks) might inform the adoption and design of video-mediated survey interviews. Bloom engages in a similar investigation for speech dialogue systems and considers how they might advance current speech-based survey systems. Johnston considers multimedia for survey data collection, such as speech, video, and gesture as well as conventional typed input all integrated in a single device; he explores the potential benefits for respondents who do not speak the language of the survey designers, or whose vision, hearing, or mobility are challenged. Cassell and Miller consider whether embodied conversational agents are viable as interviewers and how they might affect respondents' sense of rapport and the candor of their answers. Hancock discusses what is known about lying through different media, like IM and e-mail, and considers how this might affect responses to sensitive survey questions presented through different media. Person, D'Mello, and Olney explore new kinds of data that can inform the system of the user's cognitive and affective state, such as body posture as an indication of attentiveness. Fussell, Zhang, Conrad, Schober, and Setlock connect evidence from recent work on cross-cultural communication across different media (e.g., IM versus video versus face-to-face) to survey interviews across similar media. Konstan,

Rosser, Horvath, Gurak, and Edwards discuss the issues of guaranteeing privacy and communicating this to respondents in conducting web surveys about risky behavior. Marx discusses ethical concerns introduced by technologies on the horizon and lists the questions one must ask to determine if, when the technology is used to collect data, it constitutes a survey or surveillance. Finally, Graesser, Jeon, and McDeniel systematically go through the book's major ideas and lay out a research agenda; they provide important links between survey methodology, artificial intelligence, and human–computer interaction.

We gratefully acknowledge support from the National Science Foundation, and in particular Cheryl Eavey, that made possible the workshop and the activities that have led to this book. The Survey Research Center at the University of Michigan contributed additional funds and made space available, both of which were essential for the workshop to take place, and we are grateful for this support. Nick Prieur did excellent work coordinating all logistics of the workshop and book project; he was indispensable. Stephanie Chardoul gave an illuminating tour of the Survey Research Center's telephone lab, enabling communication technologists without survey experience to get a taste of the interview process. Becca Rosen and Matt Jans at the New School for Social Research and the University of Michigan, respectively, did superb work in helping to assemble the manuscript and were generous with their time.

FREDERICK G. CONRAD
Ann Arbor, Michigan

MICHAEL F. SCHOBER
New York, New York

Contributors

Anne H. Anderson, PhD, University of Dundee, Dundee, Scotland, UK

Jonathan Bloom, PhD, SpeechCycle, Inc., New York, New York, USA

Justine Cassell, PhD, Northwestern University, Evanston, Illinois, USA

Frederick G. Conrad, PhD, University of Michigan, Ann Arbor, Michigan, USA

Mick P. Couper, PhD, University of Michigan, Ann Arbor, Michigan, USA

Sidney D' Mello, PhD, University of Memphis, Memphis, Tennessee, USA

Weston Edwards, PhD, University of Minnesota, Minneapolis, Minnesota, USA

Marek Fuchs, PhD, Universität Kassel, Kassel, Hessen, Germany

Susan R. Fussell, PhD, Carnegie Mellon University, Pittsburgh, Pennsylvania, USA

Arthur C. Graesser, PhD, University of Memphis, Memphis, Tennessee, USA

Laura Gurak, PhD, University of Minnesota, Minneapolis, Minnesota, USA

Jeffrey T. Hancock, PhD, Cornell University, Ithaca, New York, USA

Keith J. Horvath, PhD, University of Minnesota, Minneapolis, Minnesota, USA

Moongee Jeon, MS, University of Memphis, Memphis, Tennessee, USA

Michael Johnston, PhD, AT&T Labs Research, Florham Park, New Jersey, USA

Joseph A. Konstan, PhD, University of Minnesota, Minneapolis, Minnesota, USA

Gary T. Marx, PhD, Massachusetts Institute of Technology, Cambridge, Massachusetts, USA

Douglas W. Maynard, University of Wisconsin–Madison, Madison, Wisconsin, USA

Bethany McDaniel, MS, University of Memphis, Memphis, Tennessee, USA

Peter Miller, PhD, Northwestern University, Evanston, Illinois, USA

Andrew Olney, PhD, University of Memphis, Memphis, Tennessee, USA

Natalie K. Person, PhD, Rhodes College, Memphis, Tennessee, USA

B. R. Simon Rosser, PhD, University of Minnesota, Minneapolis, Minnesota, USA

Nora Cate Schaeffer, PhD, University of Wisconsin–Madison, Madison, Wisconsin, USA

Michael F. Schober, PhD, New School for Social Research, New York, New York, USA

Leslie D. Setlock, MS, Carnegie Mellon University, Pittsburgh, Pennsylvania, USA

Qiping Zhang, PhD, Long Island University, Brookville, New York, USA

CHAPTER 1

Survey Interviews and New Communication Technologies

Michael F. Schober
New School for Social Research, New York, New York

Frederick G. Conrad
University of Michigan, Ann Arbor, Michigan

1.1 INTRODUCTION

How people communicate with each other today—at least those with access to new technologies—is enormously different from how they communicated with each other even ten years ago. Internet-based cell phones, desktop videoconferencing, mobile instant messaging, blogs, podcasts, and other current modes of communication would have been unimaginable, or the stuff of science fiction, for prior generations. And the pace of change only seems to be increasing, with new generations embracing and creating newer forms of communication.

The pace of this change poses a substantial challenge to one of the institutions on which modern societies rest: the collection of data in survey interviews. It is through individual survey interviews that societies track their employment, health, crime, demographic shifts, and citizens' opinions about controversies of the day, among many other social phenomena. The moment in which a survey respondent provides an answer to a survey question is thus far more consequential than it might at first seem, and what takes place at that moment is in a state of major transition. Changes in communication technology have already led to questions about the accuracy and generalizability of large-scale survey data. For example, as cell phone adoption increases,

Envisioning the Survey Interview of the Future, Edited by Frederick G. Conrad and Michael F. Schober
Copyright © 2008 John Wiley & Sons, Inc.

it is unclear whether the data collected via traditional landlines are still as reflective of the entire population, and it is unclear whether social researchers will continue to be able to sample populations in ways that reliably generalize if they cannot interview respondents' whose only phone is a mobile phone. Yet cell phones also introduce new problems, such as the effects of answering questions in public, or while driving (see Fuchs, Chapter 4 in this volume).

The urgency of understanding and anticipating how new technologies will affect respondents' willingness to participate and their ability to provide accurate answers is acute. The barrage of requests that people receive to provide information has increased precipitously, as market researchers and opinion pollsters of varying degrees of legitimacy vie with census bureaus and social scientists to gain access to respondents. For these and other reasons (e.g., see Groves and Couper, 1998; Groves et al., 2002), people are becoming increasingly reticent to participate in survey interviews, and a greater percentage are opting out altogether. The implications become disturbing if we consider the point made by Ken Prewitt (2006), political scientist and former director of the U.S. Census Bureau: if governments become unable to gather essential information about citizens through interviews, in which the citizens have consented to participate, they will start to depend on surveillance data that citizens have *not* consented to provide. It is already possible to track a great deal about people's behavior (though probably not their opinions) from credit card usage, travel reservations, police and hospital records, and so on, and about their tastes and preferences from purchase profiles. Just how far societies develop into regimes of surveillance will depend, in part, on the degree to which survey interviews continue to be the major source of societal data.

Contending with the arrival of new technologies that affect communication is not new to survey research. Historically, survey researchers have, in fact, been key innovators in creating new technologies; for example, Hollerith invented the all-important punch card at the Census Bureau in 1890. In a more general sense, the great success of telephone surveys in the 1970s, especially after the incorporation of computer support in the 1980s, has served as a model for other mass scale telephone transactions. For example, telemarketing (sidestepping for the moment an evaluation of its social impact) and reservation centers owe at least an indirect debt to the pioneers of telephone surveys.

Yet more recently, survey practitioners have tended to follow rather than lead in devising and adopting new technologies. Quite some time after desktop computers with graphical user interfaces had become ubiquitous worldwide, survey practitioners continued to report on their plans for the transition from DOS to Windows. The same is true for the introduction of laptop computers into field data collection. Cell phones are another example: many years after they have become the primary phones in many homes, survey researchers still grapple with whether and how to incorporate them into telephone surveys. While speech and handwriting recognition tools are becoming more commonplace (e.g., speech interfaces to reservation systems and in-car GPSs; handwriting recognition in hand-held computers), they still occupy only a small niche in survey methodologists' suite of tools. Although the Internet is increasingly used for web surveys, its adoption lags far behind that for other activities like e-commerce, electronic tax filing, and so on.[1] And survey researchers have hardly considered the potential use and impact of mobile instant messaging, desktop videoconferencing, animated agent (avatar) systems, and other technologies whose role in daily communication is increasing rapidly.

There are legitimate reasons for conservative adoption of new technologies by survey practitioners. Given the extensive infrastructure costs, survey practitioners may seek to avoid moving to less proven approaches from technologies that are working reliably and that produce data that fulfill clients' needs. Survey practitioners are rightly concerned about the quality of the statistical estimates produced by surveys. If the incomplete penetration of a technology or the lack of complete sample frames leads to unknown error, adopting the technology could be risky.

Despite these legitimate concerns, we have reached the point of needing to be more proactive in trying to understand how current communication technologies, and those on the horizon, are used by humans and how their use in surveys would affect data quality. Because preferred ways of communicating are changing on a large scale, adhering to tried and true methods for survey interviewing is likely to lead to loss of participants and data quality; some potential respondents may already be unwilling to participate with today's standard technology (e.g., landline telephones). For example, people who spend much of the day multitasking on their smart phones may be more likely to interact with an interviewer via instant messaging than by talking. The point is that we can no longer afford *not* to consider the communicative properties of new technologies. This will require developing theories that allow us to exploit the characteristics of new media, rather than simply emulate familiar practice with new tools.

We also will need to move beyond assuming that design principles from one medium are the right principles for another. So, for example, rather than simply applying the design principles developed for paper questionnaires in web-based instruments, because these are well known and vetted (e.g., see Dillman, 2000), we propose that we should capitalize on the unique properties of web interaction. For example, one can imagine web questionnaire interfaces that adapt or tailor themselves to respondents' behavior, that diagnose respondents' need for clarification, and that allow respondents to request clarification (Conrad, Schober, and Coiner, 2007; Schober, Conrad, Ehlen, Lind, and Coiner, 2003). Other visual and interactive aspects of web surveys are also exploitable, with systems that detect respondents' lack of effort, that instantly check some answers against electronic records, and other sorts of features on the horizon.

Because what is cutting edge today will not be so for long, what is even more essential is developing principles and theories that remain useful and applicable irrespective of what technology is on the horizon at any point in time. This will require far more extensive connection and conversation between communication technologists and survey researchers than currently occurs.

---

[1] Of course, the barriers to widespread deployment of web surveys have at least as much to do with incomplete access to the Internet within the population as they do with the technology itself (see Chapters 3, 4, and 12 in this volume).

This chapter, and this volume, represent an attempt to foster this conversation. For survey researchers, the conversation requires learning a bit more not only about features of technologies, but also the principles and taxonomies that guide the development of those technologies. For communication technologists, the conversation requires learning a bit more about the survey enterprise, the nature of interviewing, and what survey designers would need to be assured of before being willing to adopt a new technology. Here we will present brief introductions to what each community should know in order to engage in the conversation. We also raise a set of questions that need to be addressed when considering whether to adopt a new communication technology. Some of these questions are relatively straightforward to answer, but answering others will require knowing the results of research that hasn't yet been carried out.

We should also be clear about what we are *not* doing. Although there are various important technological developments that peripherally affect the survey interview, this chapter and volume will focus on communication technologies in the survey interview itself. Thus, we will not focus on new technologies for constructing samples; non-interview methods for collecting data (e.g., passive measurement devices that eliminate respondents' self-report, like devices that detect cigarette smoke for a survey on smoking); or questionnaire construction, question development, and survey translation (into different languages), which happen before the interview. All these issues deserve serious attention, but, again, the focus here is on the interview itself.

1.2   SURVEY INTERVIEWS: THE STATE OF THE ART TODAY

A primary distinction in how surveys are administered today is between interviewer- and self-administration (see Couper, Chapter 3 in this volume). As we will argue, for considering the next generation of survey data collection, this distinction is becoming hazier.

The most frequent medium in which interviewer-administered surveys occur today is via landline telephone calls; face-to-face interviews still occur, but they are not the norm as they were fifty years ago. Typically, interviewers call from call centers where they have been carefully trained, and their performance is often monitored, either live or through random recording. Interviewers are most often trained to implement some version of standardized interviewing (see Fowler and Mangione, 1990), in which interviewers read scripted questions exactly as worded and make sure not to deviate from their script. In the strictest forms of standardized interviewing, interviewers leave the interpretation of questions entirely up to respondents; they do not define words in questions even if respondents ask for clarification, because departing from the script for some respondents but not others could lead to different interviewers affecting the answers differently. These days interviewers most often sit in front of computer screens (computer-assisted telephone interviewing, or CATI), so that a program prompts them with the questions they are to deliver, fills in specifics like the respondent's name, collects the answers they enter during the interview, and can do some error-checking and skip to relevant next questions based on answers to prior questions.

The energy and expenditure that go into maintaining call centers, and into selecting, training, and monitoring interviewers, are enormous. In fact, interviewer costs are often the greatest expense for carrying out a survey (although this is not to suggest that interviewers, many of whom work part time, are highly paid). Finding interviewers who sound natural on the phone, who are good at enticing respondents to participate, who manage to follow the script while maintaining rapport, and who are careful and meticulous enough to treat the data with care even at the end of a long shift is not a trivial task. And, while inferable (from audio) characteristics of interviewers like their age, gender, dialect, or ethnicity can affect respondents' willingness to participate and the answers that respondents give (Ehrlich and Riesman, 1961; Hatchett and Schuman, 1975; Kane and Macaulay, 1993; Schuman and Converse, 1971), it is not usually within the control of survey centers what the characteristics of their interviewers turn out to be.

The other major form of survey data collection these days is self-administration, via paper and pencil in mail-in surveys like the U.S. Census, but more and more via text-based web surveys. From survey designers' point of view, there are a number of advantages and disadvantages to self-administration. Advantages include both the cost savings of not needing interviewers and the fact that, if there are no interviewers, then different interviewers can't have differential effects on the data. Respondents can also choose to respond when it is convenient for them, and without time pressure that might distort their responses. Disadvantages include the greater difficulty of persuading respondents to participate, as well as the impossibility of establishing rapport with respondents, which might convince them to consider response options and complete the survey carefully.

Newer forms of administration begin to blur the lines between interviewer- and self-administration. Consider Interactive Voice Response surveys, where respondents listen to recorded voices on their telephone and speak or keypad their answers (see Bloom, Chapter 6 in this volume). On the one hand, from the survey designers' perspective this is clearly self-administration, in that no live interviewers are involved and the interface is perfectly standardized. From respondents' perspective, however, the interaction with the survey instrument unfolds over time in ways that are more similar to interviewer-administration, with spoken questions that sound as they would if a live interviewer were reading them, and with a turn-taking structure akin to survey discourse. Or consider Audio Computer-Assisted Self-Interviewing (A-CASI), in which a face-to-face interviewer turns the laptop over to the respondent for sensitive or threatening portions of interviews; the respondent listens to audiorecorded questions over headphones and types answers into the laptop. This provides privacy, so that other household members and even the interviewer don't know what the answers are, not to mention the domain of questioning; it can lead respondents to answer more honestly (e.g., Tourangeau and Smith 1996). Again, survey designers classify this as self-administration, but the interactive dialogue from the respondent's perspective bears notable similarity to interviewer-administration.

These more ambiguous cases only hint at the level of complexity that is on its way. Self-administered interviewing is likely to continue to add more and more features of human interviewing, turning the distinction between self- and interviewer-administration into a continuum or even, eventually, making the distinction obsolete.

Interviews with human interviewers that are mediated by videoconferencing certainly include interviewers, but they are removed from full physical copresence. Textual web interfaces and speech interfaces that allow respondents to request clarification by clicking or asking are under development (e.g., Conrad, Schober, and Coiner, 2007; Conrad et al., 2006; Ehlen et al., 2007); these could incorporate a level of interactivity that goes beyond what live interviewers trained in the strictest form of standardization are currently able to deliver. Interviews by talking head animated agents, akin to the University of Memphis' functioning AutoTutor tutoring system, have demonstrated effects of the ethnic appearance of the animated tutor (see Person, D'Mello, and Olney, Chapter 10 in this volume; and Cassell and Miller, Chapter 8 in this volume) that suggest the interaction may include more aspects of interviewer-administration than self-administration. Self-administration seems to lead to more candid responses to sensitive questions, but it is unclear at this point whether respondents will consider an interviewing system that includes an animated interviewing agent to be self- or interviewer-administered.

A next generation of interviewing systems is likely to make use of "paradata" (process measures) from respondents during the interaction, as a way of diagnosing when help might be needed so that the system can intervene appropriately to collect high quality data (see Couper, Chapter 3 in this volume; and Person, D'Mello, and Olney, Chapter 10 in this volume). One could imagine making use of respondents' typing errors, changed answers, response times, speech disfluencies, or facial expressions to assess their confidence in their answers or their likelihood of having misunderstood. Designers of interviewing systems will have to make decisions about when and how an interviewing agent should respond to this kind of user (respondent) behavior. Should the system respond during or after answers? What kind of audio or visual evidence should a computerized "interviewer" display? Should the interviewing agent say "okay" or "uh-huh" like human interviewers do? Should the interviewing agent smile or nod in response to an answer? When should the interviewing agent look at the respondent?

Considerations like these make it clear that survey researchers don't yet have the full set of computational and theoretical tools that it will take to make informed choices about the next generation of interviewing systems. And designers of other sorts of systems don't understand the unique constraints and requirements of survey interviewing. In the sections that follow, we present brief (and nonexhaustive) descriptions of what survey researchers need to know about what is going on in communication technology research and development and what communication technologists need to know about surveys.

1.3 WHAT SURVEY RESEARCHERS NEED TO KNOW ABOUT COMMUNICATION TECHNOLOGIES

People interact differently, in principled and characterizable ways, in different modes or media of communication. Scholars in a wide array of disciplines, including subareas of psychology, computer science, linguistics, communication, media studies, human–computer interaction, management studies, information science, and sociology (among others), have proposed distinctions, observed behavior, and carried out experiments that are potentially useful for survey researchers who want to consider adopting a new communication technology.

The existing empirical base has, thus far, rarely come from studying survey interaction, but rather from examining people's behavior on a variety of real-world and laboratory-created tasks. These can range from studies of group decision-making in mediated business communication, to comparing people's instructions and references about maps or abstract shapes when they are videomediated versus interacting face to face or via instant messaging, to careful analyses of people's interactions with working animated agents in intelligent tutoring systems.

The extent to which existing work is relevant for survey interaction thus depends on how similar and different those tasks are from survey interviews. (This is why clearer characterization of today's survey interaction is extremely important as we move forward; see Schaeffer and Maynard, Chapter 2 in this volume, and papers in Maynard, et al., 2002). It also will depend on how convinced survey designers are about the generalizability of work that comes from case studies and samples of participants smaller than the large samples that surveys rest upon. Given that there are almost no large-sample studies available (carrying them out would be prohibitively expensive, and not clearly any more generalizable), survey designers will need a bit of interdisciplinary flexibility in order to extrapolate from the existing work.

Survey designers should know that communications theorists do not have a single agreed-upon taxonomy of the features that are relevant in understanding a communicative situation, although a number of important distinctions have been proposed, including:

- *Communicative Goals.* Interactive goals can be oriented toward completing the task (*task-oriented* communication) or toward the relationship with the conversational partner (*socioemotional* or *affective* communicative goals), or both.
- *Initiative.* Who "drives" the interaction?

  Task versus *Conversational* Initiative. Different parties in an interaction may be in charge of the joint task (e.g., interviewers are responsible for collecting the data) and in charge of what is happening in the conversation at the moment (a respondent may take the conversational initiative by requesting that the question be re-read).

  Single-Agent versus *Mixed* Initiative. In some interactions the task or conversational initiative comes from one party only (e.g., in standardized survey interviews the interviewer owns the task initiative), but in other kinds of interactions the initiative can pass back and forth (e.g., in potential interactive interviewing systems of the future).

  Stakes. Some interactions are high stakes, where getting the details right matters (e.g., air-traffic control communication, negotiating child visitation rights), and others have lower stakes (e.g., small talk at a party, and maybe, from survey respondents' perspectives, their answers during interviews).

Synchrony. In *synchronous* interactions, there is virtually no lag between conversational turns (e.g., on the telephone), but in *asynchronous* interactions there can be an indefinite lag (e.g., postings to blogs, voice mail messages, e-mail).

We see a number of areas of research on interaction in communication technologies that are potentially important for survey designers to know more about. This isn't the place for a detailed or exhaustive review, but here is some of the existing work, along with references for where to look further.

1.3.1   Comprehension in Different Media

People in conversation make sure they understand each other by confirming ("uh-huh," "okay," nodding) and questioning ("huh?", "what was that?", looking confused) during and after each other's utterances (see Clark, 1992, 1996; Schober and Brennan, 2003, among many others). The form of this "grounding" differs in different communication media, which *afford* different costs and constraints (see Clark and Brennan, 1991, and Whittaker, 2003, for useful and reader-friendly overviews). For example, whether participants talk, type, or click; whether they can see each other's faces or hear each other's voices, and whether the communication leaves a reviewable record are all affordances of a medium and can all affect how people ground their understanding (e.g., clicking is easier than typing; remembering is harder than being able to review). This works out differently for different kinds of communicative tasks. The kinds of findings in this area include evidence that over textual chat people coordinate more efficiently when using an explicit turn marker (like "o" for "over") than when such a marker is unavailable (Hancock and Dunham, 2001), and evidence that task efficiency can be relatively unaffected by availability of facial information (e.g., as afforded by video) when people are directing each other on maps (Anderson et al., 1997).

Beyond the verbal aspects of grounding understanding, nonverbal communicative displays are an important part of communication. Research and theorizing have focused on two main kinds of displays, which are differently available in different media. *Paralinguistic* aspects of communication are tied to the linguistic aspects; they include speakers' tone of voice, intonational contours, their fluency and disfluency, and pausing (for reviews and discussion, see Clark, 1996; Levelt, 1989). *Visual* displays are less directly connected with the language, although they are deeply intertwined; these include gesture, gaze, nods, and facial expressions (see Goodwin, 1991; Krauss et al., 1996; McNeil, 1992, among many others). Paralinguistic and visual displays can affect turn taking, referring, attention, and interpersonal impressions; this has mostly been demonstrated in comparisons of face-to-face and telephone conversations, but a growing literature is demonstrating how access to and lack of access to visual and audio cues can affect the nature of interaction in newer media (see Whittaker, 2003).

These cues are all potential paradata for survey researchers to make use of. Little is yet known about the reliability and utility of facial expressions and paralinguistic displays in survey interviews, and there have been few comparisons across modes, but some work suggests that paralinguistic cues can be valid indicators of need for clarification. So, for example, survey respondents who pause longer and are more disfluent (*um*ming more) before answering are more likely to be presenting answers that are inaccurate (Schober and Bloom, 2004); survey respondents who avert their gaze longer while answering questions in face-to-face interviews are more likely to be giving unreliable answers (Conrad, Schober, and Dijkstra, 2008).

1.3.2   Least Effort Strategies and Satisficing

As has been argued at least since Herbert Simon's classic work (1957), people tend to satisfy their problem-solving goals, including their interactional goals, with the least effort possible given their processing constraints. This phenomenon has been called "satisficing"—doing what it takes to get something done well enough, rather than working harder to maximize the outcome. The basic idea is that humans lack the cognitive resources (particularly working memory capacity) to consider all relevant information in making a decision and so make decisions that are suboptimal but often good enough. While the notion was developed to account for organizational decision making, satisficing and similar mental shortcuts have been observed across a vast range of human behavior.

In human–computer interaction, interface design has been shaped by the recognition that users will not engage in much of what they consider to be unnecessary action. For example, it has been shown (Gray and Fu, 2005) that when information that is needed to complete a task is displayed on the screen, people will avoid even a small eye movement, let alone physical interaction with the system, in order to obtain this information by *recalling* it—even though what they recall is less accurate than what is on the screen. A version of the satisficing idea has been applied to survey interviews (e.g., see Krosnick, 1991): a respondent can conceive of his/her goal as finishing the interview or questionnaire as quickly as possible rather than providing the most accurate data. Respondents might, for example, select the first adequate response option even though it isn't the best one, or provide similar answers to all questions in a grid (nondifferentiation), or respond before they have thought about an issue for very long. This obviously can lead respondents to provide suboptimal answers and is not what survey designers hope respondents will do.

Satisficing strategies have been shown to differ depending on what information is afforded by different communication media. For example, survey respondents are differentially likely to endorse earlier and later response options in a list when they are reading (earlier) rather than listening (later) (Krosnick and Alwin, 1986). Certainly whether questions are delivered as text or speech and whether the respondent controls the onset and timing of the question delivery will become only more important as new interviewing technologies are developed.

This suggests that, when survey researchers consider whether to adopt a new communication technology, they would do well to carry out the kind of task analysis advocated by Simon, thinking through how and when respondents can satisfice in that medium. Media that promote additional satisficing are likely to be at odds with the goals of survey researchers; designs that reduce the opportunities for satisficing are likely to produce higher quality responses—though may increase non-response if short-cuts are not possible.

1.3.3   Deception and Self-Disclosure Differences

There is evidence that when people lie to one another they do so differently in different media, with different proportions of and different kinds of lies in e-mail, telephone, face-to-face, instant messaging, and so on. For example, participants in a study by Hancock et al. (2004) reported being more likely to lie on the phone than face-to-face, and still less in instant messages and e-mail messages. Lies about people's actions were most common on the phone, and lies about feelings were most common face-to-face; false justifications and reasons were most common in e-mail. A range of differences in what a medium affords (whether there's a reviewable trace that could lead to discovery) and people's purposes in using a medium (small talk versus business) are likely to play a role.

Surveys often ask respondents for sensitive information, and their level of honesty and how much they disclose may well be affected by their disclosure experiences in other communicative situations (see Hancock, Chapter 9 in this volume). Survey modes that create a sense of privacy seem to lead to more honest reporting of information that might embarrass or harm the respondent. Privacy is usually created by removing the interviewer from the data collection process, and so computerized self-administration (e.g., CASI and A-CASI) is now widely used to ask people about their sexual behavior, drug and alcohol use, criminal behavior, and so on. But interviewing systems of the future challenge these traditional distinctions. Consider a text chat interview. This is a mode that clearly involves real-time interaction with another person—an interviewer—but nothing is communicated about the interviewer's appearance or voice. And consider an animated agent interviewer. In this mode, all of the agent's facial and vocal characteristics are evident to the respondent but the agent is clearly not a living human. On intuitive grounds both text chatting and animated agent interviewing seem somewhere between interviewer- and self-administration in terms of the degree to which respondents are likely to truthfully disclose sensitive information, but there isn't empirical evidence yet one way or another.

1.3.4   Social Presence

Face-to-face communication and remote video have high social presence (Short, William, and Christie, 1976) because they allow the exchange of rich interpersonal cues. In comparison, text has low social presence. Thus, we would expect interviewing systems that use video to mediate between respondents and interviewers or that display an animated computer face in a web browser to elicit a greater sense of social presence than a text chat interview. However, people sometimes attribute human-like characteristics to very *in*animate user interfaces, even line drawings of people, reacting socially and emotionally (Reeves and Nass, 1996). So it could be that interfaces that communicate any animacy at all, including live text and animated interviewing agents, will create enough social presence that respondents will shade the truth as they sometimes do when human interviewers ask sensitive questions.

How the literature on social presence of different media applies to the survey response task is not yet clear. Although Reeves and Nass (1996) observed social responses were produced by line drawings in a nonsurvey task (rating personality), Tourangeau, Miller, and Steiger (2003) found very little effect of including a still photograph of the researcher in a web survey interface. Perhaps survey respondents really do know the difference between animate and inanimate interface objects and when it is to their advantage to interact with an interviewing system as just a piece of technology—because, for example, they wish to be truthful if the costs are not too great—then they are able to do so. More evidence that survey respondents can ignore or compartmentalize social cues when it is to their advantage to do so is the apparent sense of privacy that is created by listening to a recording of an interviewer's question via headphones, even though it adds voice—a particularly human attribute—to the text of conventional self-administration. Respondents must understand at a fundamental level that a recorded voice is inanimate.

1.3.5　User Modeling and Adaptive Interfaces

An intense and controversial area of activity in communication technologies has involved "user modeling"—designing interfaces that are tailored to different cultures, groups, or individuals (e.g., see Fink and Kobsa, 2000; Fink et al., 1998; Kay, 1995; Kobsa, 1990, 2001; Rich, 1983, among many others).

In general, there is agreement that making an interface seamless and easy to use is paramount. Where there is disagreement on whether a one-size-fits-all approach can sufficiently satisfy all users, and whether explicitly modeling individual users or groups of users (e.g., experts versus novices) is effective in all domains. The evidence is mixed and varies for different applications and task domains. But there are domains in which user modeling seems to work quite well. One example is intelligent tutoring systems that maintain individual student models, presenting different exercises for different students to work on depending on which skills need practice (e.g., see Corbett et al., 1997). Another is Rich's (1979) literature recommendation system, which demonstrated that a model of a group of users (or "stereotype model") can improve the system's performance and the user's experience relative to a system that uses no model. The system recommends novels to users that it believes they might like based on descriptions of the novels and the stereotype to which users belong, based on very brief self-descriptions. Users rated the system's recommendations as far more successful when they were based on the stereotype model than when they were made generically.

For survey technologies, an additional twist is the longstanding and deeply held view that a survey is standardized only if every single respondent experiences precisely the same stimulus (cf. Fowler and Mangione, 1990). So a debate on tailored interfaces must consider what counts as "the same stimulus"—must it look the same and use the same wording? Or, in the interest of standardizing the user/respondent's interpretation of survey questions, should interfaces be individualized? We will return to this issue, which we see as at the heart of the conversation between survey researchers and communication technologists, at the end of the chapter.

1.3.6 Exploiting Mode Differences

A final point that survey designers should consider as they approach the scholarship on communication technologies: communication technologists have tended to exploit the information available in different media, and (as a general trend) to create technologies that make use of all available channels of information (e.g., consider measures of vocal stress in automated speech interfaces that can diagnose user frustration as discussed by Person, D'Mello, and Olney in Chapter 10 in this volume). In contrast, the survey world has differing tendencies in thinking about "mode effects" (differences in the quality of data collected, e.g., on the telephone versus face-to-face versus web surveys). One tendency is to see mode effects as undesirable irritants that must be eliminated, as evidenced in the years of concern during the transition from face-to-face to telephone interviewing. Another has been to exploit what a new mode has to offer so as to maximize data quality.

So survey designers thinking about new technologies have a choice. They can either work hard to keep data quality comparable across modes, even if this means that some exploitable features of a mode won't be used. (For example, assuming respondents' facial information is useful to diagnosing need for clarification, this would mean *not* using facial information if a survey is also being administered on an audio-only telephone.) The alternative choice is to exploit different modes to maximize data quality, even if this leads to noncomparable data across different modes of administration. Getting used to consciously choosing which strategy to take is one of the challenges that survey designers will face as they consider adopting new communication technologies.

1.4 WHAT COMMUNICATION TECHNOLOGISTS NEED TO KNOW ABOUT SURVEY INTERVIEWS

The survey enterprise is vast and far more complex than most communication technologists are likely to realize. An astonishing amount of thought and care go into question design, sampling, interviewer training and monitoring, so that the resulting data are as unbiased and accurate as possible. This thought and care have resulted in a specialized vocabulary and a set of acronyms[2] that can be mysterious to the uninitiated. Communication technologists who want to collaborate with survey designers—and the size

---

[2] To give a flavor of a few of the acronyms, survey researchers distinguish between CATI (Computer-Assisted Telephone Interviewing—human interviewer on telephone is prompted by computer and enters answers into computer), CAPI (Computer-Assisted Personal Interviewing—human interviewer face-to-face is prompted by computer and enters answers into computer), SAQ (Self-Administered Questionnaire—respondent fills out paper form with interviewer present), A-CASI (Audio-Computer-Assisted Self-Interviewing—respondent hears questions over headphones and enters into laptop, with interviewer

and impact of the survey arena should make this an attractive proposition—will need to become familiar with some basic distinctions.

In thinking through whether to adopt a new communication technology for interviewing, survey designers will want to be confident that if they adopt it:

- Data quality (the quality of the estimates about the population that result from the interviews) is no worse with a new than existing technology. Data quality is measured by a number of indicators. This includes data reliability (the extent to which you get the same estimates when you redo the survey, or compare the answers of similar subgroups in the sample), data validity (the extent to which you are measuring what you claim to be measuring), coverage of the population (the extent to which the sample frame—the list of names or addresses from which the sample is selected—reflects the full population), and nonresponse rates (see below).
- Costs for developing the survey, administering it, training interviewers (if there are any), and data analysis are no greater, or at least remain in the realm of possibility.
- Respondents find the interaction no more irritating or difficult, that is, *respondent burden* of various kinds doesn't increase.
- Respondents are no less likely to agree to participate (*response rates* do not decrease), no more likely to hang up or abort their participation (*completion rates* do not decrease, *break-off* rates do not increase), and no more or less likely to participate or complete the survey if they come from different subpopulations (which would lead to *nonresponse error* if different subpopulations answer differently).

Here is a very rudimentary and oversimplified set of basic facts and distinctions about the survey process and what survey designers worry about along the way. Obviously the list isn't exhaustive, and the distinctions aren't exactly as clear as we're making them sound. A number of reader-friendly introductions to surveys and interviewing present these issues with more precision, including Beimer and Lyberg (2003) and Groves et al. (2004).

1.4.1 Major Distinctions

A primary distinction in survey research concerns whether the respondents report about themselves and the other people with whom they live (*household surveys*) or the organizations to which they belong, usually companies but also schools, farms, hospitals, prisons, and so on (*establishment surveys*). Household surveys are typically

present), and PAPI (Paper and Pencil Interviewing—human interviewer, face-to-face, enters answers on paper). The acronyms result from different historical moments and trajectories rather than a unified labeling scheme, and can thus be less than transparent. Readers who want to familiarize themselves with these terms are recommended to consult the definitions provided in Couper et al. (1998) and to some degree in Chapter 3 in this volume.

quite different from establishment surveys in many ways including the interviewing techniques. For example, the principles of standardized interviewing are far less prevalent in establishment than household surveys, and the responses come from records like the payroll data base in contrast to household survey where they tend to be self-reports (e.g., based on respondent's memory). The particular technologies that might be feasible in the two kinds of surveys are likely to differ not only because of differences in interviewing and the types of data but also because businesses are likely to adopt new technologies sooner than households. For example, web surveys were used regularly in some establishment surveys well before they were used in household surveys primarily because businesses were likely to have high speed Internet access and computers on all desktops prior to households.

Another major distinction is often drawn between surveys that interview respondents just once or on multiple occasions. The former is a *cross-sectional* survey (it takes measures at one slice in time); the latter is a type of *longitudinal* survey (it enables comparisons not only between respondents but within respondents over time). In this kind of longitudinal survey, respondents are considered to be in a *panel,* a group of people brought together to provide data across multiple waves of interviewing. Panel surveys may create unique opportunities to use interviewing technologies that adapt to respondents. Across waves, quite a lot of information becomes known about respondents. For example, if a respondent's attributes, e.g. race or gender or health status, become known in one interview, an interviewing agent might be configured to resemble (or not) the respondent in subsequent interviews.

Yet another major distinction concerns the kinds of questions asked of respondents, specifically whether they are asked for their opinions or about their behaviors. The process of answering questions is similar across these distinctions (see Tourangeau et al., 2000) but there are some differences that could affect the kinds of interviewing technologies one would develop. For example, the answers to behavioral questions can in principle be verified; there is a right or wrong answer to a question about whether one voted in the election. But answers to opinion questions, like how well elected officials are doing at their jobs, really can't be verified. Thus, one can imagine an interviewing technology that is able to confirm certain answers in real time by comparing against online databases (see Marx, Chapter 13 in this volume, for a discussion about the ethics of such technologies). But it's hard to imagine an analogous procedure when the survey concerns opinions.

A final distinction that communication technologists should be aware of is that between closed- and open-ended questions. Closed questions provide a limited number of response options and require respondents to select from the set, as in requiring a respondent to select "yes" or "no"; requiring a number for answers to numerical questions (e.g., number of people living in your house); or requiring answers on continuous or discrete rating scales (e.g., answering on a scale of 1 to 5 where 1 means "strongly agree" and 5 means "strongly disagree"). Open questions give respondents free rein in answering; what they report verbally is usually transcribed by interviewers (in paper self-administered questionnaires, handwritten open text is clerically entered into the database, and in web-based questionnaires the text is directly entered by the respondent). The text is coded, usually by people with help from computers, so similar responses can be aggregated and tallied. Closed format questions are far more common than open responses primarily because of the extra processing effort associated with open text.

1.4.2   Major Players

A basic issue for communication technologists to be aware of is that the content of a survey, and the politics and funding of its sponsoring agency, can affect how a communication technology gets adopted. A major arena of survey research is government surveys that produce official statistics in domains like employment, crime, health, business, energy, and demographics. Governments publish official statistics on a cyclical basis (e.g., monthly unemployment statistics) and so rely on well established administrative and operational machinery to collect and analyze survey data. Official statistics tend to be based on answers to behavioral questions, but occasionally opinion questions are asked in government surveys. Some official statistics are produced on an international scale, such as the Eurobarometer, which involve collaboration between government agencies, designed to allow cross-national comparisons.

Another set of scientific surveys, measuring both opinions and behaviors, often over time, are carried out by academic survey centers affiliated with universities. Large-scale academic surveys share much with government surveys but differ in important ways. One key difference is that academic surveys are typically designed to produce data that will allow social scientists to test theories with multivariate methods. As a result, such surveys often present batteries of related questions to respondents so that the researchers can construct scales by combining answers. It is also the case that academic surveys and their sponsors do not have the deep pockets of many national government survey organizations and so may be less able to bear the financial costs of adopting new technologies than would a government agency.

Another important sector in the survey research industry—in some ways the most visible face of the industry—is the corporate entities that specialize in measuring public opinion. These surveys obviously ask questions that are primarily about opinions, but they also collect information about respondents' backgrounds, for example, about their education and income. Finally, market research involves vast amounts of survey research. This work is often done on very tight schedules and one-time surveys are the norm. The lion's share of market research is now done online, in contrast to how it is conducted in the other sectors we have mentioned.

The costs of adopting a new technology are greater for long-running, large-scale surveys than for smaller scale and less entrenched surveys. Researchers are reluctant to change the administrative machinery that supports ongoing surveys and are loath to compromise time series by changing the mode of data collection. In favor of adopting new technology is that the most high-impact surveys have the most to lose by failing to keep up with new communication technologies. Changes in how the population communicates will require increased use of mobile phones, the web, and chat methods in order to reach respondents.

In any case, understanding the administrative and financial obstacles to expanding how a survey does its business will be essential for communication technologists interested in collaborating with survey researchers. The incentives and obstacles to implementing new communication technologies for survey interviewing are likely to vary in these different settings.

1.5   WHAT HAPPENS IN SURVEY INTERVIEWS

1.5.1   Sequence of Events

The sequence of events in an interview actually includes a few steps before the questions are asked. First, there is a contact initiation phase—a telephone call to a household, an e-mail solicitation, a postal mail request. In a telephone call to a household, a respondent must be selected from among possible household members; not all respondents are interchangeable or even appropriate for particular kinds of questions. An important task for interviewers is the process of what is called *refusal conversion*: convincing reluctant respondents to participate. This isn't done to annoy respondents, but rather because the sampling procedures that have led to the selection of the household and respondent require that a reasonable percentage of the selected respondents participate to increase the chances that the resulting population estimates are accurate.[3] After obtaining agreement to participate, interviewers work to present standardized interviews and maintain rapport such that the interview is appropriately completed.

Note that each of these steps—contact initiation, respondent selection, refusal conversion, and interview completion—can look different with different modes of interviewing. For example, all of the activities until the first question is asked are largely unscripted, but once the interview begins, interviewers typically are far more limited in the words they are empowered to use. Thus, technologies used in the early stages of interaction will need to be capable of supporting exchanges in which the particular words and conversational turns have not been planned, whereas the technology that supports the actual interview may not need to be so flexible, depending on the interviewing approach. Any new communication technology that is considered for survey interviews will have to be able to allow each of these steps to be carried out adequately, and this may require multiple technologies for the different stages.

---

[3] Survey organizations expend the effort to persuade initial non-respondents to participate because higher response rates are often assumed to be associated with lower non-response bias, i.e. the difference between the answers of respondents and non-respondents when the answers of the latter group are somehow knowable. In fact response rates seem not to be a good predictor of non-response bias (e.g. Keeter, et al., 2000). Yet refusal conversion continues to be a key activity of survey interviewers.

1.5.2 Features of Interviews

Interviews themselves are quite an unusual form of interaction when compared with other kinds of conversations, and self-administered surveys are different from other kinds of interactions with computers. Their distinctive features will need to be considered by communication technologists working on survey applications. First, in interviewer-administered surveys, the two parties in the interaction are in unusual roles (e.g., see, Clark and Schober, 1991; Schaeffer, 2002; Schober, 1999; Schober and Conrad, 2002; Suchman and Jordan, 1990) compared with roles in question-asking in spontaneous conversations. Standardized interviewers follow a script of questions and probes written by others, and so they are really acting as intermediaries for the survey designers who wrote the questions. When an interviewer reads a question, she isn't formulating it with the same freedom and responsivity to prior talk that spontaneous conversationalists engage in; when she refuses to explain what a question means (as she should if she is carrying out a strictly standardized interview), she is behaving in ways that would be at least impolite, if not bizarre, in everyday interactions (and which potentially harm data quality: see Conrad and Schober, 2000; Suchman and Jordan, 1990; Schober and Conrad, 1997). Whether or not respondents know it, there is an invisible third agent—the survey designer—involved. This can affect respondents' cognitive processes and emotional reactions to the interaction in odd ways: if respondents fail to recognize that the interviewer's behavior is constrained, they can interpret the interviewer's robotic behavior as uncooperative or even hostile, and they can interpret the interviewer's scripted "okay" as denoting real approval when it may not (Schober and Conrad, 2002). Savvy respondents who understand the game are operating by somewhat different conversational rules than they would be under other circumstances.

Computer-assisted interviews add an extra element to the situation (see Couper, Chapter 3 in this volume). Unlike a piece of paper with only static instructions to the interviewer to read the next question, a computer program is active and potentially leads the interviewer down different paths depending on respondents' prior answers (e.g., skipping all the questions about job satisfaction if the respondent reported being unemployed). There is thus an additional attention-demanding agent in the situation, or, to put it another way, the survey designer structures (intrudes into?) the interview in a new way through computer-assisted interviews.

1.5.3 Features of Self-administered Surveys

As interviews differ from other kinds of human conversation, interactions with self-administered survey systems are different from other kinds of human–computer interaction (Schober, Conrad, Ehlen, and Fricker, 2003). For one, the system initiates the interaction rather than the user; more often computer users go online to obtain information, but in online surveys they provide it. The location of the information being transferred is different: it comes from the user's memory rather than being something that can be found in existing documents. And, as we have demonstrated (Schober, Conrad, Ehlen, and Fricker, 2003), respondents to web surveys are more likely to (mistakenly) assume that they understand what words in surveys mean than to assume that their search queries are interpretable by a search engine.

This is not to suggest that online textual surveys are entirely different from other kinds of human–computer interaction, but simply that they differ in important and systematic ways that may be useful to consider when thinking through new communication technologies for surveys.

We should note that new technologies for surveys that blur the distinction between interacting with an interviewer and interacting with a computer are blurring the distinction between CMC (computer-mediated communication between two people) and HCI (human–computer interaction). Agent-like interviewing systems, for example, already have some features of human interviewers (spoken voice, dialogue responsivity, facial movement), and it is not yet clear whether principles and experiences from HCI or CMC are the more appropriate to bring to bear.

1.6 COMPARABILITY OF QUESTIONS ACROSS MODES

Ideally, survey questions are designed to produce comparable answers no matter what the medium of communication. Years of testing compared the answers produced via landline telephone with those produced face to face, with the general consensus that the answers were sufficiently comparable to adopt the new medium (see de Leeuw and van der Zouwen, 1988). On the other hand, we now know that self-administered interviews can lead to greater (presumably more honest) reporting of sensitive behaviors than face-to-face interviews (Tourangeau & Smith, 1996). So comparability of questions across modes—where comparability involves wording as well as technological and social context—is important for communication technologists to think about.

Considering that the mode of administration can affect the comparability of questions, there are some dangers in adopting a new technology for administering an existing questionnaire. For example, survey researchers often pretest questions to identify problems in interviewer delivery and respondent answering, but if a questionnaire is inherited from an earlier survey in a different mode, it is not clear that problems identified earlier necessarily recur with the new technology and it is also possible that new problems will occur when the new technology is introduced. This will depend on how similar the old and new modes are in these *affordances* (see page 8.) For example, moving from telephone to IM means that there is a reviewable record of the earlier exchanges between interviewer and respondent and an opportunity to reread questions that might have been too long to grasp in a single interviewer reading over the phone; but there is still the temporal pressure of a real-time interview, which could limit "the respondent's likelihood of" reviewing text in contrast to a self-administered questionnaire on the web.

It remains to be seen whether new media for survey interviewing will require new kinds of pretesting, or whether the kinds of problems uncovered via today's means will be observed with the some pretesting methods.

1.7   CONTENT AND FORMAT OF QUESTIONS

The impact of a new communication technology may depend on both the content and format of questions.

1.7.1   Content

The opinions that respondents report can be affected by interviewers' observable (or inferable) attributes (e.g., race, gender, age) if these are relevant to the question topic. This general phenomenon is well illustrated by a survey of black households (Schuman and Converse, 1971) in which either white or black interviewers asked respondents a variety of questions. On many questions, there was virtually no difference due the interviewers' race. When the questions concerned race, however, respondents expressed different opinions to black than white interviewers. For example when a black interviewer asked these black respondents if they could trust white people, 7% said "yes." When white interviewers asked the same question, 35% said "yes." These sorts of findings suggest that new communication technologies that highlight an interviewer's (or interviewing agent's) race, gender, or age group could affect answers to questions about these topics differently from technologies that do not (although it is also not clear whether one can devise truly neutral interviewing agents!). Cassell and Miller (Chapter 8 in this volume) discuss this issue in more detail.

As we have noted earlier, another way that impact of question content may differ by interviewing technology concerns sensitive questions. If a new technology reduces people's comfort enough that they stop reporting embarrassing behaviors, survey researchers will be wary of adopting it. Thus, the social presence afforded by different media will be a key consideration in the use of new media: streaming video seems likely to produce effects similar to what is observed in face-to-face interviews, but it is unclear to what degree virtual animated interviewers, IM, or speech dialogue systems create a sense of social presence for respondents.

1.7.2   Format

The format of response options to questions can affect the kinds of answers that respondents give, and these different formats will be differently possible to implement in different media. Closed options are already implemented quite differently in today's technologies; when an interviewer or a recorded voice speaks the response options, depending on the implementation the respondent must either listen to the entire list or can interrupt before the full set of options is listed. In a text-based self-administered interview, the respondent can see all the options at once and may even consider them before reading the question (or may ignore them—see Graesser et al., 2006).

Closed responses have been the norm in today's surveys, as they are far easier to capture and analyze than open-ended answers. But new communication technologies are opening up possibilities for systematic analysis of open-ended answers that go far beyond what is currently possible. The AutoTutor program, for example, uses Latent Semantic Analysis to allow students in tutoring situations to type natural-language answers, and it then classifies responses, with notable accuracy, so as to engage in fairly realistic dialogue. As such technologies improve, there is no reason to imagine that they won't be implementable in survey situations for dealing with written and spoken answers in far more naturalistic ways.

1.8   ERROR IN SURVEYS

As noted earlier, survey researchers will be unwilling to adopt a new communication technology if it is shown to increase survey error. There are a number of extensive treatments of different kinds of survey error (Groves, 1989), but the main areas that need to be considered include the following:

> *Coverage error.* Coverage error derives from a mismatch between those people or organizations in the sampling frame–the list of addresses, phone numbers, email addresses that represents the population to be surveyed–and those people or organizations who are actually in the population. This mismatch produces coverage error when those not in the frame differ on the attributes of interest from those in the frame. For example, if the sampling frame does not include people whose only telephone is a mobile device and these mobile-only users differ on the attributes being measured (e.g. political preferences) this would produce coverage error. The larger the undercoverage (the number of cell-only respondents in the population), the greater the total impact of coverage error on the survey results.

> *Sampling Error.* Sampling error occurs when the people one selects from a sampling frame don't accurately represent the population from which they are selected (e.g., report different opinions and behaviors than would the entire population if everyone were interviewed) and to which one wants to generalize. No sample will ever be a perfect microcosm of the population and different samples will almost surely lead to slightly different results; a certain amount of error is just inherent in the sampling process. In general, the larger the sample size the smaller the sampling error.
>
> There is a vast amount of theoretical and practical work aimed at understanding and reducing sampling error. Because the current volume is concerned with collecting data from respondents after they have been selected in a sample, this source of error is unlikely to interact with the choice of interviewing technology. However, a communication researcher working in the survey domain needs to recognize that survey designers will be concerned about how adopting the technology affects the sampling.

> *Nonresponse error.* Nonresponse error occurs when the people in the sample who didn't participate in the survey differ in the opinions or behaviors being measured than those who did participate. If a new communication technology selectively irritates or scares or confuses certain subgroups in the sample enough that they don't participate in or finish the survey, and they would have answered the questions differently than the groups with higher participation and

completion rates, this could lead to nonresponse error. As with coverage error, the overall impact of this discrepancy depends on the nonresponse rate. Because interviewers are involved in recruiting respondents and keeping them on task, this type of error is quite relevant to the choice of interviewing technology.

Measurement Error. Measurement error is how far off from true values respondents' answer are. Measurement error can result from *interviewer-related* error—e.g. bias due to particular interviewers' behavior or observable attributes—or *respondent-related* error—e.g. problems in respondents' comprehension, memory, or judgment that affect answers. There is some evidence that variation in interviewer "probing" behavior (e.g., not always asking the respondent to select from the response categories) increases the impact of interviewers on answers (Mangione, Fowler, and Louis, 1992). Such variation might be substantially reduced by a speech dialogue system that is clever enough to recognize when a probe is required and administers the probes systematically. Of course, the particular voice and dialect chosen by the designers can introduce error of its own. Evidence of respondent-related measurement error would be lower reports of embarrassing behaviors with a new technology, presumably because the technology introduces social presence that was not present in earlier technology.

It is worth elaborating on respondent-related error, given the rather large survey literature giving examples (often amusing, but with disturbing consequences) of what can influence answers (e.g., see Clark and Schober, 1991; Schuman and Presser, 1981; Sudman, Bradburn, and Schwarz, 1996; Tourangeau et al., 2000, among others). Seemingly minor changes in wording, in question order, and in response alternatives can (though they don't always) dramatically affect the percentages of respondents providing particular answers. These phenomena are known as *response effects*, and they can occur at a number of different levels.

To give a flavor, consider Loftus's (1975) comparison of answers to these two questions:

1. Do you get headaches frequently, and if so, how often?
2. Do you get headaches occasionally, and if so, how often?

Respondents to the first question reported an average of 2.2 headaches per week, while those for the second reported an average of 0.7 headaches per week. This presumably results from what kinds of headaches respondents consider they are being asked about (see Clark and Schober, 1991).

Or consider the study by Schwarz, Hippler, Deutsch, and Strack (1986) which asked Germans about their television-watching habits. When asked how many hours of television they watched each week, respondents given a set of response alternatives ranging from "less than $1/2$ hour" to "more than $2^1/2$ hours" provided notably lower estimates than respondents given a set of alternatives ranging from "less than $2^1/2$ hours" to "more than $4^1/2$ hours." Extrapolating from these data to the population would have provided an estimate of 16% of the population watching more than $2^1/2$

hours of television per week with the first response scale, but 38% of the population with the second response scale. In general, the nature of response alternatives (Is there a "don't know" option? Must respondents choose from existing alternatives, or should they provide open-ended responses?) dramatically affect in what answers respondents give.

So far, the story on whether these response effects work the same way with new interviewing technologies isn't entirely clear. Comparisons of current modes reveal some response effects that occur both in interviews and self-administered question-naires as well as self-administered questionnaires. So, for example, just as in telephone interviews, respondents to self-administered paper and pencil questionnaires are more likely to answer that they "don't know" when a question's response options explicitly include "don't know" (Schuman and Presser, 1981).

On the other hand, some response effects definitely work differently in different interviewing media. For example, recall that when an interviewer presents response options, respondents are more likely to go with the most recent one. But when they answer the same question on paper, they are more likely to go with the first response option (Krosnick and Alwin, 1987). As another example (Schwarz, Strack and Mai 1991; Strack, Martin and Schwarz, 1987), question order in interviews can make a big difference: people who first report how satisfied they are with their romantic life and then answer about how happy they are in general tend to answer in the same direction, i.e. their answers are correlated; their answers are uncorrelated when the questions are asked in the reverse order (general and then specific). But on paper (self-administered), this order effect is attenuated, presumably because people can look back and forth and change their answers.

As we see it, the moral is that new communication technologies for surveys will need to make sure they don't create new kinds of response effects.

1.9 ADOPTING NEW TECHNOLOGIES

Our assumption is that the communication technologies emerging today are the tip of the iceberg, and that the case studies in this volume may not be timely in future. Nonetheless, we believe that considering these examples helps raise sets of questions that will continue to be important as new technologies are developed and become options for use in survey interviewing.

Adopting current and upcoming communication technologies for survey inter-viewing could lead to a number of benefits, assuming the technologies have the right properties. One could imagine reducing interviewer-related error by making entirely standardized interviewing agents (although this isn't quite as straightforward as it seems). There are certainly potential cost savings from automation, e.g., in not hiring as many human interviewers, or increasing the interviewing pool by making inter-viewers less location-bound. Multimodal technologies might increase accessibility for respondents with sensory and motor deficits, or for people who are immobile or unreachable. Collecting paradata from respondents might allow better interpretation of responses and better diagnosis of respondents' needs.

But the potential benefits for any particular technology may just as easily be outweighed by the downsides, and the ones that come to mind are not trivial. As we have collected the work presented in this volume, a number of practical and ethical questions, to which we do not have answers, have come to the fore. As we see it, survey researchers need principles to guide decision-making about adopting new communication technologies in surveys in at least the following areas:

1. Should a new technology only be adopted for a survey once enough people in the population being measured have access to it? Once enough people use it frequently in their own communication, and so are used to using it? What counts as enough?

2. What are the costs of *not* adopting new technologies? For example, will a survey using an old technology (e.g., paper, or face-to-face interviews), or not allowing alternative modes for answering, seem antiquated or dull or nonscientific?

3. How connected does the new technology have to be to what people already know? WebTV assumes people know how to use a TV; agent technology assumes people already know how to interact face to face; IM assumes respondents can key in text. Just how different can a technology be from what a respondent already knows to allow accurate survey measurement (with all this entails— appropriate coverage, response rates, completion, satisfaction, validity)?

4. Is there a gold standard to which all new technologies should be compared? Is a face-to-face interview administered by an experienced personable interviewer the best, and a new technology should only be adopted to the extent that the interaction—or, to complicate matters, the data quality—mimics what happens in that sort of interview? Or might mediated communication potentially lead to improved data quality in some domains by reducing the social presence produced by a live interviewer as with A-CASI?

5. Survey designers have tried to increase rapport between interviewers and respondents. Communication technologists have tried to increase usability with computer interfaces. As survey interfaces have the potential to become more anthropomorphic, does rapport become part of usability? Should it?

6. How do we conduct interviews when more and more members of a population don't share the same first language, or the same set of cultural beliefs? How are interviews different when interviewers in Bangalore, rather than Michigan, are calling U.S. households? Should a remote video-mediated or animated survey interviewer look and sound and dress like the respondent? Should a web interface tailor interviews to local background cultural beliefs (e.g., about whether the number of children you report having includes ones who didn't survive)?

7. If new technologies allow interviews to be tailored to characteristics of individual respondents, will this infringe upon privacy, or is such infringement a necessary feature of a society that collects data about its citizens? Will respondents' privacy have been violated if survey questions are tailored to what is known about them in databases from prior survey responses, monitored web use, or linked datasets? The flip side: If you know something about a respondent

and don't divulge that, or don't rely on it as you ask questions, is it ethical to waste their time asking again? Is it ethical not to confirm and to just assume from what you collected via surveillance?

8. New communication technologies increase the number of ways that respondents can be contacted and surveilled and surveyed; along with this the opportunities to hassle and burden people increase. What level of intrusion is acceptable?

9. Informed consent: Is it ethical to capture paradata (e.g., typing errors, changed answers, response times, speech disfluencies, facial expressions) without explicit consent? When people agree to participate are they aware that this is being captured? Should they be?

10. There is an obvious technical survey coverage issue when not everyone has access to the same technologies. Is there an ethical issue, above and beyond this, about haves and have-nots? Or is it fine for some respondents to have access to technologies that others don't as long as everyone is able to participate somehow?

1.10   THE THEORETICAL HORIZON

The issues raised by contemplating new communication technologies for survey interviewing are serious in both practical and political terms, as sorting out what comes next will affect how we think of the role of consent in data-gathering about citizens, and ultimately what kinds of societies we become.

The issues also raise some basic theoretical questions about the nature of interaction and about standardization, and the answers are not straightforward.

1.10.1   Interaction

As we see it, the new communication technologies currently on the horizon all fall somewhere on the continuum between face-to-face interaction and written asynchronous communication. Videomediation keeps most of the features of face-to-face interaction while removing physical copresence; textual web interfaces like IM add a back-and-forth turn-taking structure into the "dialogue" that could allow clarification dialogue (Conrad, Schober, and Coiner, 2007); survey systems that diagnose facial paradata from respondents are emulating parts of what human interviewers can do. The same is also true for features of interviewers/interviewing agents: interviewers have a number of social identities, some of which are differently evident or inferable depending on the medium of communications. For example, accent cues are unavailable under textual mediation but can lead to all sorts of attributions about interviewers on the phone, and appearance, dress, and facial cues are unavailable on the phone but can lead to all sorts of attributions face to face. Thus, the range of behavioral and social cues from interviewers (and, in the other direction, respondents) give degrees of anthropomorphism to the interfaces (Schober and Conrad, 2007), with variability on the dimensions of dialogue capability, perceptual realism, and intentionality.

What is new, both practically and theoretically, is that interviewing systems will not only allow but require *conscious choice* about which elements of anthropomorphism are implemented. If one considers videomediated interviewing, then visible characteristics of the interviewer are of necessity part of the equation, with all the attendant benefits and baggage. If one considers building an automated animated agent, that agent must have an appearance and resulting inferable social identity; the survey researcher will now have to *choose* whether to create only one interviewer with one appearance for all respondents, or to create alternate interviewers for different respondents, or to allow respondents to choose their interviewer's appearance. One can imagine changing the appearance and voice of an interviewer mid-interview once the topic of questioning changes; one can imagine changing the interviewing agent's responsivity for different questions, with friendly "okays" for nonsensitive questions, but much more measured responses that don't suggest approval or disapproval for sensitive questions. One can imagine changing the interviewer's responsivity depending on the system's measurement of the respondent's reactions as the interview goes along; for example, if the respondent becomes irritated or agitated the system might loosen some of the constraints of standardization.

Having this kind of choice at our disposal is something brand new. It means that aspects of interviewer-related influence that have always been assumed to be error in the data may not only be controllable, but even exploited, if it turns out that this is helpful. It also means that survey designers will need to have far greater knowledge about interactive processes from basic social sciences, perhaps even more than the social scientists currently know. Questions that survey researchers need to know the answers to could thus drive new basic research in the social sciences, with the outcome that we will end up understanding where survey interviews of different kinds fit into a grander taxonomy of tasks.

To complicate things even further, new technologies may allow for access to information about respondents that goes beyond what face-to-face interviewers have available to them. Automated analysis of respondents' language, for example, could potentially let an interviewer, or an interviewing system, know when a respondent is likely to be under- or overreporting a behavior, and the system could intervene to probe further. We fully recognize the disturbing implications of following this line of thinking, and do not mean to suggest that we advocate building Brave New World systems of this sort. But given that technologies for detecting and augmenting evidence of one's partner's mental and physical processes are in development for use in other arenas [we are thinking of "augmented reality" systems for benign purposes like having beyond-visibility access to, say, your chamber music collaborator's breathing or bowing (Schober, 2006)], the space of technological applications to surveys may range beyond the continuum from face-to-face to text.

1.10.2 Standardization

The facts of new interviewing technologies complicate survey researchers' notions of standardization to a new degree. Current standardized methods arose in reaction to a long history of missteps and errors, in response to evidence that different interviewers were affecting the quality of answers differentially (see Beatty, 1995). The solution was to standardize question wording and interviewer delivery of those questions, such that all respondents receive precisely the same wording, and interpretation of questions is left entirely to respondents.

We have argued elsewhere, along with others (e.g., Conrad and Schober, 2000; Schober and Conrad, 1997, 2002; Suchman and Jordan, 1990), that although the goals of standardization are laudable, the logic of this solution is problematic. As Suchman and Jordan put it, it is perhaps more consistent with the goals of standardization to standardize the *meaning* of each question for respondents, rather than the wording, even if this leads to different respondents being presented with different words. So, for example, if a respondent has a different interpretation of what it means to live in a household than the government agency asking the question, it might be useful for the interview to allow the kind of idiosyncratic interaction that would allow this discrepancy to be uncovered and for the respondent's report of how many people live in his house to conform with other respondents' interpretation. We have shown that data quality can be improved, although interviews take longer, on the telephone (Conrad and Schober, 2000; Schober and Conrad, 1997; Schober, Conrad, and Fricker, 2004), face to face (Schober et al., 2006), and in automated textual (Conrad, Schober, and Coiner, 2007) and speech (Ehlen, Schober, and Conrad, 2007; Schober et al., 2000) interviewing systems that allow nonstandardized clarification dialogue to take place.

New interviewing technologies that allow survey researchers to choose features of anthropomorphism complicate the logic even more. Following the logic of strict standardization, a survey interview of the future will be most standardized if all the available features of the interviewer or interviewing system are exactly the same: if it looks the same, sounds the same, and responds the same. Thus, any variation in answers will not be attributable to variation in the interviewer. (This makes such an interviewing system a better approximation of strict standardization than is possible in any current call center.) But extrapolating from the findings on standardizing wording versus meaning, perhaps what is more desirable is that *data quality* be comparable across all respondents. By this logic, any aspect of an interviewer or interviewing agent that potentially gets in the way of data quality—say, race of interviewer for certain subpopulations for certain kinds of questions, responsivity of the interviewer on sensitive questions—should be tailored for the particular respondent population, to meet the goal of standardizing the respondent's *experience* of the interview. So whatever it takes for the interview to feel nonthreatening, while asking embarrassing questions, should be implemented, and if that means making the interviewer look and sound motherly for one sort of respondent but very distant and computer-like for another sort, then that will satisfy the goals of standardization.

Current data do not allow advocacy of one position over the other, or of another position entirely. But this is the sort of very basic question that thinking about new communication technologies for surveys will require us to deal with.

REFERENCES

Anderson, A. H., et al. (1997). The impact of VMC on collaborative problem-solving: an analysis of task performance, communication process and user satisfaction. In K. Finn, A. Sellen, and S. Wilbur (Eds.), *Video-Mediated Communication* (pp. 133–155). Mahwah, NJ: Lawrence Erlbaum, Inc.

Beatty, P. (1995). Understanding the standardized/non-standardized interviewing controversy.*Journal of Official Statistics, 11*, 147–160.

Biemer, P. P., and Lyberg, L. E. (2003), *Introduction to Survey Quality*, New York: Wiley.

Clark, H. H. (1992). *Arenas of Language Use* (pp. 176–197). Chicago: University of Chicago Press.

Clark, H. H. (1996). *Using Language.* Cambridge: Cambridge University Press.

Clark, H. H., and Brennan, S. E. (1991). Grounding in communication. In L. B. Resnick, J. M. Levine, and S. D. Teasley (Eds.), *Perspectives on Socially Shared Cognition* (pp. 127–149). Washington, DC: APA.

Clark, H. H. and Schober, M. F. (1991). Asking questions and influencing answers. In J. M. Tanur (Ed.), *Questions About Questions: Inquiries into the Cognitive Bases of Surveys* (pp. 15–48). New York: Russell Sage Foundation.

Conrad, F. G. and Schober, M. F. (2000). Clarifying question meaning in a household telephone survey. *Public Opinion Quarterly, 64*, 1–28.

Conrad, F. G., Couper, M. P., Tourangeau, R., and Peytchev, A. (2006). Use and non-use of clarification features in web surveys. *Journal of Official Statistics, 22*, 245–269.

Conrad, F. G., Schober, M. F., and Coiner, T. (2007). Bringing features of human dialogue to web surveys. *Applied Cognitive Psychology, 21*, 165–188.

Conrad, F. G., Schober, M. F., and Dijkstra, W. (2008). Cues of communication difficulty in telephone interviews. In J. M. Lepkowski et al. (Eds.), *Advances in Telephone Survey Methodology* (pp. 212–230). Hoboken, NJ: John Wiley & Sons.

Corbett, A. T., Koedinger, K. R., and Anderson, J. R. (1997). Intelligent tutoring systems (Chapter 37). In M. G. Helander, T. K. Landauer, and P. Prabhu (Eds.), *Handbook of Human–Computer Interaction*, 2nd ed. Amsterdam, The Netherlands: Elsevier Science.

Couper, M. P., et al, (Eds.). (1998). *Computer Assisted Survey Information Collection*. Hoboken, NJ: John Wiley & Sons.

de Leeuw, E. D., and van der Zouwen, J. (1988). Data quality in telephone and face to face surveys: a comparative meta-analysis. In R. M. Groves, P. P. Biemer, L. E. Lyberg, J. T. Massey, W. L. Nicholls, and J. Waksberg (Eds.), *Telephone Survey Methodology* (pp. 283–300). Hoboken, NJ: John Wiley & Sons.

Dillman, D. A. (2000). *Mail and Internet Surveys: The Tailored Design Method.* Hoboken, NJ: John Wiley & Sons.

Ehlen, P., Schober, M. F., and Conrad, F. G. (2007). Modeling speech disfluency to predict conceptual misalignment in speech survey systems. *Discourse Processes*, 44.

Ehrlich, J., and Riesman, D. (1961). Age and authority in the interview. *Public Opinion Quarterly, 24*, 99–114.

Fink, J. and Kobsa, A. (2000). A review and analysis of commercial user modeling servers for personalization on the World Wide Web. *User Modeling and User-Adapted Interaction, 10*(3–4), *Special Issue on Deployed User Modeling*, 209–249.

Fink, J., Kobsa A., and Nill, A. (1998). Adaptable and adaptive information provision for all users, including disabled and elderly people. *New Review of Hypermedia and Multimedia, 4,* 163–188.

Fowler, F. J., and Mangione, T. W. (1990). *Standardized Survey Interviewing: Minimizing Interviewer-Related Error.* Newbury Park, CA: SAGE Publications.

Goodwin, C. (1991). *Conversational Organization: Interaction Between Speakers and Hearers.* New York: Academic Press.

Graesser, A. C., Cai, Z., Louwerse, M., and Daniel, F. (2006). Question understanding aid (QUAID): a web facility that helps survey methodologists improve the comprehensibility of questions. *Public Opinion Quarterly, 70,* 3–22.

Gray, W. D., and Fu, W. T. (2005). Self-constraints in interactive behavior: the case of ignoring perfect knowledge-in-the-world for imperfect knowledge-in-the-head. *Cognitive Science, 28,* 359–382.

Groves, R. M. (1989). *Survey Errors and Survey Costs.* Hoboken, NJ: John Wiley & Sons.

Groves, R. M., and Couper, M. P. (1998). *Nonresponse in Household Interview Surveys.* New York: Wiley.

Groves, R. M., Dillman, D. A., Eltinge, J. L., and Little, R. J. A. (Eds.). (2002). *Survey Nonresponse.* Hoboken, NJ: John Wiley & Sons.

Groves, R. M., Fowler, F. J. Jr., Couper, M. P., Lepkowski, J. M., Singer, E., and Tourangeau, R. (2004). *Survey Methodology.* New York: Wiley.

Hancock, J. T., and Dunham, P. J. (2001). Language use in computer-mediated communication: The role of coordination devices. *Discourse Processes, 31,* 91–110.

Hancock, J. T., Thom-Santelli, J., and Ritchie, T. (2004). Deception and design: the impact of communication technologies on lying behavior. *Proceedings of the Conference on Computer Human Interaction,* Vol. 6, pp. 130–136. New York: ACM.

Hatchett, S., and Schuman, H. (1975). White respondents and race of interviewer effects. *Public Opinion Quarterly, 39*(4), 523–528.

Kane, E., and Macaulay, L. (1993). Interviewer gender and gender attitudes. *Public Opinion Quarterly, 57,* 1–28.

Kay, J. (1995). Vive la difference! Individualized interaction with users. In C. S. Mellish (Ed.), *Proceedings of the 14th International Joint Conference on Artificial Intelligence* (pp. 978–984). San Mateo, CA: Morgan Kaufmann.

Keeter, S., Miller, C., Kohut, A., Groves, R., and Presser, S. (2000). Consequences of reducing nonresponse in a national telephone survey. *Public Opinion Quarterly, 64,* 125–148.

Kobsa, A. (1990). User modeling in dialog systems: potentials and hazards. *AI and Society 4*(3), 214–240.

Kobsa, A. (2001). Generic user modeling systems. *User Modeling and User-Adapted Interaction, 11*(1–2), 49–63.

Krauss, R. M., Chen, Y., and Chawla, P. (1996). Nonverbal behavior and nonverbal communication: What do conversational hand gestures tell us? In M. Zanna (Ed.), *Advances in Experimental Social Psychology* (pp. 389–450). San Diego, CA: Academic Press.

Krosnick, J. A. (1991). Response strategies for coping with the cognitive demands of attitude measures in surveys. *Applied Cognitive Psychology, 5,* 213–236.

Krosnick, J., and Alwin, D. (1987). An evaluation of a cognitive theory of response order effects in survey measurement. *Public Opinion Quarterly, 51,* 201–219.

Levelt, W. J. M. (1989). *Speaking: From Intention to Articulation.* Cambridge, MA: MIT Press.

Loftus, E. F. (1975). Leading questions and the eyewitness report. *Cognitive Psychology, 7,* 560–572.

Mangione, T. W., Fowler, F. J., and Louis, T. A. (1992). Question characteristics and interviewer effects. *Journal of Official Statistics, 8,* 293–307.

Maynard, D., Houtkoop-Steenstra, H., Schaeffer, N. C., and van der Zouwen. (Eds.). *Standardization and Tacit Knowledge: Interaction and Practice in the Survey Interview.* New York: John Wiley & Sons.

McNeil, D. (1992). *Hand and Mind. What Gestures Reveal About Thought.* Chicago: University of Chicago Press.

Prewitt, K. (2006). Keynote address at Second International Conference on Telephone Survey Methodology, Miami, FL.

Reeves, B., and Nass, C. (1996). *The Media Equation.* Stanford, CA: Center for the Study of Language and Information.

Rich, E. (1979). User modeling via stereotypes. *Cognitive Science, 3,* 329–354.

Rich, E. (1983). Users are individuals: individualizing user models. *International Journal of Man-Machine Studies, 18,* 199–214.

Schaeffer, N. C. (2000). Conversation with a purpose – or conversation? Interaction in the standardized survey interview. In D. Maynard, H. Houtkoop-Steenstra, N. C. Schaeffer, and van der Zouwen, J. (Eds.), *Standardization and Tacit Knowledge: Interaction and Practice in the Survey Interview.* (pp. 95–124). New York: John Wiley & Sons.

Schober, M. F. (1999). Making sense of questions: an interactional approach. In M. G. Sirken, D. J. Hermann, S. Schechter, N. Schwarz, J. M. Tanur, and R. Tourangeau (Eds.), *Cognition and Survey Research* (pp. 77–93). Hoboken, NJ: John Wiley & Sons.

Schober, M. F. (2006). Virtual environments for creative work in collaborative music-making. *Virtual Reality, 10*(2), 85–94.

Schober, M. F., and Bloom, J. E. (2004). Discourse cues that respondents have misunderstood survey questions. *Discourse Processes, 38,* 287–308.

Schober, M. F., and Brennan, S. E. (2003). Processes of interactive spoken discourse: the role of the partner. In A. C. Graesser, M. A. Gernsbacher, and S. R. Goldman (Eds.), *Handbook of Discourse Processes* (pp. 123–164). Mahwah, NJ: Lawrence Erlbaum Associates.

Schober, M. F., and Conrad, F. G. (2002). A collaborative view of standardized survey interviews. In D. Maynard, H. Houtkoop-Steenstra, N. C. Schaeffer, and J. van der Zouwen. (Eds.), *Standardization and Tacit Knowledge: Interaction and Practice in the Survey Interview* (pp. 67–94). Hoboken, NJ: John Wiley & Sons.

Schober, M. F., and Conrad, F. G. (1997). Does conversational interviewing reduce survey measurement error? *Public Opinion Quarterly, 61,* 576–602.

Schober, M. F., and Conrad, F. G. (2007). Dialogue capability and perceptual realism in survey interviewing agents. Paper presented at the 62nd Annual Conference of the American Association for Public Opinion Research, Anaheim, CA.

Schober, M., Conrad, F., Ehlen, P., and Fricker, S. (2003). How web surveys differ from other kinds of user interfaces. *Proceedings of the American Statistical Association, Section on Survey Research Methods.* Alexandria, VA: American Statistical Association.

Schober, M. F., Conrad, F. G., Ehlen, P., Lind, L. H., and Coiner, T. (2003). Initiative and clarification in web-based surveys. *Proceedings of American Association for Artificial Intelligence Spring Symposium: Natural Language Generation in Spoken and Written Dialogue* (pp. 125–132). Menlo Park, CA: American Association for Artificial Intelligence.

Schober, M. F., Conrad, F. G., and Fricker, S. S. (2004). Misunderstanding standardized language in research interviews. *Applied Cognitive Psychology, 18,* 169–188.

Schuman, H., and Converse, J. M. (1971). The effects of black and white interviewers on white respondents in 1968. *Public Opinion Quarterly*, *35*(1), 44–68.

Schuman, H., and Presser, S. (1981). *Questions and Answers in Attitude Surveys: Experiments in Question Form, Wording, and Context.* New York: Academic Press.

Schwarz, N., Hippler, J., Deutsch, B., and Strack, F. (1985). Response scales: effects of category range on reported behavior and comparative judgments. *Public Opinion Quarterly*, *49*, 388–385.

Schwarz, N., Strack, F., and Mai, H.-P. (1991). Assimilation and contrast effects in part-whole question sequences: a conversational logic analysis. *Public Opinion Quarterly, 55,* 1–25.

Short, J.A., Williams, E., and Christie, B. (1976). *The social psychology of telecommunications.* New York: John Wiley & Sons.

Strack, F., Martin, L. L., and Schwarz, N. (1988). Priming and communication: The social determinants of information use in judgments of life-satisfaction. *European Journal of Social Psychology, 18*, 429–42.

Simon, H. A. (1957). *Models of Man.* Hoboken, NJ: John Wiley & Sons.

Suchman, L., and Jordan, B. (1990). Interactional troubles in face-to-face survey interviews. *Journal of the American Statistical Association*, 85, 232–241.

Sudman, S., Bradburn, N., and Schwarz, N. (1996). *Thinking About Answers: The Application of Cognitive Processes to Survey Methodology.* San Francisco, CA: Jossey-Bass.

Tourangeau, R., Couper, M. P., and Steiger, D. M. (2003). Humanizing self-administered surveys: experiments on social presence in web and IVR surveys. *Computers in Human Behavior, 19*, 1–24.

Tourangeau, R., and Smith, T. W. (1996). Asking sensitive questions: the impact of data collection mode, question format and question content. *Public Opinion Quarterly*, 60, 275–304.

Tourangeau, R., Rips, L., and Rasinski, K. (2000). *The Psychology of Survey Response.* Cambridge: Cambridge University Press.

Whittaker, S. (2003). Theories and methods in mediated communication. In A. C. Graesser, M. A. Gernsbacher, and S. R. Goldman (Eds.), *Handbook of Discourse Processes* (pp. 243–286). Mahwah, NJ: Lawrence Erlbaum, Inc.

CHAPTER 2

The Contemporary Standardized Survey Interview for Social Research[*]

Nora Cate Schaeffer and Douglas W. Maynard
University of Wisconsin–Madison, Madison, Wisconsin

2.1 INTRODUCTION

Investigators in many fields have long relied on research interviews conducted as part of sample surveys as a major source of data. This chapter introduces readers to the justification for the standardized survey interview, summarizes the rules that standardized interviewers follow, and describes some features of interaction in the interview. We briefly enumerate some actions in which interviewers and respondents engage and describe how the respondent's answer sometimes displays conversational practices learned in other social situations. We then present a detailed analysis of reports, a type of answer to a survey question that is not in the format projected by the question, which illustrates one mechanism by which respondents display problems

*Some of the research reported here was supported in part by the National Institute of Child Health and Human Development (Center Grant R24 HD047873), by the National Institute on Aging (Center Grant P30 AG017266 and the Wisconsin Longitudinal Study: Tracking the Life Course P01 AG021079), by the University of Wisconsin Center for Demography and Ecology, by the University of Wisconsin Center for Demography of Health and Aging, by grants from the Graduate School Research Committee to Schaeffer and Maynard, and by the University of Wisconsin Survey Center (UWSC). Nora Cate Schaeffer and Douglas W. Maynard are coinvestigators on the WLS Supplement "Cognition and Interaction in Surveys of Older Adults." Collaborators are Jennifer Dykema (UWSC), Dana Garbarski (Department of Sociology), and Cabell Gathman (Department of Sociology).

Envisioning the Survey Interview of the Future, Edited by Frederick G. Conrad and Michael F. Schober
Copyright © 2008 John Wiley & Sons, Inc.

in answering survey questions.[1] In our discussion we consider some implications of such studies of interaction for survey interviews in the future.

2.2   THE SURVEY RESEARCH INTERVIEW

The goal of most social research surveys is to describe a population. Like other research goals this one entails certain requirements. One requirement is that units of observation must be selected with known, nonzero probabilities of selection; this is part of what makes the data collection effort a "survey." In addition, the data collected from the units in the sample, the respondents, must be comparable. For the data to be logically comparable across units, all respondents must be asked the same questions (with the same meaning across respondents), offered the same exhaustive and mutually exclusive categories, and answer all questions. The rationale for these logical conditions is straightforward: if, for example, respondents are asked different questions (or questions with different meanings), then variation in their answers could be due to variation in the wording of the question, not to differences in the underlying concept being measured. If one interviewer asks, "Do you favor or oppose Governor Doyle's veto of this bill?" and another interviewer asks, "Do you favor Governor Doyle's veto of this bill?" the answers are not logically comparable. If all respondents are not given the same opportunities to provide each item of information, then the absence of a response becomes difficult to interpret: Did the question not apply, or was it not asked?

Survey researchers think of the questions asked in a survey interview as measurement tools, and survey interview protocols are commonly referred to as "instruments." Survey methodologists use two criteria derived from measurement theory to evaluate the quality of survey questions. The criterion of validity addresses whether the question measures the concept it is intended to measure. The criterion of reliability addresses whether a question would be answered in the same way by other respondents with the same true value (or by the same respondent if she or he were asked again and the underlying true value had not changed). Because assessing the reliability or validity of a survey item is itself a substantial undertaking, researchers often rely on surrogate assessments—indicators that are known or presumed to be associated with validity or reliability—to evaluate survey questions. These indicators include, for example, the results of cognitive interviews, a method of investigating cognitive processing that has been adapted for testing survey questions, in which subjects think out loud as they answer the survey questions (see Presser et al., 2004 for examples). Similarly, researchers have coded the interaction between the interviewer and the respondent to see whether the behaviors of the actors suggests that the questions cannot

---

[1] "Reports" refer to a class of answers that do not match the format projected by a question. For example, if a question projects a "yes" or "no" answer (e.g., "Can you read ordinary newsprint without glasses?"), and the respondent gives an answer other than "yes" or "no" (e.g., "Only reading glasses"), the answer does not match the format projected by the question. We use "reports" to refer to this type of answer, and not to "self-reports" more generally. (See Moore, 2004; Schaeffer and Maynard, 1995).

be administered in a standardized way or that the answers may have measurement error (e.g., see Draisma and Dijkstra, 2004; Schaeffer and Dykema, 2004). Measurement error may be invalidity (measuring something other than the target concept) or unreliability (variability in the answers that is due to something other than the concept being measured).

Because questions in surveys are instruments for measurement, survey researchers attempt to improve their validity and reliability in a variety of ways. For example, aids such as calendars or questions that provide memory cues may be used to improve the accuracy of answers. Researchers also attempt to write questions that will be interpreted consistently by respondents and thus provide more reliable data. Because surveys often use many interviewers to collect data from thousands of respondents, researchers have long been concerned with the effect different interviewers might have on the answers that are recorded (Hyman, 1954/1975). The practices of standardization have been developed to attempt to hold the effect of the interviewer constant and thereby reduce unreliability due to variation in the behavior of interviewers across respondents.

2.3   STANDARDIZATION IN THE SURVEY INTERVIEW[2]

Standardization refers to a set of practices that have developed over many years, and that are defined and implemented in different ways at different institutions (Viterna and Maynard, 2002). In *Standardized Survey Interviewing*, Fowler and Mangione (1990) codified a comprehensive set of rules that were grounded in the rationale for standardization and refined through many years of practice and observation. Fowler and Mangione (1990, p. 35) describe four principles of standardized interviewing:

1. Read questions as written.
2. Probe inadequate answers nondirectively.
3. Record answers without discretion.
4. Be interpersonally nonjudgmental regarding the substance of answers.

Although survey researchers would probably universally accept these principles, researchers and staff charged with supervising field operations at various centers have, over the years, developed their own specifications for training interviewers, sometimes importing practices from other centers (Viterna and Maynard, 2002, p. 393). Although the 12 academic survey centers that Viterna and Maynard examine vary considerably in their practices, "all centers in this study require all questions

---

[2]This section is based on Douglas W. Maynard and Nora Cate Schaeffer (2002), standardization and its discontents. In *Standardization and Tacit Knowledge: Interaction and Practice in the Survey Interview* (pp. 3–45), edited by Douglas W. Maynard, Hanneke Houtkoop-Steenstra, Johannes van der Zouwen, and Nora Cate Schaeffer. Hoboken, NJ: John Wiley & Sons. Copyright © 2002 by John Wiley & Sons, Inc. used by permission of John Wiley & Sons, Inc.

to be read verbatim in the order they appear" (Viterna and Maynard, 2002, p. 394), a practice that thus might be considered fundamental to standardization, and which appears even in Schober and Conrad's (1997) flexible or conversational interviewing. The version of standardization that Fowler and Mangione systematized is among the most rigorous and allows interviewers less autonomy than the practices used at most of the centers that Viterna and Maynard describe.

A comparison between the version of standardization articulated by Fowler and Mangione and the rules assembled by Brenner (1981a, pp. 19–22; 1981b, p. 129) illustrates how methods of standardization vary. Brenner allows interviewers to "show an interest in the answers given by the respondent," to "volunteer" clarification "when necessary," and to "obtain an adequate answer by means of nondirective probing,[3] repetition of the question or instruction, or nondirective clarification" when the respondent gives an inadequate answer. If a respondent asks for clarification, an interviewer must provide it, but nondirectively, by using "predetermined clarifications." The predetermined clarifications described by Brenner appear to be similar to the information that some survey centers routinely provide to interviewers in "question-by-question specifications" or that Schober and Conrad (1997; Conrad and Schober, 2000) provided to the interviewers in the experiments on "flexible" or "conversational" interviewing described later. Such practices contrast with Fowler and Mangione's (1990) recommendations for severely restraining the discretion of interviewers.

The variability in methods of standardization can also be illustrated by considering the practice of "verification," which some survey organizations, such as the Census Bureau, appear to consider compatible with Fowler and Mangione's first principle. Verification is a practice that acknowledges that respondents sometimes provide information before the interviewer asks for it, and that if an interviewer ignores the fact that she has already heard the answer to a question, the interaction may be awkward for both the interviewer and respondent. For example, an interviewer who has already been told the respondent's age in one context, instead of asking a later, scripted question about how old he or she is, might say, "I think you said earlier that you were 68 years old. Is that correct?" (Kovar and Royston, 1990, p. 246).

The second principle, to probe nondirectively, can also be implemented in different ways. For example, Fowler and Mangione (1990, pp. 39–40) require interviewers to repeat all the response categories when they probe closed questions. In support of this rule, there is research showing that the meaning of any individual category depends on the entire set of categories the respondent considers (Schaeffer and Charng, 1991; Smit et al., 1997). But there is variation in how this principle is implemented. For example, van der Zouwen and Dijkstra (2002) describe a practice called "tuning" in which the interviewer needs to repeat only the categories that appear to be in the vicinity of the respondent's answer. Tuning is itself similar to a technique called "zeroing in" that Fowler and Mangione (1990, p. 41) describe for numerical answers.

Survey practitioners recognize that there are situations in which the behavior of the interviewer cannot be standardized completely. For example, when investigators train

---

[3] A "directive probe," sometimes called a leading probe, is one that directly or indirectly offers a candidate answer; for example, "So would you say yes to that one?"

interviewers using "question-by-question specifications" of question objectives, the investigators recognize that the interviewer may need this information, the application of which cannot be prespecified, to diagnose and correct respondents' confusions and misunderstandings. Interviewers become less standardized when they actually apply the information provided to them during training, but that is part of the tension latent in traditional styles of standardization. Another example of a situation that cannot be highly standardized is probing, for example, of "Don't know" answers. Probing by the interviewer is associated with increased interviewer variance (Fowler and Mangione, 1990, p. 44; Mangione et al., 1992); that is, some of the variation in the answers is due to variation in the behavior of the interviewer rather than to variation in the concept being measured.

The rationale for standardization is that standardization decreases behaviors of interviewers that could give rise to variance in the answers that is not due to the concept being measured; reducing variation in the behavior of the interviewers should reduce error variance. In principle, when standardization is successful, it makes the effect of the interviewer in the aggregate very small, in the specific sense that it reduces the interviewers' contribution to variance. Standardization should also reduce bias by reducing the number of opportunities for the interviewer's expectations or opinions to intrude on the process by which the respondent's answer is generated, interpreted, or recorded. Studies in centralized telephone facilities where standardization is practiced, for example, find that the component of variance due to differences in the behavior of the interviewer across respondents (interviewer variance) is usually quite small for most types of survey items (Groves and Magilavy, 1986; Mangione et al., 1992). This is so even though there are many lapses in achieving the ideals of standardization (Oksenberg et al., 1991).

Standardization does not eliminate the effect of the individual interviewer on the individual respondent. An interviewer who follows the rules of standardization might react to a respondent's ambiguous answer by repeating all the response categories, for example. From the point of view of standardization, this is preferable to alternative behaviors such as the interviewer's using her beliefs about the respondent's likely answer to choose which response categories to repeat. But the standardized repetition of the response categories influences how respondents express their answers, and, over the course of the interview, probably has the desirable (from the point of view of facilitating the standardized interview) effect of training respondents to choose a category from among those offered. An answer is clearly an effect of the individual interviewer on the individual respondent. Nevertheless, if standardization were comprehensive and perfectly implemented, we could say that the effect of any interviewer would be the same as the effect of any other interviewer. That is, standardized interviewers are interchangeable.

This chapter does not examine the impact of standardizing the interviewer's behavior on the reliability or validity of the resulting data. Studies of interviewer variability are difficult to execute and are small in number as a result. A study comparing interviewer variability for standardized and unstandardized interviewing faces the challenge of selecting an appropriate comparison: Does one compare different varieties of standardized interviewers to each other or standardized to unstandardized interviewers, and, if the latter, how would those unstandardized interviewers be trained to behave, how would their answers be recorded, and how should outcome variables be specified? Sites within standardized interviews where standardization breaks down—for example, because of poor question design or poor training—provide a glimpse of the impact of standardization on reliability and validity. Collins (1980) described an instrument that omitted critical filter questions (questions that determine whether or not a target question should be asked); because of this design flaw, interviewers had to decide when target questions should be asked, and the result was a large component of interviewer variance in the answers. Interviewer effects were also present for the number of mentions to open questions and when probing was involved (see also Mangione et al., 1992). With respect to validity, a record-check study that compared respondents' reports about legal custody of their children to court records found that nonparadigmatic question-and-answer sequences in which the interviewer engaged in follow-up behaviors were less likely to yield accurate answers than sequences without follow-up by the interviewer (Schaeffer and Dykema, 2004). (However, we note that it is as plausible that a respondent's difficulty in answering occasioned the follow-up as that an interviewer's follow-up behavior somehow led to the inaccuracy.)

Because demands on interviewers—such as determining which questions to ask, asking open questions, probing, and defining complex concepts—are greater when interviewing is not standardized, many researchers are reluctant to experiment in loosening the rules of standardization. One such experiment is the event history calendar (EHC), and the initial experience reported by Dijkstra, Smit, and Ongena (forthcoming) illustrates the challenges of preparing interviewers for nonstandardized interviewing. In the versions of the EHC developed by Belli and colleagues (Belli, 1998; Belli et al., 2001; Belli and Callegaro, forthcoming), interviewers do not have to ask a series of questions in a particular order, and they are trained in procedures designed to improve recall of autobiographical events. When they evaluated a paper-and-pencil EHC administered by telephone versus a traditional standardized interview (using reports of events that occurred two years before the interview as a criterion) Belli, Lee, Stafford, and Chou (2004) did not find heightened levels of interviewer variance in the EHC; but preliminary analyses of life course reports gathered via CATI interviewing indicate that many items appear to have slightly higher interviewer variance in EHC interviews (Belli, personal communication, 2007). Even if EHC methods slightly increased interviewer variance, however, the total response error of EHC methods compared to traditional standardized interviewing could be smaller if EHC methods also reduced response bias.

It is plausible that the various components of standardization have different effects on the quality of data and that the size and direction of effects depend on the type of question (e.g., events, social categories, evaluations). Consider findings about the impact of reading the questions as worded. This practice appears to increase reliability, which is consistent with a reduction in the interviewer component of variance (Hess et al., 1999). And accurate reading of questions sometimes yields more accurate answers (Schaeffer and Dykema, 2004), but sometimes does not (Dykema et al., 1997). In a study that compared survey answers to court records to assess their accuracy, Dykema (2005) examined the impact of interaction on the accuracy of answers about characteristics of court orders about child support and legal custody (seven variables), the years of several salient events such as marriage and paternity establishment (four variables), and the amount of child support exchanged (one variable). Whether or not the interviewer made major changes in the wording of the question had no significant impact in nine cases and was associated with less accurate answers in two cases (whether the visitation arrangement was legalized and whether the parents had joint legal custody) and with more accurate answers in one case (amount of child support exchanged). When a cumulative pattern of behavior by the interviewers was assessed (errors in reading questions up to the question being evaluated), the pattern of interviewers' making major changes was associated with less accuracy for five of the seven questions about the court order and two of the four questions about years of events; as before, the pattern was reversed for the amount of child support exchanged. Such different results could occur because of variation in the difficulty of the task, because of problems in the design of the particular questions, or because of variation in the quality of the criterion.

Research suggests that other departures from standardization can also affect the quality of the resulting answers. For example, as already mentioned, when interviewers follow up by repeating only some response options or express their opinions, the response distribution of answers is affected (Smit et al., 1997), and when interviewers (who are standardized in that they are trained to read the question as worded) diagnose and correct comprehension problems and instruct respondents to ask for clarification, respondents give answers more consistent with survey concepts (Conrad and Schober, 2000; Schober and Conrad, 1997).

2.4   DATA

Up to this point, we have presented a description of the standardized interview in an idealized form. It is useful to temper this view by examining some examples from standardized interviews as they exist in practice. In the next sections we illustrate features of standardized survey interviews with excerpts from digitally recorded interviews from the 2004 wave of the Wisconsin Longitudinal Study (WLS). These observations were made in the course of developing a system for coding the interaction in the WLS interview. The health section, from which these excerpts are taken, comes early in the interview and has several types of survey item. The WLS began with a sample of 10,317 men and women, one-third of those who graduated from Wisconsin high schools in 1957; sample members have been interviewed several times in the intervening decades. The 2004 interviews were digitally recorded (with the permission of the respondents, who almost always consented), so that the interaction between interviewer and respondent is preserved along with the answers to questions. The WLS sample is structured in independent random replicates, or random subsamples. For our detailed examination of the interaction, we identified all cases in replicate two that were conducted by a single interviewer. From this group of cases we randomly selected 50 interviewers and sampled one interview from each interviewer, discarding cases in which the respondent did not give permission for the recording of their interview to be used in research. We made detailed transcriptions using conversation analytic conventions.[4]

2.5 PARADIGMATIC AND NONPARADIGMATIC QUESTION–ANSWER SEQUENCES

Because the question–answer sequence in which a question is asked as written, the respondent provides an answer in the format proposed by the question, and the interviewer acknowledges the answer is the same across interviewers, it could be considered the paradigmatic question–answer sequence of standardized interviewing (Schaeffer and Maynard, 1996). This paradigmatic sequence is illustrated by Excerpt 1.

```
Excerpt 1:    WLS Case 10, Q38

 1  FI:  # (0.2) .hh (0.1) A:n how would you duhscribe yer
 2       abi:lity tuh ruhmember things? (0.1) du:rrin' tha
 3       pa:ss fou:r wee↑:ks (0.1) .hh Were you a:ble tuh
 4       ruhmember mo:st thi↑:ngs (0.1) .h so:mewhat fergetful
 5       (0.3) ve:ry fergetful (0.4) or unable tuh ruhmember
 6       a:nythin' at a:ll.
 7       (0.3)
 8  FR:  Mo:hss thi:ngs.
 9       (1.3)
10  FI:  °Nnka:y?°
```

In Excerpt 1 the interviewer reads the question as worded (line 1), the respondent identifies and clearly selects one of the offered response options ("most things," line 8), and the interviewer acknowledges that selection ("okay," line 10). This sequence presents the interviewer with a routine situation: she simply records the response option identified by the respondent. The paradigmatic sequence can take some embellishment, as in Excerpt 2, and still be recognizable:

```
Excerpt 2: WLS Case 15, Q38

 1  FI:  Oka:y? (0.1) .h Wood yuh duhscribe ye:r- (0.1) ho:w
 2       wood yuh duhscribe yer ability tuh reme:m↑ber
 3       thi:ngs durin' tha pa:ss four wee:ks? .hh Wuh you
 4       able tuh remember mo:st thi:ngs, somewhat fergetfu:l
 5       very fergetfu:l er u:nable tuh
 6       [remember a:nthin' at a:ll.]
 7  MR:  [Yeh mo:st thi:ngs I pritty] much r'member
 8       °evrythin' pritty we:ll.° hh (0.4)
```

---

[4] Transcription conventions are described in the Appendix.

The response in Excerpt 2 is more complex than that in Excerpt 1: the initial "yeh" (line 7) is not a response to the survey question and cannot be coded by the interviewer, but it is followed by "most things," a phrase that appears in only one of the response options and can be heard as selecting that response option. The respondent then describes his situation as "I pretty much remember everything pretty well," a type of comment that we label a consideration. Even though "most things" is only a partial repetition of the response option and is preceded by an uncodable "yeh" and followed by a consideration, the skeleton of the paradigmatic sequence is recognizable, and the interviewer proceeds to the next question (not shown).

Paradigmatic sequences probably constitute a large majority of question–answer sequences (they do in the study examined here, for example), particularly if the questions are well designed and the interviewers well trained. The smoothness of the paradigmatic sequence does not in itself, however, indicate that either the question or the answer is of high quality. The respondent may interpret a question in a way different from the way the investigator intended without experiencing or giving evidence of any difficulty. Similarly, a paradigmatic answer may be a routine socially acceptable answer (e.g., "of course I do not cheat on my income taxes") or express an estimate as more certain than it is.

Thus, the fact that a sequence is paradigmatic does not necessarily indicate that a particular answer is valid or reliable, only that the respondent found a way to complete the task. Nor do deviations from the paradigmatic sequence necessarily indicate that data are of lower quality. It is easy to find examples in which a respondent's answer is not in the format requested by the survey question, the interviewer records the answer in the response categories without probing, and it appears that the quality of the data has not been compromised (Hak, 2002; Schaeffer and Maynard, 1996). [However, Schaeffer and Dykema (2004) found that in the case of legal custody, answers that interviewers treated as implicitly codable in this way were no more accurate than uncodable answers.] When the respondent asks for clarification, the question–answer sequence is no longer paradigmatic, but we have already noted that such requests sometimes are associated with more accurate answers (Schober and Conrad, 1997). Nevertheless, some deviations from the paradigmatic sequence, such as qualifications, are repeatedly associated with reduced accuracy (Draisma and Dijkstra, 2004; Dykema, 2005; Schaeffer and Dykema, 2004).

Paradigmatic sequences provide little information about the understandings, misunderstandings, thinking, accommodations, or adjustments that contributed to the respondent's answer. That is, paradigmatic sequences provide little information about the interpretive practices (or the cognitive process) by which they are constructed. Nonparadigmatic sequences can be informative in several ways. For example, nonparadigmatic sequences may display the respondent's understanding of the question, understanding of the task, level of uncertainty, facts the respondent is considering or rejecting, and so forth. These and other features of nonparadigmatic sequences have usefully been interpreted as evidence of problems in question design (Fowler, 1992; Oksenberg et al., 1991; van der Zouwen and Smit, 2004) or the quality of the resulting data (Draisma and Dijkstra, 2004; Dykema et al., 1997; Hess et al., 1999; Schaeffer and Dykema, 2004). Nonparadigmatic sequences also often constitute occasions on which the interviewer may—or must—react in some way that cannot be completely prescribed by the practices of standardization. These sequences thus provide sites to observe the participants deploying conversational practices learned in other social situations within the structure of the standardized interview. Actual practices reflect the "tacit knowledge" (Maynard and Schaeffer, 2000) that interviewers and respondents use to make the interview happen. One way in which the paradigmatic sequence breaks down occurs when the respondent fails to provide a properly formatted answer, and later we examine some of the ways in which this can happen and the interactional skills that participants use to conduct the task at hand.

2.6 CONVERSATIONAL PRACTICES WITHIN THE STANDARDIZED INTERVIEW

An early view of the survey interview characterized it as a "conversation with a purpose" (Bingham and Moore, 1924, cited in Cannell and Kahn, 1968), and this view of interviews as conversations was later echoed in the description of survey interviews as "conversations at random" (Converse and Schuman, 1974). In contrast to these informal characterizations of the survey interview stand the formal rules and constraints of standardization that we have just described. Someplace between a "conversation with a purpose" and a perfectly implemented standardized interview are the actual practices of interviewers and respondents.

Survey methodologists have studied the interaction between the interviewer and respondent for many years (Cannell et al., 1968). Most examinations of interaction in the survey interview have used standardization as a starting point and focused on how successfully standardization has been implemented, for example, by examining whether interviewers read questions as worded or respondents provide adequate answers (Brenner, 1981b), although exactly how these actions are defined varies. However, as researchers have looked more closely at what interviewers and respondents do, they have noticed and described in more detail how the participants import into the survey interview conversational practices from other contexts (Maynard and Schaeffer, 2006; Schaeffer, 1991; Schaeffer and Maynard, 2002; Schaeffer et al., 1993; Suchman and Jordan, 1990). As such observations have accumulated, they provide material for considering what these conversational practices accomplish, how they might support or undermine the goals of valid measurement within the survey interview, and how they might subvert or support the practices of standardization.

2.7 ACTIONS OF INTERVIEWERS AND RESPONDENTS

It is possible to examine nonparadigmatic sequences from a stance that attempts to characterize the utterances of interviewers and respondents and their sequential organization into actions. The actions of interviewers and respondents can be described at several levels. At the most general level, the participants ask and answer survey questions. At an atomic level, they deploy particles, talk, or remain silent. At intermediate

TABLE 2.1   Some Utterances and Actions of Interviewers

Utterances	Actions
Read survey item verbatim	Perform question from interview script
Modify survey item	Train respondent
Repeat survey item verbatim	Follow up to correct comprehension
Repeat response options verbatim	problems
Repeat answer	Follow up to correct inadequacies in the
Propose formatted answer	answer
Accept or code answer	Motivate respondent
Signal return to script	Maintain interaction
Request clarification	Record answer
Evaluate respondent's performance	Compliment
Say "sorry"	Reassure or encourage
Laugh	Apologize
Comment	Affiliate
Digress	

levels are utterances such as repeating questions, repeating answer categories, paraphrasing questions, and so on, and actions such as diagnosing problems or training respondents. One way of thinking about these utterances and actions is illustrated in Tables 2.1 and 2.2. Actions are accomplished as sequences of utterances, some of them prescribed by standardization, others improvised. Some of the utterances and actions that a description of interviewing must account for, such as reports and immediate coding of answers, have been described before (Hak, 2002; Schaeffer and Maynard, 1995, 2002), but others, such as encouragement and compliments, have received little attention (Gathman et al., 2006).

The lists of utterances and actions in Tables 2.1 and 2.2 are not comprehensive, and the distinction between utterances and actions is blurred in some cases. Nevertheless, these lists display some features of interaction that must be accounted for in

TABLE 2.2   Some Utterances and Actions of Respondents

Utterances	Actions
Repeat question or part of question	Interrupt
Repeat or paraphrase one or more response	Display understanding
options	Display retrieval of information
Use affirmative	Display consideration
Use negative	Display uncertainty or difficulty
Report with consideration	Display construction of answer
Report with hesitation or other marker	Request clarification
Report with a quantification	Provide answer
Ask question	Show a stance or position
Laugh	
Comment	
Digress	

a description of the survey interview. What some of these features look like, and the way that utterances build toward actions, can be shown with a few illustrations. In the following excerpt, the respondent interrupts the interviewer's initial reading of the response options:

```
Excerpt 3: Case 49, Q39

 1   FI:   .hh A:n u:m- ho:w wood you duhscribe yer ability tuh
 2         thi:nk an so:lve- day tuh day pro:bl'ms, u:m durin'
 3         tha pa:ss four wee:ks, .hh a↑gai:n were you a:ble
 4         tuh think clearly 'n' solve probl'ms, hadduh little
 5         di:fficulty:, had [so:me-  ]
 6   MR:                    [A:h clea]:rly.
 7   FI:   .h Oka:y, I'm sorry I'm juss required tuh read
 8         a:llree- all thee o:ptions fuh these. .hh U:m had
 9         so:me di:fficulty:, (0.1) hadduh great deal of
10         di:fficulty:, o~:r u:nable tuh thi:nk, °o~:r so:hlve
11         pro:bl'ms.°
12         (0.1)
13   MR:   Na::h. (0.2) (uh)
14         (0.1)
15   FI:   Uh=whi:sh [o:ne,] >did you w'nt-<
16   MR:             [Put- ]
17   MR:   Put clea:rly.
18   FI:   Oka:y, tha firs' one? Thi[:n-]
19   MR:                            [Ye ]:ah.
20   FI:   Oka:y, (0.1)
```

In response, the interviewer's next utterance announces that she will read all the options, which she then does. The respondent's "nah" (line 13) is insufficient to identify which option he is choosing, and the interviewer's next utterance explicitly asks for that identification. The interviewer's utterances roughly meet the requirements of standardization outlined earlier. Because of their sequential placement, the interviewer's utterances constitute follow-ups or probes. At a more general level the probes are an interactional method of training the respondent that he or she must provide an answer that is formally as well as substantively adequate (Moore and Maynard, 2002).

In Excerpt 4, the interviewer's intervention comes before the respondent speaks:

```
Excerpt 4: Case 49, Q40

 1   FI:   .hhh (0.1) A:nd have you ha:d any trouble with
 2         pai::n or disco:mfort durin' tha pa::ss fou:r
 3         wee:ks?
```

```
4                (1.0)
5      MR:       U:::h hhhh
6                (1.6)
7      FI:       (hh)
8                (0.8)
9      FI:       An this is u:m- (0.5) completely up tuh ye:r
10               (incare-) interpreta:tion of i~t.=
11     MR:       =I guess ri:lly no:t. h
12               (.)
13     FI:       O⁻ka:y yuh want me tuh put no: ↑fuh that izzat
14               righ'?=
15     MR:       =Ye:ah.=
16     FI:       =°Okay,°
```

In Excerpt 4, the interviewer treats the respondent's particle and the silences (lines
4–8) as potentially indicating a problem, and her utterance proposes what might be
a simplification of the response task ("and this is um… up to your… interpretation
of it," lines 9–10). The interviewer's utterance here is a preemptive probe (Schaeffer
and Maynard, 1995, 2002), which responds to the delays in lines 4–8 by reducing the
difficulty of the task.

2.8   REPORTS IN STANDARDIZED INTERVIEWING

One way in which the paradigmatic sequence breaks down occurs when the respondent provides a "report." In Drew's (1984) description of reports, the recipient of an
invitation (e.g., "Do you want to have dinner on Sunday?") did not reply directly, but
instead provided information relevant to a refusal of the invitation (e.g., "Isn't that
the night of the opera?"), and left the upshot to the person issuing the invitation. In
survey interviews, what we label a report is an utterance by the respondent that is
provided in the position where an answer (to the original reading of a question or to
a probe) is expected, that does not use the format projected by the question (e.g., the
utterance is not "yes" or "no" to a "yes/no" question), but that provides information
relevant to the process of answering the question (i.e., it is not a digression). Thus,
the report provides an opportunity for the interviewer to gather or propose the upshot
of the report and to record an answer or take some other action (see Moore, 2004;
Schaeffer and Maynard, 1996, 2002).

  This basic structure can be seen in the following excerpts:

```
Excerpt 5: WLS Case 10, Q2

1     FI:    # (0.3) .t .hh Du:rrin' tha pa:ss fou:r wee↑:ks
2            have you: been able tuh see well enu↑:ff tuh read
3            o:rdinary newsprin' .hh (0.1) withou:t gla:sses er
4            contac' le:nses?
             (1.2)
```

```
 6   FR:   A:h- (0.1) Ju:ss rea:din' gla:sses?
 7         (0.7)
 8   FI:   Oka:y? I'm 'ust gonna reread tha que:stion? .hh=
 9   FR:   =O↑:h [o  ]ka:(h)y h (0.1) .hh=
10   FI:         [uh-]
11         Du:rrin' tha past fou:r wee↑:ks (0.1) .hh have you
12         been able tuh see well enu:ff (0.1) tuh read
13         o:rdinary news↓print (0.1) .hh with↑ou:t (0.2)
14         gla:↓ss~es (0.1) o:r co:ntac' le:ns~es.
15         (0.3)
16   FR:   A:~h- (0.3) No↓:. hh
17   FI:   Oka:y?
```

Excerpt 6: WLS Case 1, Q7

```
 1   FI:   Yer doin' be~tter than I a(h):(h)m. (0.1) .hh (0.1)
 2         Without a hearing aid an while in a group
 3         conversa:↑tion with at least three other pee↑pu:ll
 4         have you been able tuh hear what is sai:d?
 5         (1.7)
 6   MR:   Not very goo:d. hh[h]
 7   FI:                     [h]hh Would you say no:?
 8         (0.6)
 9   MR:   No[:.]
10   FI:     [°iz]zat-° (0.1) .hh (A:hem) what about wi:th a
11         hearing aid.
12         (1.5)
13   MR:   U:m be:tter cha:nce the:n.
14   FI:   °Kay.° (0.2) Wou:ld you say yes er no in ge:nrull.
```

Excerpts 5 and 6 present examples in which the respondent does not provide a formatted answer but instead states relevant information without providing an upshot. In Excerpt 5 the interviewer responds to the report ("Ah. . . just reading glasses," at line 6) by announcing that she is going to reread the question (line 8) and then doing so (lines 10–14). Two additional examples of reports and the way interviewers respond to them are presented in Excerpt 6, and we begin with the second example in that excerpt. When answering the second question (lines 10–11) about whether he can hear with a hearing aid, the respondent provides the report that there is a "better chance then" (line 13). In line 14 the interviewer acknowledges the answer ("okay") and uses a standardized probing technique to ask the respondent to select a response category ("yes or no"). But she also seems to treat the report as indicating a trouble in answering; she downgrades her request with a phrase, "would you say," which we refer to as a distancing phrase, and simplifies the task ("in general"). In both cases, the utterance of the interviewer, in effect, returns to the respondent the task of producing the upshot of the report. As actions, the interviewers' utterances are follow-ups that support the practices of standardization by instructing the respondent to attend to details of the question when formulating an answer, but in neither of these cases does the interviewer suggest an answer.

In the first example of a report in Excerpt 6 ("not very good," line 6), the interviewer displays another method that interviewers commonly deploy for responding to a report: proposing a candidate answer ("Would you say no?" line 7) in a directive probe. This sequence is similar to sequences that Moore and Maynard (2002) examine in which the answer is "formally" (which they contrast with "substantively") inadequate. In their corpus of 3880 sequences from five surveys, Moore and Maynard (2002, p. 304) observe that a directive probe is much more likely when the respondent gives an answer that is formally inadequate than when the answer is substantively inadequate. As in this case, however, it is not always clear when the inadequacy of an answer should be considered "formal" rather than substantive, and the requirements of standardized measurement in this respect may be more constraining than those of other social contexts. If the inadequacy of the answer is truly simply formal, the quality of the resulting data is unlikely to be affected by this sort of directive probe; the difficulty lies in determining when the inadequacy is only formal. (We return to this example in our discussion of mitigators that quantify.)

The respondent in Excerpt 5 presents relevant facts or considerations in her report ("just reading glasses," line 6), and we use the term "considerations" to describe this type of report. These and other types of reports (described later) show respondents taking a step toward fitting their experience into the terms of the survey question and failing to do so. Reports present occasions on which measurement error is revealed to be present (e.g., by the respondent's uncertainty) or perhaps introduced (e.g., by the interviewer's response). In addition, our examples show interviewers using their tacit knowledge to alternate between the formal practices of standardized interviewing and conversational practices as they attempt to negotiate a complete formatted answer (Maynard and Schaeffer, 2000).

2.9   REPORTS THAT MITIGATE: UNCERTAINTY AND GENERALIZATION

In addition to presenting considerations, reports can also display doubt or uncertainty:

```
Excerpt 7: WLS Case 30, Q12

1    MI:  .hh (0.1) A:n- (0.1) hev pee↑pull who do not kno:w
2         you: understood you completely when you spea:k?
3         (0.6)
4    MR:  Ah thi:nk so:. (0.1) [#] (0.1)
5         I still gotta little bit of uh
6         w- Wis[co]:nsin accent, (0.1) but [(these fo:lks, ]
7    MI:         [# ]                       [HH hu(h) hu(h)h]
8         (0.2)
```

```
9    MI:   HU~ (H)H [.HH ]
10   MR:              [They] j'ss put that asi:de,
11         (0.1)
12   MI:   O(h)ka(h):y, (0.2) U:h- wood you say like- (0.2)
13         ye:h on that o:ne (0.1) iz[zah c'rreh?]
14   MR:                              [Ye(h):       ](h)h.
15         (0.2)
16   MI:   O(h)ke(h):(h)h.  .hh
```

In Excerpt 7, the respondent provides a conventional expression of uncertainty ("I think so," line 4) and follows that with a self-deprecating joke or humorous offering (lines 5–6, 10). The answer is not properly formatted and leaves the upshot to the interviewer. The interviewer treats the respondent's answer as requiring only confirmation and produces a directive probe (lines 12–13). This type of report (like others, such as giving ranges) embodies practices that participants engage in to express uncertainty or otherwise mitigate answers, and interviewers commonly confront such answers as they conduct standardized interviews.

Another type of report represents a practice that may be more common when answering questions about threatening topics, that is, questions that ask for information that might be embarrassing or incriminating (Schaeffer, 2000):

Excerpt 8: WLS Case 4, Q35

```
1    MI:   Oka:y, (0.1) A:n durin' tha pa:st four wee:ks,
2          didjuh ever fee:l fre:tfu:l, a:ngry, irrituhbull
3          ankshuss er duhpressed?
4          (0.7)
5    MR:   O:ah I think evrybody gets i:rrituhbull,
6          (0.3)
7    MI:   °Mm↑hmm° So: I should sa::y- ye:s t'that
8          que:ss[ch'n?]
9    MR:         [Ye   ]:ss.=
10   MI:   =O:ka:y, (0.2) .hh (0.1)
```

In answering the question about his recent emotional experiences, the respondent does not describe his own behavior. Instead, the report provides a speculative generalization (line 5), which, by referring to "everybody," can be heard as including the respondent without his having to make a direct admission. As with other reports, this one provides an opportunity for the interviewer to gather or propose an upshot, which he does with a directive probe that is crafted as a request for confirmation (lines 7–8). In this case, the interviewer's directive probe can be heard as relieving the respondent of the obligation to initiate a potentially embarrassing admission about himself.

2.10   REPORTS THAT QUANTIFY

In the next example, the respondent, rather than using the dichotomous response options, produces an answer that suggests a scale or continuum:

```
Excerpt 9: WLS Case 28, Q32

1   FI:   HH=O(h):ka:y? .HH Durin' tha pa:st four ↑wee:ks .hh
2         have you been feelin' ha:p↑py or u:nhappy.
3         (3.1)
4   MR:   O:↑::hhh I guess I'm- (0.2) feelin' pritty ha:ppy.
5         (0.5)
6   FI:   A::ri:ght.
```

Offered a choice between happy and unhappy, the respondent in Excerpt 9 displays a stance of uncertainty (with the mitigating phrase, "I guess," line 4), incorporates a word from the question ("feeling"), and volunteers a response that was not offered as a category ("pretty happy"). By using a modifier ("pretty"), the respondent displays an orientation to a scale with gradations of happiness. An early scaling study indicated that "pretty" happy is less happy than "happy" (Cliff, 1959), and conversation analytic research suggests that a phrase such as "pretty good" can be a "qualified response to a how-are-you inquiry" (Schegloff, 1986, p. 130). Because survey measurement requires classification in the offered categories, "pretty happy" presents the problem that it is not clear whether it should be classified as "happy" or "unhappy." The interviewer solves this difficulty by classifying the respondent as "happy" (this can be determined by seeing which question is asked next), thus engaging in a practice that has been called immediate coding (Hak, 2002), without confirming her action with the respondent. The conversational practice of treating the upshot of answers such as "pretty happy" may be so commonsensical and routine that the interviewer may not notice that she recorded the answer in a form different from that in which the respondent expressed it.

In the next two excerpts, the respondents' answers may display that they are experiencing difficulty in translating a value into the two implicit response options, yes and no. The question in Excerpt 10 and Excerpt 11 (which reproduces part of Excerpt 6) asks the respondents to classify themselves with respect to several conditions: without a hearing aid, in a group conversation, with at least three other people, and able to hear what is said. For the last condition, in particular, it may not be clear how to determine when the condition is satisfied:

```
Excerpt 10: WLS Case 28, Q7

1   FI:   A::↑ri:gh(t) .hh Wwithou:t a hea:rin' ai:d, an while
2         in grou:p conversa:↑tion with at least three other
3         pee↑pu:ll .hh have you been able tuh hea:r what is
4         sai:d?
5         (1.8)
```

```
6   MR:   Mo:stuh tha ti:me.
7         (1.2)
8   FI:   I:zzat a ↑ye:s zen e[:r-]
9   MR:                     [Ye ]:s
10        (0.1)
11  FI:   O[ka:y?]
```

Excerpt 11: WLS Case 1, Q7

```
1   FI:   Without a hearing aid an while in a group
2         conversa:↑tion with at least three other pee↑pu:ll
3         have you been able tuh hear what is sai:d?
4               (1.7)
5   MR:   Not very goo:d. hh[h]
6   FI:                     [h]hh Would you say no:?
7               (0.6)
8   MR:   No[:.]
```

Each of these reports displays an orientation to a continuous dimension: the first focuses on a frequency dimension (how often), the second on an intensity dimension (how well). It is not clear whether being "able to hear what is said" means that the respondent is able to hear what is said "ever" or "always" or with some other frequency. Similarly, being able to hear what is said could mean that one is able to hear "even a little bit" or "extremely well" or somewhere in between. The question wording provides no guidance to the respondent or the interviewer in deciding where the threshold for saying "yes" lies. The reports of these respondents, like that of the respondent in Excerpt 9, implicitly poses to the interviewer the question of where on the continuum lies the boundary between the offered response options, and each interviewer responds with a directive probe.

   In the following excerpt, a similarly complex question that nevertheless asks for a yes or no answer is also met with a response that refers to a continuum, although in a humorous way:

Excerpt 12: Case 29, Q17

```
1   MI:   Have you been able to be:nd (0.1) li:ft
2         (0.1) ju:mp a:nd ru↑:n without di:fficul↑ty:
3         (0.1) an without he:lp or equip↑ment of any
4         ki:nd? .hhh
5               (1.2)
6   FR:   A:↓:h no:t too much ju:mpin'
7         he[(h)hhe(h)h]he(h)hhe[(h)h↓he(h)h]
8   MI:     [O:kay?    ]        [so the    ]:n I shud
9         put no:the:n fuh tha:t one [is that c'r↑reh-]
10  FR:                              [A:h tha:t's     ]
11        c'rre:ct ye:s.
```

DISCUSSION   **49**

The answer "not too much jumping" (line 6) addresses only one of the conditions mentioned in the question, and it does so in a way that may suggest that there is "no jumping," "a little" jumping, or possibly the ability to jump but not the occasion to do so. In addition, the respondent appends laughter to her response (line 7), thereby displaying "troubles resistance," and invites the interviewer to laugh in response (Jefferson, 1984; Lavin and Maynard, 2001). The interviewer delivers a directive probe, declining the invitation to laugh, ignoring many possible lines of inquiry, and moving the interview on to the next question.

2.11 DISCUSSION

In this chapter we introduced the standardized interview, the paradigmatic question–answer sequence, and the prescriptions of standardization that guide interviewers in handling routine and unexpected situations. We have also briefly described some of the ways that standardization can break down when respondents import common conversational practices into the standardized interview. In addition to reports, which we have examined in detail here, such practices include producing qualified responses to questions that ask for a response option, joking and laughing, and others (Lavin and Maynard, 2001; Schaeffer and Maynard, 1995, 2002). Faced with reports, interviewers may use canonical practices of standardized interviewing such as rereading the survey question. Or, and perhaps more commonly, they may routinely use their tacit knowledge of conversational practices, for example, to interpret reports as codable or not.

The reports that we have examined use the slot after the survey question to display a number of things. The considerations recited by respondents appear to describe aspects of their experience that they are using to construct an answer or that pose some problem as they attempt to find an appropriate response option. Reports display other translation problems as well, as when the respondent elects a response option that was not offered or returns to the interviewer the task of deciding where on an underlying continuum the threshold for "yes" lies. Finally, reports display answers that are mitigated in some way (e.g., by uncertainty) although interviewers routinely treat these answers as codable. One characterization of reports is that they display a lack of fit between the survey question and the experience that the respondent is describing. The lack of fit may originate in uncertainty about the facts (state uncertainty) or difficulty determining which of the offered options represents the respondent's true value (task uncertainty) (Schaeffer and Thomson, 1992), or somewhere else. Evidence has accumulated that some features of interaction—like reports, delays in answering, or qualification in answering—indicate measurement error (unreliability or invalidity) or problems in question design (Draisma and Dijkstra, 2004; Dykema et al., 1997; Hess et al., 1999; Mathiowetz, 1998, 1999; Schaeffer and Dykema, 2004; van der Zouwen and Smit 2004).

The examples that we have presented used conventional methods of interviewing: a human interviewer talking with a respondent over the telephone. The survey interview of the future is likely to increasingly rely on modes that turn tasks previously performed by interviewers over to a computer and that are self-administered (such as computer-assisted self-interviewing, CASI, or web surveys). Because self-administered instruments eliminate the interviewer component of variance (the part of unreliability that is due to variation in the behavior of the interviewer), the data they obtain are potentially more reliable than data an interviewer can acquire (O'Muircheartaigh, 1991). In addition, data from self-administered instruments may sometimes be more valid, for example, for threatening questions (Tourangeau and Smith, 1998). However, simple technological enhancements alone are unlikely to lead to substantial improvements in the accuracy of data; for example, adding audio to computer-assisted self-interviewing (audio-CASI) may not improve the validity of data more than CASI alone (Couper et al., 2003). Moreover, in the absence of an interviewer, when respondents encounter problems in answering survey questions, those problems may be more likely to lead to item nonresponse or abandoning the task altogether.

Future enhancements to self-administered instruments are likely to be of several types: the development of animated agents who might conduct the interview, the addition of some of the diagnostic skills of an interviewer to a CASI instrument, and the ability to accept spoken input from the respondent. Although self-administered applications in which the machine presents a human agent visually would still be considered relatively exotic, video interviewers have been used in some survey experiments (Krysan and Couper, 2003), and animated agents are increasingly feasible. The literature on interviewer effects suggests that the characteristics of animated agents or voices could have an impact on the data if those characteristics were relevant to the topic of inquiry (e.g., "race" of the agent might affect answers if the topic were racial attitudes). However, the physical features of an agent could provide investigators the ability to control interviewer characteristics (e.g., race of "interviewer" could easily be crossed with race of respondent), hold them constant, or randomize them. The copresence of even an animated agent has the potential to be experienced as motivating by the respondent; one effect could be to reduce item nonresponse, for example. Like other SAQs, such an animated agent would be almost perfectly standardized; that is, it would expose all respondents to the same stimulus, increasing the reliability of data compared to interviewer-administered instruments. In addition, an animated agent could slow the pace of the interview, which could improve comprehension and provide more time for retrieval. However, experience with audio-CASI, during which many respondents turn off the audio if that option is available, suggests that enforcing a slow pace could irritate the respondent.

Increasing the validity of data, however, could require using other methods to determine the requirements of enhanced automated interviewing and the "skills" that it would require—with or without an animated agent. Ethnomethodology (EM) and conversation analysis (CA), whose substantive inquiries involve embodied forms of practical knowledge and talk-in-interaction, represent tools for understanding human–machine interaction.[5] Our partial inventory of utterances and actions in the interview

---

[5] For a review of EM and CA see Maynard and Clayman (1991); for a recent overview of studies concerning technology and work see Heath and Button (2002); and for examples of EM and CA studies of human–machine interaction see Suchman (1987), Maynard and Schaeffer (2000), Heath and Luff (2002), and Whalen et al. (2002).

and our extended discussion of reports illustrates the sorts of phenomena that ethnomethodology and conversation analysis can identify, details that are informative in themselves and that also serve as the starting point for other types of inquiry (such as systematic coding and quantitative analysis).

We have not attempted to describe the full range of utterances and actions that occur within the survey interview. But we have used reports to illustrate a type of information that respondents provide in interviews that is not generally encouraged or captured by self-administered instruments. Such information is potentially informative to survey methodologists, for reasons already described, and also to analysts. Although research has not yet identified what meta-information (like that which the respondent provides in a report) is most useful to preserve, or how to incorporate it in analytic models to control for the effects of measurement error, the studies that have been done (some listed earlier) show the promise of such investigations. A challenge faced by an automated interviewing system would be to correctly interpret such reports—for example, that the respondent was expressing some doubt but was clearly leaning in one direction—and to then fashion an appropriate response—for example, to code "probably" as "yes." Automated interviewing systems have faced such challenges, for example, by treating a pause before entering an answer as a sign of trouble and spontaneously offering help or deciding when the length of a pause before answering is a sign of a problem in comprehension (Ehlen et al., 2005; Schober and Bloom, 2004; Schober, et al., 2000). Other chapters in this volume illustrate that the success of automated interviewing and other innovations in the future will depend in part on how well we describe and understand the person-to-person interviews of the present.

APPENDIX: TRANSCRIBING CONVENTIONS[6]

---

1. Overlapping speech A: Oh you do? R[eally] B: [Um hmmm]	Left-hand brackets mark a point of overlap, while right-hand brackets indicate where overlapping talk ends.
2. Silences A: I'm not use ta that. (1.4) B: Yeah me neither.	Numbers in parentheses indicate elapsed time in tenths of seconds.
3. Missing speech A: Are they? B: Yes because ...	Ellipses indicate where part of an utterance is left out of the transcript.
4. Sound stretching B: I did oka::y.	Colon(s) indicate the prior sound is prolonged. More colons, more stretching.
5. Volume A: That's where I REALLY want to go.	Capital letters indicate increased volume.

6. Emphasis
A: I do <u>not</u> want it.

Underline indicates increased emphasis.

7. Breathing
A: You didn't have to
worry about having the
.hh hhh curtains closed.

The ''h" indicates audible breathing.
The more ''h's" the longer the breath.
A period placed before it indicates
inbreath; no period indicates outbreath.

8. Laugh tokens
A: Tha(h)t was really
neat.

The ''h" within a word or sound indicates
explosive aspirations; e.g., laughter,
breathlessness.

9. Explanatory material
A: Well ((cough)) I don't
know

Materials in double parentheses indicate audible phenomena other than actual
verbalization.

10. Candidate hearing
B: (Is that right?) ( )

Materials in single parentheses indicate
that transcribers were not sure about
spoken words. If no words are in parentheses, the talk was indecipherable.

11. Intonation
A: It was unbelievable. I
↑ had a three point six?
I ↓ think.
B: You did.

A period indicates fall in tone, a comma
indicates continuing intonation, a question mark indicates increased tone. Up
arrows (↑) or down arrows (↓) indicate
marked rising and falling shifts in intonation immediately prior to the rise or
fall.

12. Sound cut off
A: This- this is true

Dashes indicate an abrupt cutoff of sound.

13. Soft volume
A: °Yes.° That's true.

Material between degree signs is spoken
more quietly than surrounding talk.

14. Latching
A: I am absolutely
sure.=
B: =You are.
A: This is one thing
[that I=
B: [Yes?
A: =really want to do.

Equal signs indicate where there is no gap
or interval between adjacent utterances.

Equal signs also link different parts of a
speaker's utterance when that utterance
carries over to another transcript line.

15. Speech pacing
A: What is it?
B: >I ain't tellin< you

Part of an utterance delivered at a pace
faster than surrounding talk is enclosed
between ''greater than'' and ''less
than'' signs.

[6] Adapted from Gail Jefferson (1974), Error correction as an interactional resource. *Language in Society*, 2, 181–199.

REFERENCES

Belli, R. F. (1998). The structure of autobiographical memory and the event history calendar: potential improvements in the quality of retrospective reports in surveys. *Memory, 6*(4), 383-406.

Belli, R. F., Lee, E. H., Stafford, F. P., and Chou, C-H. (2004). Calendar and question-list survey methods: association between interviewer behaviors and data quality. *Journal of Official Statistics, 20*(2): 143–84.

Belli, R. F., Shay, W. L., and Stafford, F. P. (2001). Event history calendars and question list surveys: a direct comparison of interviewing methods. *Public Opinion Quarterly, 65*(1): 45–74.

Belli, R. F., and Callegaro, M. (forthcoming). The emergence of calendar interviewing: a theoretical and empirical rationale. In R. F. Belli, F. P. Stafford, and D. F. Alwin (Eds.), *Calendar and Time Diary Methods: Measuring Well-being in Life Course Research.* Thousand Oaks, CA: Sage.

Bingham, W., and Moore, B. (1924). *How to Interview.* New York: Harper & Row.

Brenner, M. (1981a). Aspects of conversational structure in the research interview. In P. Werth (Ed.), *Conversation and Discourse* (pp. 19–40). London: Croom Helm.

Brenner, M. (1981b). Patterns of social structure in the research interview. In M. Brenner (Ed.),*Social Method in Social Life* (pp. 115–158). London: Academic Press.

Cannell, C. F., and Kahn, R. L. (1968). Interviewing. In G. Lindzey and E. Aronson (Eds.), *The Handbook of Social Psychology* (Vol. 2, pp. 526–595). Reading, MA: Addison-Wesley.

Cannell, C. F., Fowler, F. J. Jr., and Marquis, K. H. (1968). *The Influence of Interviewer and Respondent Psychological and Behavioral Variables on the Reporting in Household Interviews.* Washington DC: U.S. Department of Health and Human Services.

Cliff, N. (1959). Adverbs as multipliers. *Psychological Review, 66*(1), 27–44.

Collins, M. (1980). Interviewer variability: a review of the problem. *Journal of the Market Research Society, 22*(2), 77–95.

Conrad, F. G., and Schober, M. F. (2000). Clarifying question meaning in a household telephone survey. *Public Opinion Quarterly, 64*(1), 1–28.

Converse, J. M., and Schuman, H. (1974). *Conversations at Random: Survey Research as Interviewers See It.* Hoboken, NJ: John Wiley & Sons.

Couper, M. P., Singer, E., and Tourangeau, R. (2003). Understanding the effects of audio-CASI on self-reports of sensitive behavior. *Public Opinion Quarterly, 67*(3), 385–395.

Dijkstra, W., Smit, J., and Ongena, Y. (forthcoming). An evaluation study of the event history calendar. In R. F. Belli, F. P. Stafford, and D. F. Alwin (Eds.), *Calendar and Time Diary Methods: Measuring Well-being in Life Course Research.* Thousand Oaks, CA: Sage.

Draisma, S., and Dijkstra, W. (2004). Response latency and (para)linguistic expression as indicators of response error. In S. Presser, J. M. Rothgeb, M. P. Couper, J. T. Lessler, E. Martin, J. Martin, and E. Singer (Eds.), *Methods for Testing and Evaluating Survey Questionnaires* (pp. 131–148). New York: Springer-Verlag.

Drew, P. (1984). Speakers reportings in invitation sequences. In J. M. Atkinson and J. Heritage (Eds.), *Structures of Social Action: Studies in Conversation Analysis* (pp. 129–151). Cambridge: Cambridge University Press.

Dykema, J. (2005). An investigation of the impact of departures from standardized interviewing on response errors in self-reports about child support and other family-related variables. Paper presented at the annual meeting of the American Association for Public Opinion Research, Miami Beach, FL.

Dykema, J., Lepkowski, J. M., and Blixt, S. (1997). The effect of interviewer and respondent behavior on data quality: analysis of interaction coding in a validation study. In L. Lyberg, P. Biemer, M. Collins , E. de Leeuw, C. Dippo, N. Schwarz, and D. Trewin (Eds.), *Survey Measurement and Process Quality* (pp. 287–310). Hoboken, NJ: Wiley-Interscience.

Ehlen, P., Schober, M. F., and Conrad, F. G. (2005). Modeling speech disfluency to predict conceptual misalignment in speech survey interfaces. *Proceedings of the Symposium on Dialogue Modeling and Generation, 15th Annual Meeting of the Society for Text & Discourse.* Vrije Universiteit, Amsterdam.

Fowler, F. J. Jr. (1992). How unclear terms affect survey data. *Public Opinion Quarterly, 56*(2), 218–231.

Fowler, F. J., Jr., and Mangione, T. W. (1990). *Standardized Survey Interviewing: Minimizing Interviewer-Related Error.* Newbury Park: Sage.

Gathman, E. H. C., Maynard, D. W., and Schaeffer, N. C. (2006). The respondents are all above average: compliment sequences in a survey interview. Unpublished manuscript, University of Wisconsin, Madison.

Groves, R. M., and Magilavy, L. J. (1986). Measuring and explaining interviewer effects in centralized telephone surveys. *Public Opinion Quarterly, 50*(2), 251–266.

Hak, T. (2002). How interviewers make coding decisions. In D. W. Maynard, H. Houtkoop-Steenstra, J. van der Zouwen, and N. C. Schaeffer (Eds.), *Standardization and Tacit Knowledge: Interaction and Practice in the Survey Interview* (pp. 449–470). Hoboken, NJ: John Wiley & Sons.

Heath, C., and Button, G. (2002). Editorial introduction to special issue on workplace studies. *British Journal of Sociology, 53*, 157–161.

Heath, C., and Luff, P. (2000). *Technology in Action.* Cambridge: Cambridge University Press.

Hess, J., Singer, E., and Bushery, J. M. (1999). Predicting test–retest reliability from behavior coding. *International Journal of Public Opinion Research, 11*(4), 346–360.

Hyman, H. H. (1954/1975). *Interviewing in Social Research.* Chicago: The University of Chicago.

Jefferson, G. (1984). On the organization of laughter in talk about troubles. In J. M. Atkinson and J. Heritage (Eds.), *Structures of Social Action* (pp. 346–69). Cambridge: Cambridge University Press.

Kovar, M. G., and Royston, P. (1990). Comment on "Interactional troubles in face-to-face survey interviews." *Journal of the American Statistical Association, 85*(409), 246–247.

Krysan, M., and Couper, M. P. (2003). Race in the live and virtual interview: racial deference, social desirability, and activation effects in attitude surveys. *Social Psychology Quarterly, 66*(4), 364–383.

Lavin, D., and Maynard, D. W. (2001). Standardization vs. rapport: respondent laughter and interviewer reaction during telephone surveys. *American Sociological Review, 66,* 453–479.

Mangione, T. W., Fowler, F. J., Jr., and Louis, T. A. (1992). Question characteristics and interviewer effects. *Journal of Official Statistics, 8*(3), 293–307.

Mathiowetz, N. A. (1998). Respondent expressions of uncertainty: data source for imputation. *Public Opinion Quarterly, 62,* 47–56.

Mathiowetz, N. A. (1999). Expressions of respondent uncertainty as indicators of data quality. *International Journal of Public Opinion Research, 11*(3), 289–296.

Maynard, D. W., and Clayman, S. E. (1991). The diversity of ethnomethodology. *Annual Review of Sociology, 17,* 385–418.

Maynard, D. W., and Schaeffer, N. C. (2000). Toward a sociology of social scientific knowledge: survey research and ethnomethodology's asymmetric alternates. *Social Studies of Science, 30*(3), 323–370.

Maynard, D. W., and Schaeffer, N. C. (2002). Standardization and its discontents. In D. W. Maynard, H. Houtkoop-Steenstra, J. van der Zouwen, and N. C. Schaeffer (Eds.), *Standardization and Tacit Knowledge: Interaction and Practice in the Survey Interview* (pp. 449–470). Hoboken NJ: John Wiley & Sons.

Maynard, D. W., and Schaeffer, N. C. (2006). Standardization-in-interaction: the survey interview. In P. Drew, G. Raymond, and D. Weinberg (Eds.), *Talk and Interaction in Social Research Methods* (pp. 9–27). London: Sage.

Moore, R. J. (2004). Managing troubles in answering survey questions: respondents' uses of projective reporting. *Social Psychology Quarterly, 67*(1), 50–69.

Moore, R., and Maynard, D. W. (2002). Achieving understanding in the standardized survey interview: respondents' and interviewers' uses of next-position repair techniques. In D. W. Maynard, H. Houtkoop-Steenstra, J. van der Zouwen, and N. C. Schaeffer (Eds.), *Standardization and Tacit Knowledge: Interaction and Practice in the Survey Interview* (pp. 281–312). Hoboken, NJ: John Wiley & Sons.

Oksenberg, L., Cannell, C. F., and Kalton, G. (1991). New strategies for retesting survey questions. *Journal of Official Statistics, 7*(3), 349–365.

O'Muircheartaigh, C. A. (1991). Simple response variance: estimation and determinants. In P. P. Biemer, R. M. Groves, L. E. Lyberg, N. A. Mathiowetz, and S. Sudman (Eds.), *Measurement Errors in Surveys* (pp. 551–576). Hoboken, NJ: John Wiley & Sons.

Presser, S., Rothgeb, J. M., Couper, M. P., Lessler, J. T., Martin, E., Martin, J., and Singer, E. (Eds.) (2004). *Methods for Testing and Evaluating Survey Questionnaires.* Hoboken, NJ: John Wiley & Sons.

Schaeffer, N. C. (1991). Conversation with a purpose—or conversation? Interaction in the standardized interview. In P. P. Biemer, R. M. Groves, L. E. Lyberg, N. A. Mathiowetz, and S. Sudman (Eds.), *Measurement Errors in Surveys* (pp. 367–392). Hoboken, NJ: John Wiley & Sons.

Schaeffer, N. C. (2000). Asking questions about threatening topics: a selective overview. In A. A. Stone, J. S. Turkkan, C. A. Bachrach, J. B. Jobe, H. S. Kurtzman, and V. S. Cain (Eds.), *The Science of Self-report: Implications for Research and Practice* (pp. 105–122). Mahwah, NJ: Lawrence Erlbaum.

Schaeffer, N. C., and Charng, H. (1991). Two experiments in simplifying response categories: intensity ratings and behavioral frequencies. *Sociological Perspectives, 34*(2), 165–182.

Schaeffer, N. C., and Dykema, J. (2004). A multiple-method approach to improving the clarity of closely related concepts: distinguishing legal and physical custody of children. In S. Presser, J. M. Rothgeb, M. P. Couper, J. T. Lessler, E. Martin, J. Martin, and E. Singer (Eds.), *Methods for Testing and Evaluating Survey Questionnaires* (pp. 475–502). New York: Springer-Verlag.

Schaeffer, N. C., and Maynard, D. W. (1995). Occasioning intervention: interactional resources for comprehension in standardized survey interviews. Paper presented at the International Conference on Survey Measurement and Process Quality, Bristol, England.

Schaeffer, N. C., and Maynard, D. W. (1996). From paradigm to prototype and back again: interactive aspects of cognitive processing in standardized survey interviews. In N. Schwarz and S. Sudman (Eds.), *Answering Questions: Methodology for Determining Cognitive and Communicative Processes in Survey Research* (pp. 65–88). San Francisco, CA: Jossey-Bass Publishers.

Schaeffer, N. C., and Maynard, D. W. (2002). Occasions for intervention: interactional resources for comprehension in standardized survey Interviews. In D. W. Maynard, H. Houtkoop-Steenstra, J. van der Zouwen, and N. C. Schaeffer (Eds.), *Standardization and Tacit Knowledge: Interaction and Practice in the Survey Interview* (pp. 261–280). Hoboken NJ: John Wiley & Sons.

Schaeffer, N. C., Maynard, D. W., and Cradock, R. (1993). Negotiating certainty: uncertainty proposals and their disposal in standardized survey interviews. University of Wisconsin, Madison, Center for Demography and Ecology Working Paper 93–25.

Schaeffer, N. C., and Thomson, E. (1992). The discovery of grounded uncertainty: developing standardized questions about strength of fertility motivation. In P. V. Marsden (Ed.), *Sociological Methodology 1992* (Vol. 22, pp. 37–82). Oxford: Basil Blackwell.

Schegloff, E. A. (1986). The routine as achievement. *Human Studies, 9,* 111–151.

Schober, M. F., and Bloom, J. E. (2004). Discourse cues that respondents have misunderstood survey questions. *Discourse Processes, 38*(3), 287–308.

Schober, M. F., and Conrad, F. G. (1997). Does conversational interviewing reduce survey measurement error? *Public Opinion Quarterly, 61,* 576–602.

Schober, M. F., Conrad, F. G., and Bloom, J. E. (2000). Clarifying word meanings in computer-administered survey interviews. In L. R. Gleitman and A. K. Joshi (Eds.), *Proceedings of the 22nd Annual Conference of the Cognitive Science Society* (pp. 447–452). Mahwah, NJ: Lawrence Erlbaum Associates.

Smit, J. H., Dijkstra, W., and van der Zouwen, J. (1997). Suggestive interviewer behaviour in surveys: an experimental study. *Journal of Official Statistics, 13*(1), 19–28.

Suchman, L. (1987). *Plans and Situated Actions.* Cambridge: Cambridge University Press.

Suchman, L., and Jordan, B. (1990). Interactional troubles in face-to-face survey interviews. *Journal of the American Statistical Association, 85*(409), 232–253.

Tourangeau, R., and Smith, T. W. (1998). Collecting sensitive information with different modes of data collection. In M. P. Couper, R. P. Baker, J. Bethlehem, C. Z. F. Clark, J. Martin, W. L. Nicholls II, and J. M. O'Reilly (Eds.), *Computer Assisted Survey Information Collection* (pp. 431–453). Hoboken, NJ: John Wiley & Sons.

van der Zouwen, J., and Dijkstra, W. (2002). Testing questionnaires using interaction coding. In D. W. Maynard, H. Houtkoop-Steenstra, J. van der Zouwen, and N. C. Schaeffer (Eds.), *Standardization and Tacit Knowledge: Interaction and Practice in the Survey Interview* (pp. 427–448). Hoboken, NJ: John Wiley & Sons.

van der Zouwen, J., and Smit, J. H. (2004). Evaluating survey questions by analyzing patterns of behavior codes and question–answer sequences: a diagnostic approach. In S. Presser, J. M. Rothgeb, M. P. Couper, J. T. Lessler, E. Martin, J. Martin, and E. Singer (Eds.), *Methods for Testing and Evaluating Survey Questionnaires* (pp. 109–130). New York: Springer-Verlag.

Viterna, J. S., and Maynard, D. W. (2002). How uniform is standardization? Variation within and across survey research centers regarding protocols for interviewing. In D. W. Maynard, H. Houtkoop-Steenstra, J. van der Zouwen, and N. C. Schaeffer (Eds.), *Standardization and Tacit Knowledge: Interaction and Practice in the Survey Interview* (pp. 365–401). Hoboken, NJ: John Wiley & Sons.

Whalen, J., Whalen, M., and Henderson, K. (2002). Improvisational choreography in teleservice work. *British Journal of Sociology, 53,* 239–258.

CHAPTER 3

Technology and the Survey Interview/Questionnaire

Mick P. Couper
University of Michigan, Ann Arbor, Michigan

3.1 INTRODUCTION

In this chapter a review is given of some of the recent trends related to the role of technology in survey data collection, whether in interviewer-administered survey interviews or in the completion of self-administered questionnaires. The chapter begins with a brief history of technology developments and how these have changed survey data collection; then a review is given of some of the implications of these trends for the future of the survey interview.

Two related trends are shaping the world of survey data collection. One is the trend toward increased use of technology, whether to enhance and supplement the role of the interviewer with computer-assisted interviewing (CAI), or to replace the interviewer or automate self-administered questionnaires (SAQs). A parallel trend is the move toward self-administration, both as a means to reduce costs relative to interviewer administration, and as a means to control or reduce measurement errors, particularly in the reporting of sensitive topics. The latter trend has been fueled by an accumulation of research over several decades that respondents tend to give more honest answers, particularly on sensitive topics, when they do not have to verbalize these responses to an interviewer. In recent years, the acceleration in the adoption of self-administered methods has been facilitated by technological developments. The marriage of new technologies and increase in self-administration of survey questionnaires has the potential to profoundly change the nature of the survey interview.

Figure 3.1 illustrates some of the survey data collection methods that have evolved in recent years. This set is by no means exhaustive, and there are many variants

Envisioning the Survey Interview of the Future, Edited by Frederick G. Conrad and Michael F. Schober
Copyright © 2008 John Wiley & Sons, Inc.

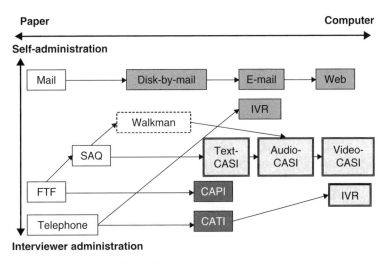

FIGURE 3.1   The evolution of survey technology.

of these main methods. The horizontal axis shows the trend from paper-based surveys to computer-based ones. The vertical axis arrays the methods from fully self-administered to fully interviewer-administered methods, with the interviewer-assisted methods in between.

Some explanation of the acronyms and coding is in order before we proceed. The white boxes represent the main paper-based methods that have been used by survey researchers for many decades. These include mail or postal surveys, face-to-face (FTF) or personal visit surveys, and telephone surveys. Note that self-administered questionnaires (SAQs) are a subset of FTF surveys and are distinct from mail surveys. The latter require no involvement of an interviewer and no direct contact with the respondent. SAQs, on the other hand, are administered as part of a face-to-face survey, with the interviewer present during completion of the SAQ.

The light grey shaded boxes are what are commonly referred to as computerized self-administered questionnaire (CSAQ) methods (e.g., see, Ramos et al., 1998). Not to be confused with SAQs, these are fully self-administered methods that are the logical extension of mail surveys to the Internet. The first attempts to introduce technology involved mailing disks to respondents (hence disk-by-mail or DBM), which they would install on their computers, run the software to complete the questionnaire, and return the disks to the survey organization. This was superseded by e-mail surveys, in which the questionnaire was included in the body of the e-mail message (e.g., see Couper et al., 1999; Schaefer and Dillman, 1998). Both these methods were short-lived and have been fully replaced by Internet or Web surveys (for a review, see Couper, 2000). This is the area of greatest growth in the survey world—and particularly in the market research sector—in recent years.

The methods in white text on dark grey are the computer equivalents of the two main interviewer-administered methods: computer-assisted personal interviewing

(CAPI) and computer-assisted telephone interviewing (CATI). In both cases it is the interviewer using the technology. For CAPI this is generally a laptop computer, while for CATI it is typically a networked desktop computer in a centralized facility. However, decentralized CATI—interviewers conducting telephone interviews from their homes, using the laptop computer—is an increasingly popular option.

The methods in the boxes with double-line borders are self-administered methods conducted in the presence of, or with the aid of, an interviewer, and as such are an outgrowth of SAQs. The three computer-assisted self-interviewing (CASI) methods are variants of self-administered surveys conducted by respondents on a laptop computer, typically in their own homes, in the presence of an interviewer, as part of a CAPI survey. The distinctions between text-, audio-, and video-CASI are described later.

The method labeled "IVR" and appearing in two places in Fig. 3.1 requires some explanation. IVR stands for interactive voice response and refers to the automated delivery of survey interviews over the telephone, in which a respondent listens to a prerecorded voice reading the survey questions, and answers either by speaking his/her answers (voice recognition) or, more commonly, by pressing numbers on the telephone keypad (touchtone data entry). IVR can be respondent-initiated, where the respondent dials a toll-free number and responds directly without interviewer intervention. IVR can also be interviewer-initiated, where the call is placed by an interviewer who administers some of the questions (using CATI) before switching the respondent to the IVR system. For a review of IVR, see Tourangeau, Steiger, and Wilson (2002) and Steiger and Conroy (in press). IVR thus belongs both as a fully self-administered approach (along with Web surveys) and as a "CASI" method with interviewer assistance. To add to the confusion, IVR has also been variously known as touchtone data entry (TDE), voice recognition entry (VRE), automatic speech recognition (ASR), and even telephone audio computer-assisted self-interviewing (T-ACASI) (see Blyth, 1998; Cooley et al., 2000; Phipps and Tupek, 1991). The latter term was developed to emphasize the link to audio-CASI, with the unfortunate effect of developing parallel literatures on the same technique. As automated telephone interviewing further evolves, we will need to make further distinctions between the use of prerecorded human voices or computer-generated (i.e., text to speech or TTS) voices (Couper et al., 2004).

Finally, the "Walkman" box in Fig. 3.1 represents a transition technology (like DBM and e-mail surveys) that could be viewed as the precursor to audio-CASI (Camburn and Cynamon, 1993; Camburn et al., 1991). In the "Walkman" approach, survey questions were prerecorded onto an audio tape. These were then played over headphones on a portable cassette player, with the respondent marking an answer sheet devoid of the survey questions (and thus not providing any context for the answers to a third party). With the development of multimedia-capable laptop computers, this approach was soon replaced by audio-CASI.

Following this brief primer on the arcane and often confusing world of survey modes and acronyms, let's turn to a broader discussion of the two main trends that are the focus of this chapter, and their implications for survey research.

3.2   THE TECHNOLOGY TREND

As is already obvious from Fig. 3.1 and the preceding discussion, survey data collection has undergone several technological transformations in recent decades. The survey profession has been relatively quick to adopt new technologies and apply them to data collection. Indeed, some of the methods described earlier even anticipated the development of suitable technology. A few examples will suffice.

The first computer-assisted telephone interviewing (CATI) systems predated the advent of the personal computers (PCs) that are the mainstay of all CATI facilities today. For example, the first nationwide telephone facility in the United States was established in 1966; the first CATI survey was conducted in 1971, using a mainframe-based system (see Couper and Nicholls, 1998). In fact, the original patent for CATI involved a wiring diagram as well as software.

The concept of conducting face-to-face interviews with portable computers, and the acronym CAPI, were well understood in the late 1970s (see Shanks, 1983), well before the advent of viable technologies for implementation. The first tests of CAPI in the United States asked interviewers to carry around 20–25 pound (9–11 kg) computer terminals, while small hand-held computers were being explored in Europe, both in the early 1980s. The first national survey to use CAPI was the Dutch Labor Force Survey, in 1987 (van Bastelaer et al., 1988), with widespread and rapid growth in several countries following that successful implementation.

Despite the early adoption, it was recognized that the portable computers available for CAPI at the time were limited by weight (up to 15 pounds or 6.8 kg), screen visibility, capacity, and battery life. Based on a series of ergonomic studies in the early 1980s, Statistics Sweden (see Lyberg, 1985) determined the ideal weight for a hand-held CAPI computer to be about 2.2 pounds (or 1 kg). Statistics Sweden unsuccessfully attempted to contract for the manufacture of such a device. In a similar study, Couper and Groves (1992) found the mean weight comfortably held by interviewers to be around 3 lb 10 oz (1.64 kg). At that time, the weight of the state-of-the-art laptop and tablet PCs evaluated ranged from 4 lb 4 oz (1.93 kg) to 7 lb 3 oz (3.26 kg). It took over a decade before tablet PCs in this weight range and running a standard operating system were available on the market.

A similar adoption trend occurred with audio-CASI. The idea for audio-CASI was percolating before the availability of suitable hardware, as evidenced in the Walkman approach described previously. The first audio-CASI systems were tested on a Mac Powerbook PC, which had sound capabilities (Johnston and Walton, 1995) and on a DOS-based laptop PC requiring a separate digital audio device connected to the PC (Cooley et al., 1996; O'Reilly et al. 1994). With the advent of multimedia capable laptops running MS Windows, audio-CASI has become entirely software driven.

The widespread use of online panels for data collection was similarly anticipated by the development of the Telepanel in The Netherlands (Saris, 1998). The first efforts to use automated systems for direct data collection from respondents date back to Clemens (1984), who used the British Videotex system to conduct surveys. The Dutch Telepanel, which initially used PCs placed in respondents' homes with dial-up connections by modem to conduct weekly surveys, has been in operation since 1986.

Early tests of Internet surveys occurred almost simultaneously with the development of the World Wide Web, and Web surveys are at least as old as e-commerce, if not older.

But there are others who argue that survey data collection is a conservative activity, being slow to adapt to changing circumstances and embrace technological developments. Those who do so generally point to large-scale government surveys, where concerns about affecting a long time series of data may lead to overly cautious adoption of new methods. This hasn't stopped agencies from exploring or evaluating new methods, but the burden of proof required for adoption tends to produce longer delays in making changes to ongoing survey processes. For each of the above examples of early adoption, one can point to an organization or agency that is still conducting surveys using paper or is clinging to DOS-based interviewing systems. Progress may be uneven, but the march toward increased technology use in survey data collection seems inexorable.

Costs and stability of the technology are limiting factors in the adoption of new technologies. An example of this is tablet PCs. The value of the tablet for CAPI applications, especially those involving doorstep interviews, has long been recognized, but the cost of tablet PCs has generally worked against their widespread adoption. Furthermore, there are potential risks of being an early adopter—for example, RTI International purchased a large number of Apple Newton hand-helds for doorstep screening, just before the Newton was discontinued.

There is a continuing tension between the adoption of technology for technology's sake, at one end of the continuum, and the desire to delay any change until all possible kinks have been worked out and sufficient evidence has accumulated for the advantage of the new technology, at the other. This tension is good for the field. Technology is not the end product of what we do—we use technology in the service of survey data collection. New methods must prove their worth by the established criteria of cost and error reduction, but openness to new methods is essential for improving the quality of survey data and the efficiency of data collection.

The ambivalence about online data collection reflects this tension. Some early adopters argued that Internet surveys would in a few short years replace all other methods of survey data collection—in much the same way that early proponents of CATI and later CAPI made similar claims (see Couper, 2000). While this has yet to happen, a great deal of research energy is being focused on Web surveys, and how best to use them in the service of high quality data collection. For many populations, and for a growing number of topics, Web surveys have indeed become the method of choice. They are also increasingly being used to supplement other modes in mixed-mode data collection (Link and Mokdad, 2005; Schonlau et al., 2003), raising fresh challenges for survey designers and methodologists. Thus, while the early claims may appear exaggerated or at least premature, online surveys are very much part of the landscape of survey data collection.

It is fair to characterize technology adoption in surveys as more evolutionary than revolutionary. Many of the current developments were anticipated in earlier technologies, or are responses to needs identified before. Only in some domains have we seen demonstrable improvements in data quality (e.g., CASI) or efficiency (e.g., Web surveys). Most of the other technologies adopted for survey data collection have brought relatively modest gains. In summarizing the extant research evidence on computer-assisted interviewing (CAI), Nicholls, Baker, and Martin (1997, p. 241) concluded that "although moving a survey from P&P to CAI generally improves its data quality, these improvements have not been as large, broad, or well documented as the early proponents of CAI methods anticipated." But, as they also note, "none of the dire predictions made about CATI and CAPI when they were first introduced proved true after they passed from feasibility testing to production." Survey complexity has indeed increased, and the back-end processing time drastically reduced. Similarly, some things are able to be done with the new technologies that couldn't be done—or were very hard to do—with paper-based methods. And technologies such as CASI, IVR, and the Web have led to a resurgence of interest in self-administered surveys. Overall, while technology has not solved the major survey problems of the day, it is nonetheless steadily transforming the nature of survey data collection and changing the character of the survey interview.

3.3   THE SELF-ADMINISTRATION TREND

It has long been recognized that self-administration improves the quality of reporting on sensitive topics (Sudman and Bradburn, 1974). The growth in self-administration has largely been fueled by the development of technological solutions to some of the problems limiting the use of self-administered methods. In this section we focus on the two self-administered modes that have received the greatest research attention in recent years, namely, Web surveys and CASI.

Web surveys are fully self-administered, in that no interviewer is present to gain cooperation from the respondent, assist with the technology or questionnaire, and influence the respondent in any way, either positive or negative. Major concerns about issues of sample representation have been—and continue to be—raised about Web surveys (e.g., see Couper, 2000; Fricker and Schonlau, 2002), but these are not the focus of this chapter. From a measurement perspective, the Web offers the ability to deliver complex survey instruments to a widely dispersed sample. The interactivity of the Web allows for real-time edit checks, routing, fills, randomization, customized feedback, and all the other features of a computer-assisted interview. The rich visual content and graphical nature of the Web allows for the delivery of survey content and ways of answering that go well beyond what is possible with traditional paper-based mail surveys (e.g., visual analog scales, images, running tallies; see Couper, 2005). The measurement possibilities are expanding, and researchers are studying the best ways to exploit this rich new measurement medium.

Turning to computer-assisted self-interviewing (CASI) in its various forms, the use of laptop computers for computer-assisted personal interviewing facilitated the rapid expansion of CASI methods, with the same hardware and software being used for both the CAPI and CASI portions of the interview, often with minor design variations. In text-CASI, the respondent reads the questions displayed on the screen and answers by typing in a number or clicking a selection with the mouse. Audio-CASI adds a prerecorded voice of an interviewer reading the survey questions, typically listened to over headphones to maximize privacy. But most audio-CASI systems use both text and audio, with a large proportion of respondents eschewing the audio channel and choosing to complete the interview (faster) using only text (Couper and Tourangeau, 2006). Video-CASI extends the multimedia capability of modern laptop computers to present full-screen video stimulus material to respondents (Farley et al., 2005; Krysan and Couper, 2003). There is no reason why similar stimulus material (text+audio, or video) could not soon be delivered over the Web. Bandwidth and compatible media formats remain obstacles, but these are likely to be overcome as the Web matures. On the other hand, thus far there is no strong evidence that adding audio to text-CASI produces sufficient gains in data quality to justify the additional cost and effort (see Couper et al., 2004; Couper and Tourangeau, 2006). While multimedia applications are certainly very valuable for some domains (e.g., advertising research), there is as yet no evidence that it is useful or necessary for all types of surveys—but Johnston (Chapter 7 in this volume) points to some of the exciting possibilities offered by multimodal interfaces.

Both online surveys and CASI methods are transforming the way surveys are conducted. Web surveys are increasingly replacing mail surveys, especially for populations with high rates of Internet penetration, such as students, members of professional associations, and the like. The exciting measurement possibilities, coupled with the rapid speed and relatively low cost with which Web surveys can be conducted, make them an attractive alternative. Similarly, CASI modules are an increasingly common part of interviewer-administered surveys, used primarily for sensitive questions, but also for any other types of questions where interviewer effects—the influence of who asks the questions on the answer obtained—may be a concern. With IVR or T-ACASI, self-administered modules are also being incorporated into interviewer-administered surveys implemented over the telephone.

The rapid rise of these methods—particularly the Web—has led some to claim that survey interviewers are becoming obsolete. However, as far back as 1984—well before the Web existed or audio files could be played with mainstream software—this question was being posed (see Clemens, 1984). For some, the elimination of interviewers from the survey process represents the inevitable progression of technology. The self-administration trend is viewed as reducing costs, improving quality, and removing interviewer effects from the equation. My view is that these methods may supplement, but not necessarily replace, traditional forms of survey interviewing. While self-administration may help reduce measurement error, there is little evidence that other sources of survey error (especially coverage and nonresponse)[1] are reduced. Indeed, some types of surveys—such as exit interviews in election studies—are hard to imagine in a fully self-administered form.

3.4 IMPLICATIONS OF THESE TRENDS

Having briefly reviewed some of the technical developments to date in survey data collection, we now turn to a discussion of the implications of these trends for how

---

[1] See Groves et al. (2004) for definitions of these terms.

surveys are designed and conducted. The focus is on three key issues: customization, control, and social presence. These are by no means exhaustive, but they are illustrative of the challenges facing survey researchers as we not only envision, but actively make progress toward, the survey interview of the future.

3.4.1   Customization

Customization involves tailoring the survey instrument to create an individualized experience for the respondent. While trained interviewers are quite adept at following complex routing instructions and customizing question wording on the fly using paper questionnaires, computerization automates this process, potentially making the question delivery process smoother and less error-prone. Customization is hard to execute in self-administered paper questionnaires, and CASI and Web surveys make this much easier.

There are different forms of customization, for example, presenting certain questions to some respondents and not to others (skips or routing), varying the order or form in which questions or response options are presented to respondents (randomization), and varying the content of the questions themselves to reflect answers to prior questions or other knowledge about the respondent (fills).

Figure 3.2 shows an example of skip instructions from the 1992 round of the National Longitudinal Survey of Labor Market Experience, Youth Survey (NLS/Y). This survey was one of the first to convert to CAPI in the United States. While interviewers generally made few skip errors on paper, the automation of these tasks reduced the error rates even further (see Baker et al., 1995. Olsen, 1992).

An example of a question using fills is shown in Fig. 3.3. This example is from the National Medical Expenditure Survey (NMES). In paper-based surveys, interviewers would be trained to adapt the question wording appropriately, whereas again in CAI, this is taken care of by the software. Self-administered paper surveys (whether mail or SAQ) generally include little or no customization, but executing a question like that in Fig. 3.3 on the Web or in a CASI instrument is a relatively easy task.

These and other forms of customization (such as dependent interviewing; see Hill, 1994; Lynn et al., 2004) are widely used in computerized instruments, whether interviewer-administered or self-administered. Currently, there is no research on whether it improves the flow of the interview or instrument. Customization requires more programming resources and a greater need for testing (see Baker et al., 2004). But there is no evidence that this investment is worth it in terms of the interview experience or data quality. In addition, are there costs associated with poorly executed customization? For example, is it better to have a nonadaptive instrument than a highly adaptive one that makes errors? Fig. 3.4 shows an example of customization—in this case, fills—that was not executed correctly in a Web survey. To what extent do respondents notice these lapses? And, if so, to what extent do they affect the answers to the questions or indeed the interview process itself?

Customization may raise issues of control, that is, who is controlling the flow of the interview, and flexibility in determining what is asked. For example, with the branching controlled by the software, neither interviewers nor respondents have the

SECTION 10:   CHILDCARE

1. | ASK WOMEN ONLY: |

2.   <u>INTERVIEWER</u>:   ARE ANY CHILDREN LISTED ON THE **CHILDREN'S RECORD FORM, <u>PART A</u>?**

   YES 1

   NO . . . .   (SKIP TO Q.27, PAGE 10-138) 0   26-27/

3.   <u>INTERVIEWER</u>:   DO WE NEED TO ASK CHILDCARE QUESTIONS FOR ANY CHILDREN
   LISTED ON THE **CHILDREN'S RECORD FORM, <u>PART A</u>?** (ARE QS. 6, 13,
   OR 20 PREPRINTED UNDER "CHILDCARE"?)

   YES . . . . ;.   (GO TO A) 1

   NO . . . .   (GO TO B) 0   28-29/

   A.   <u>INTERVIEWER</u>:   NOTE THE FOLLOWING INSTRUCTIONS BEFORE BEGINNING THE
      REST OF THE CHILDCARE QUESTIONS.

      ENTER NAME AND ID # ON TOP OF COLUMNS IN Q.4 ON PAGE 10-114
      FOR EACH CHILD LISTED ON THE **CHILDREN'S RECORD FORM,
      PART A** FOR WHOM WE NEED TO ASK CHILDCARE QUESTIONS, THEN
      GO TO B.

   B.   <u>INTERVIEWER</u>:   WAS THERE A LIVE BIRTH SINCE DATE OF 1988 OR PRIOR
      INTERVIEW?

      YES . . . .   (SKIP TO Q.4) 1

      NO . . . .   (ASK C) 0   30-31/

   C.   <u>INTERVIEWER</u>:   IS Q.3 OR Q.3B CODED 1 -- "YES"?

      YES . . . .   (GO TO Q.4) 1

      NO . . . .   (SKIP TO Q.27, PAGE 10-138) 0   32-33/

FIGURE 3.2   Example of skip instructions from NLS/Y.

opportunity to override erroneous skips, nor do they see the logic underlying the flow decision executed by the computer. This places greater onus on the designer to accommodate all eventualities and ensure correct routing through the instrument. Customization may also increase the sense of presence. The more the instrument appears to react to the answers provided by the respondent (e.g., using feedback, fills based on prior answers), the more the respondent may become immersed in the interaction, potentially creating the perception of a more animate (interactive) agent.

3.4.2   Control

The second related issue in computerized surveys is that of control. One advantage of CAI mentioned by early proponents was that greater control could be exercised over

Paper and pencil version of question:

Now I'd like to ask you some questions about [your/PERSON'S] (current/last) (main) job. [(Are/were) you/(Is/Was) (PERSON) self-employed, or (do/did) you (does/did) (he/she) work for someone else [at your/his/her) (current/last (main) job?]

How this question may appear for one respondent in CAI:

Now I'd like to ask you some questions about Mary's last job. Was Mary self-employed, or did she work for someone else at her last job?

FIGURE 3.3 Example of fills in paper and pencil and CAI surveys.

interviewers, thereby increasing standardization (see Couper and Nicholls, 1998). Flow, routing, fills, and other customization decisions are taken out of their hands, and they simply deliver the question presented on the screen and record the response, with the system taking care of the rest. Similar arguments have been made for CSAQ and CASI methods (Ramos et al., 1998; Turner et al., 1998).

As with customization, one can view the degree of control as a continuum, with paper surveys exerting little control over interviewers and respondents, while CAI instruments typically exert a great deal of control, unless the instrument has been specifically programmed to permit "nonstandard" behaviors. Respondents completing a paper survey could check multiple responses, even if instructed to select only one; they could mark outside the boxes or add additional comments; they could answer the questions in any order, ignoring branching instructions, or choosing not to answer some questions at all. Some of these behaviors are shown in the excerpt from a mail survey shown in Fig. 3.5.

These behaviors are often not permissible in a Web survey, for example, unless explicitly allowed for in the design. Typically, a set of radio buttons would be used to force the selection of one and only one response. If this were a Web survey, the

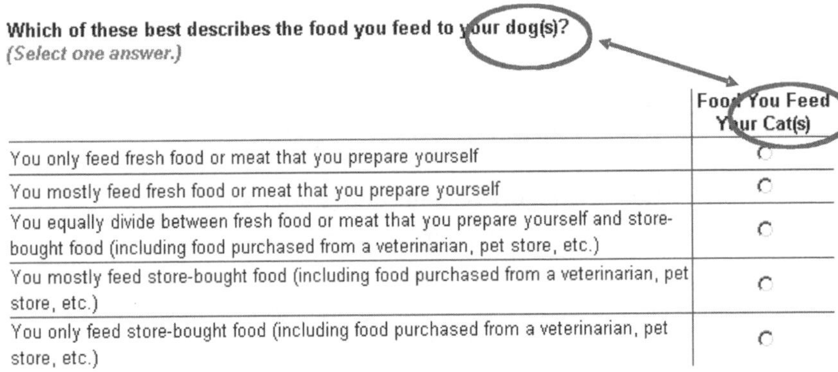
Which of these best describes the food you feed to your dog(s)?
(Select one answer.)

	Food You Feed Your Cat(s)
You only feed fresh food or meat that you prepare yourself	○
You mostly feed fresh food or meat that you prepare yourself	○
You equally divide between fresh food or meat that you prepare yourself and store-bought food (including food purchased from a veterinarian, pet store, etc.)	○
You mostly feed store-bought food (including food purchased from a veterinarian, pet store, etc.)	○
You only feed store-bought food (including food purchased from a veterinarian, pet store, etc.)	○

FIGURE 3.4 Example of customization error.

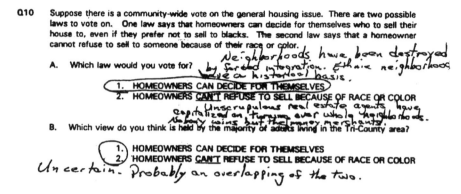

Q10 Suppose there is a community-wide vote on the general housing issue. There are two possible laws to vote on. One law says that homeowners can decide for themselves who to sell their house to, even if they prefer not to sell to blacks. The second law says that a homeowner cannot refuse to sell to someone because of their race or color.

Neighborhoods have been destroyed by forced integration. Ethnic neighborhoods have a historical basis.

A. Which law would you vote for?

1. HOMEOWNERS CAN DECIDE FOR THEMSELVES
2. HOMEOWNERS CAN'T REFUSE TO SELL BECAUSE OF RACE OR COLOR

Unscrupulous real estate agents have capitalized on turning over whole neighborhoods. Nobody wins but the money merchants.

B. Which view do you think is held by the majority of adults living in the Tri-County area?

1. HOMEOWNERS CAN DECIDE FOR THEMSELVES
2. HOMEOWNERS CAN'T REFUSE TO SELL BECAUSE OF RACE OR COLOR

Uncertain. Probably an overlapping of the two.

FIGURE 3.5   Paper questionnaires have low control.

respondent in Fig. 3.5 would be forced to choose one of the existing responses or abandon the survey, and would likely not have an opportunity to elaborate on the response chosen. CASI instruments similarly constrain respondents' behaviors to those permitted by the designer.

Some may argue that constraining the respondent is a good thing, producing cleaner data. But is there any downside to such control? There is some evidence that missing data and errors of omission and commission are reduced with automated instruments (Nicholls et al., 1997; Olsen, 1992). But other types of control may have detrimental effects on the quality of the answers obtained. Fig. 3.6 shows a Web survey with a high degree of control, forcing the respondent to provide an answer. This may lead to a power struggle between instrument and respondent.

You may not leave this answer blank.

In which zip code do you live?

You may not leave this answer blank.

Which category best describes your age?

You may not leave this answer blank.

FIGURE 3.6   High control in a Web survey.

However, control is not an intrinsic feature of the technology. For example, Web surveys can be designed to resemble paper surveys, permitting respondents to answers questions in any order (Peytchev et al., 2006). Item nonresponse can be dealt with by requiring an answer (as in Fig. 3.6), by prompting a respondent to provide an answer, or by permitting a question to be skipped. Indeed, deRouvray and Couper (2002) demonstrated that a Web survey that emulates the behavior of an interviewer—prompting once for an answer, then accepting the nonresponse—may be the optimal strategy for reducing missing data. In addition, optional comment fields could be added to each item. The degree of control exercised in various Web surveys reflects more the design philosophies of the organization that developed the software and/or implemented the survey than the inherent properties of the medium.

An important research question when thinking about the future of survey interviewing is what effect various types and levels of control have on the quality of survey data. The survey interview is a structured interaction, and not a conversation. But it is also not an interrogation. How do we use technology to achieve a goal of standardized survey interviewing while still allowing respondents the flexibility to express their views or provide information in a form most appropriate to their own circumstances? How we design computerized survey instruments may reflect our beliefs about the nature of survey interviews. The instruments should be viewed as tools by which we achieve our goals, rather than as determinants of the design. In other words, the hardware and software we use should not dictate the design of computerized survey instruments. Furthermore, the degree of control exercised over a trained interviewer may differ from that exercised over a respondent completing a self-administered survey, and may vary according to the nature of the questions and the complexity of the survey.

3.4.3 Social Presence

The third ramification of the technology and self-administration trends is the potential for increased social presence (see Hancock, Chapter 9 in this volume). A key reason for the development of self-administered methods (other than cost) is to remove the interviewer from the equation, presumably to minimize the (negative) effects of interviewer presence on survey responses. There is evidence in the literature that interviewer attributes such as race and gender affect responses to questions related to those attributes (Kane and Macauley, 1993; Schaeffer, 1980). Furthermore, the presence of an interviewer has been found to produce less honest reporting of socially sensitive information, hence the development of CASI methods (Tourangeau and Smith, 1996; Turner et al., 1998). But as automated self-administered questionnaires (whether Web, IVR, or CASI) increasingly use humanizing cues, such influences may well reappear. For example, the addition of a human voice to audio-CASI or IVR may trigger response effects such as socially desirable responding, acquiescence, or deference[2] (see Krysan and Couper, 2005). Web surveys increasingly make use of images and interactive elements that may connote a sense of presence, whether in

---

[2] For a discussion of acquiescence and deference, see Schuman and Presser (1981, Chapter 8).

the sense of a human somewhere else (computer-mediated communication or CMC) or directly with the computer itself (human–computer interaction or HCI) (e.g., see Tourangeau et al. 2003).

The "computers as social actors" (CASA) paradigm, most closely associated with Clifford Nass at Stanford, suggests that even minimal social interface cues can engender presence. For example, Nass, Moon, and Green (1997, p. 874) write that "the tendency to gender stereotype is deeply ingrained in human psychology, extending even to machines. . . . when voice technology is embedded in a machine interface, voice selection is highly consequential" (see also Nass et al., 1997, 2001, 2003). In contrast to this view, Turner et al. (1998) write in the audio-CASI context that "even in sex surveys, the gender of the voice is unimportant." Which of these is true for survey interviews?

We have conducted several studies on social presence in surveys, focusing on new technology and the measurement of sensitive issues. For example, we have examined social presence in CASI surveys, focusing on interviewer voice in audio-CASI (Couper et al., 2003) and using "virtual interviewers"—videos of interviewers reading survey questions, played on a computer screen—to explore race-of-interviewer effects (Krysan and Couper, 2003). We have also examined gender-of-voice effects in IVR surveys (Couper et al., 2004; Tourangeau et al., 2003), and various manipulations of social presence on the Web using images and feedback (Krysan and Couper, 2005; Tourangeau et al., 2003). An example of social presence from the latter study is shown in Fig. 3.7.

Generally, our results on the effect of presence on the responses to sensitive questions have been quite underwhelming, suggesting that social presence may not be of as much concern in automated surveys as implied by the work of Nass and his colleagues. One reason for the failure to replicate their findings may lie in the different

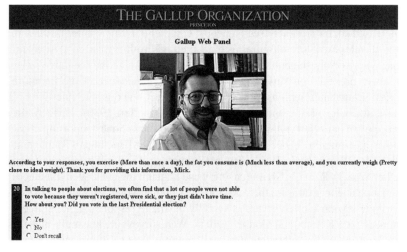

FIGURE 3.7   Social presence in a Web survey.

demands of laboratory-based experiments and surveys—the degree of cognitive and emotional engagement may differ in the two settings, and a variety of other (potentially stronger) cues may overwhelm the effect of the presence manipulation in the less-controlled survey setting. Nonetheless, this remains an intriguing area for further research to figure out why the laboratory studies do not appear to replicate in the survey setting, or to understand the conditions under which social presence may be triggered in computer-assisted surveys. As animated agents become increasingly more life-like, and as their use in surveys becomes more widespread, concerns about the reintroduction of "interviewer" effects may be raised (see Cassell and Miller, Chapter 8 in this volume).

3.5 DISCUSSION

Three possible implications of the increasing use of computer technology and the trend toward self-administration in surveys have been highlighted. These relate to customization, control, and social presence, which are interrelated issues.

For example, social presence may interact with customization. Will we be more forgiving if an automated agent makes an error if it is less human? In other words, the more "human" the agent, and the greater social presence conveyed, the greater the expectations we may have of the agent. Alternatively, we may assume that machines are less fallible than humans (to err is human, after all), and we may be more tolerant of human than machine errors. This is an area ripe for future research.

Similarly, social presence may also affect how respondents react to control. Again, will respondents be more tolerant of highly controlled interactions when politely informed that an action is not permissible by an animated agent, or whether the software rejects an answer with a terse message? Will the increasing "humanness" of computer interfaces raise our expectations about the ability to negotiate an answer?

These three dimensions may involve trade-offs. Increasing customization comes at the cost of increased time and effort, and increased risk of an error. Increased control may produce more "complete" data, but it may not necessarily be more accurate, and it may frustrate respondents to the point of abandoning the survey. Increased social presence may make the interview a more interesting and enjoyable experience, but may lengthen the survey interview, and may engender some of the very effects we are trying to reduce with self-administration.

As we continue to explore the new technologies discussed in this volume, and begin to implement these in our surveys, we need to examine how they change the very nature of what constitutes a survey "interview." The distinction between interviews and questionnaires is becoming blurred. In the lexicon of survey research, an interview generally implies an interaction with another person (the interviewer). The term "questionnaire" is used to refer either to a self-administered instrument or to the instrument an interviewer uses to conduct an interview. Does an interaction with an avatar count as an interview? Does the person have to be live (i.e., is audio-CASI with a prerecorded human voice an interview)? How interactive does an automated instrument need to be, and what degree of "presence" is needed before we consider it akin to an interview? This is not just a question of terminology or labels, but about what those labels imply in terms of the nature of the interaction or task.

With the rise in self-administered surveys in recent years, research on interviewers seems to have faded into the background. The introduction of virtual interviewers, animated agents, and the like may well rekindle research interest in the role of the interviewer (whether present or not, real or otherwise) in affecting how respondents answer survey questions. Not only does this raise new questions about "interviewer effects," but it also presents research opportunities in the sense that interviewer behavior can be more tightly controlled (from an experimental viewpoint) and elements of the interviewer's role or task examined in greater detail.

In my view, we should not rush to adopt new technology simply for the sake of doing so. We are first and foremost survey researchers. We do best when we use the technology to address particular challenges in survey data collection. For example, the Knowledge Networks approach of giving respondents Web access in exchange for participation in surveys was an attempt to address the coverage problems associated with Web surveys (see Couper, 2000). In like fashion, audio-CASI was developed to address concerns about both privacy and literacy, in an effort to improve reporting on sensitive issues. Technologies are being considered to solve many of the other pressing challenges of the day, from coverage and nonresponse, through to measurement error and the need to improve efficiency of survey data collection.

While many exciting new technologies are appearing, we should carefully consider their use before wholesale adoption. The challenge is to find better ways to use the technology in the service of better quality data (for us and our clients), a better experience (for our respondents), and greater efficiency (both for us and for our respondents). To do this, we need to better understand what the desirable elements of a survey interview are, and to learn how to replicate them in an increasingly automated, self-administered data collection world. Moving from interviewer-administration to self-administration may eliminate or reduce some of the "bad things" that interviewers may do (such as leading questions[3], directed probes[4], and social desirability effects) but it may also reduce some of the "good" things interviewers do (such as persuading, motivating, probing, and clarifying). In order to use the new technologies to improve the survey interview, we first need to have a better sense of what is "good" and what is "bad" about the interview process.

We are up to the challenge. As long as we have survey researchers who are willing to take risks, and as long as we have funders who are tolerant and supportive of such innovation and exploration, whether separate from or integrated with ongoing survey efforts, the profession will continue to develop and implement ways of using technology to improve survey data collection.

---

[3] Questions that suggest a preferred answer.
[4] See Schaeffer and Maynard, Chapter 2 in this volume.

REFERENCES

Baker, R. P., Bradburn, N. M., and Johnson, R. A. (1995). Computer-assisted personal interviewing: an experimental evaluation of data quality and costs. *Journal of Official Statistics, 11*(4), 415–431.

Baker, R., Crawford, S., and Swinehart, J. (2004). Development and testing of Web questionnaires. In S. Presser, J. Rothgeb, M. P. Couper, J. Lessler, E. A. Martin, J. Martin, and E. Singer (Eds.), *Methods for Testing and Evaluating Survey Questionnaires* (pp. 361–384). Hoboken, NJ: John Wiley & Sons.

Blyth, W. G. (1998). Current and future technology utilization in European market research. In M. P. Couper, R. P. Baker, J. Bethlehem, C. Z. F. Clark, J. Martin, W. L. Nicholls II, and J. O'Reilly (Eds.), *Computer Assisted Survey Information Collection* (pp. 563–581). Hoboken, NJ: John Wiley & Sons.

Camburn, D., and Cynamon, M. L. (1993). Observations of new technology and family dynamics in a survey of youth. *Proceedings of the American Statistical Association, Social Statistics Section* (pp. 494–502). Alexandria, VA: ASA.

Camburn, D., Cynamon, D., and Harel, Y. (1991). The use of audio tapes and written questionnaires to ask sensitive questions during household interviews. Paper presented at the National Field Technologies Conference, San Diego, CA.

Clemens, J. (1984). The use of View Data Panels for data collection. In *ESOMAR Seminar on Are Interviewers Obsolete? Drastic Changes in Data Collection and Data Presentation* (pp. 47–65). Amsterdam: ESOMAR.

Cooley, P. C., Turner, C. F., O'Reilly, J. M., Allen, D. R., Hamill, D. N., and Paddock, R. E. (1996). Audio-CASI; hardware and software considerations in adding sound to a computer-assisted interviewing system. *Social Science Computer Review, 14*(2), 197–204.

Cooley, P. C., Miller, H. G., Gribble, J. N., and Turner, C. F. (2000). Automating telephone surveys: using T-ACASI to obtain data on sensitive topics. *Computers in Human Behavior, 16*, 1–11.

Couper, M. P. (2000). Web surveys: a review of issues and approaches. *Public Opinion Quarterly, 64*(4), 464–494.

Couper, M. P. (2005). Technology trends in survey data collection. *Social Science Computer Review, 23*(4), 486–501.

Couper, M. P., Blair, J., and Triplett, T. (1999). A comparison of mail and e-mail for a survey of employees in federal statistical agencies. *Journal of Official Statistics, 15*(1), 39–56.

Couper, M. P., and Groves, R. M. (1992). Interviewer reactions to alternative hardware for computer-assisted personal interviewing. *Journal of Official Statistics, 8*(2), 201–210.

Couper, M. P., and Nicholls, W. L. II (1998). The history and development of computer-assisted survey information collection. In M. P. Couper, R. P. Baker, J. Bethlehem, C. Z. F. Clark, J. Martin, W. L. Nicholls II, and J. O'Reilly (Eds.), *Computer Assisted Survey Information Collection* (pp. 1–21). Hoboken, NJ: John Wiley & Sons.

Couper, M. P., Singer, E., and Tourangeau, R. (2003). Understanding the effects of audio-CASI on self-reports of sensitive behavior. *Public Opinion Quarterly, 67*(3), 385–395.

Couper, M. P., Singer, E., and Tourangeau, R. (2004). Does voice matter? An interactive voice response (IVR) experiment. *Journal of Official Statistics, 20*(3), 551–570.

Couper, M. P., and Tourangeau, R. (2006). Taking the audio out of audio-CASI. Paper presented at the European Conference on Quality in Survey Statistics, Cardiff.

deRouvray, C., and Couper, M. P. (2002). Designing a strategy for capturing "respondent uncertainty" in Web-based surveys. *Social Science Computer Review, 20*(1), 3–9.

Farley, W. R., Krysan, M., and Couper, M. P. (2005). Racial residential segregation in rust belt metropolises: traditional and new approaches to studying its causes. Paper presented at the annual meeting of the American Sociological Association, Philadelphia, PA.

Fricker, R. D., and Schonlau, M. (2002). Advantages and disadvantages of Internet research surveys: evidence from the literature. *Field Methods, 14*(4), 347–365.

Groves, R. M., Fowler, F. J. Jr., Couper, M. P., Lepkowski, J. M., Singer, E., and Tourangeau, R. (2004). *Survey Methodology.* Hoboken, NJ: John Wiley & Sons.

Hill, D. H. (1994). The relative empirical validity of dependent and independent data collection in a panel survey. *Journal of Official Statistics, 10*(4), 359–380.

Johnston, J., and Walton, C. (1995). Reducing response effects for sensitive questions: a computer-assisted self interview with audio. *Social Science Computer Review, 13*(3), 304–309.

Kane, E. W., and Macauley, L. J. (1993). Interviewer gender and gender attitudes. *Public Opinion Quarterly, 57*(1), 1–28.

Krysan, M., and Couper, M. P. (2003). Race in the live and virtual interview: racial deference, social desirability, and activation effects in attitude surveys. *Social Psychology Quarterly, 66*(4), 364–383.

Krysan, M., and Couper, M. P. (2005). Race-of-interviewer effects: What happens on the Web? *International Journal of Internet Science, 1*(1) 5–16.

Link, M. W., and Mokdad, A. H. (2005). Alternative modes for health surveillance surveys: an experiment with Web, mail, and telephone. *Epidemiology, 16*(5), 701–704.

Lyberg, L. E. (1985). Plans for computer assisted data collection at Statistics Sweden. *Proceedings of the 45th Session, International Statistical Institute*, Book III, Topic 18.2.

Lynn, P., Jäckle, A., Jenkins, S. P., and Sala, E. (2004). The impact of dependent interviewing on measurement error in panel survey measures of benefit receipt: evidence from a validation study (paper 2004-28). Colchester: University of Essex.

Nass, C., Isbister, K., and Lee, E. J. (2001). Truth is beauty: researching embodied conversational agents. In J. Cassell, J. Sullivan, S. Prevost, and E. Churchill (Eds.), *Embodied Conversational Agents* (pp. 374–402). Cambridge, MA: MIT Press.

Nass, C., Moon, Y., and Green, N. (1997). Are machines gender neutral? Gender-stereotypic responses to computers with voices. *Journal of Applied Social Psychology, 27*(10), 864–876.

Nass, C., Moon, Y., Morkes, J., Kim, E. Y., and Fogg, B. J. (1997). Computers are social actors: a review of current research. In B. Friedman (Ed.), *Human Values and the Design of Computer Technology.* Stanford, CA: CSLI Press.

Nass, C., Robles, E., Bienenstock, H., Treinen, M., and Heenan, C. (2003). Speech-based disclosure systems: effects of modality, gender of prompt, and gender of user. *International Journal of Speech Technology, 6*, 113–121.

Nicholls, W. L. II, Baker, R. P., and Martin, J. (1997). The effect of new data collection technologies on survey data quality. In L. Lyberg, P. Biemer, M. Collins, E. deLeeuw, C. Dippo, N. Schwarz, and D. Trewin (Eds.), *Survey Measurement and Process Quality* (pp. 221–248). Hoboken, NJ: John Wiley & Sons.

Olsen, R. J. (1992). The effects of computer-assisted interviewing on data quality. In *Working Papers of the European Scientific Network on Household Panel studies* (paper 36). Colchester, England: University of Essex.

O'Reilly, J. M., Hubbard, M., Lessler, J., Biemer, P. P., and Turner, C. F. (1994). Audio and video computer assisted self-interviewing: preliminary tests of new technologies for data collection. *Journal of Official Statistics, 10*(2), 197–214.

Peytchev, A., Couper, M. P., McCabe, S. E., and Crawford, S. (2006). Web survey design: paging versus scrolling. *Public Opinion Quarterly, 70*(4), 596–607.

Phipps, P. A., and Tupek, A. R. (1991). Assessing measurement errors in a touchtone recognition survey. *Survey Methodology, 17*(1), 15–26.

Ramos, M., Sedivi, B. M., and Sweet, E. M. (1998). Computerized self-administered questionnaires (CSAQ). In M. P. Couper, R. P. Baker, J. Bethlehem, C. Z. F. Clark, J. Martin, W. L. Nicholls II, and J. O'Reilly (Eds.), *Computer Assisted Survey Information Collection* (pp. 389–408). Hoboken, NJ: John Wiley & Sons.

Saris, W. E. (1998). Ten years of interviewing without interviewers: the Telepanel. In M. P. Couper, R. P. Baker, J. Bethlehem, C. Z. F. Clark, J. Martin, W. L. Nicholls II, and J. O'Reilly (Eds.), *Computer Assisted Survey Information Collection* (pp. 409–429). Hoboken, NJ: John Wiley & Sons.

Schaefer, D. R., and Dillman, D. A. (1998). Development of a standard e-mail methodology: results of an experiment. *Public Opinion Quarterly, 62*(3), 378–397.

Schaeffer, N. C. (1980). Evaluating race-of-interviewer effects in a national survey. *Sociological Methods and Research, 8*(4), 400–419.

Schonlau, M., Asch, B. J., and Du, C. (2003). Web surveys as part of a mixed-mode strategy for populations that cannot be contacted by e-mail. *Social Science Computer Review, 21*(2), 218–222.

Shanks, J. M. (1983). The current status of computer assisted telephone interviewing: recent progress and future prospects. *Sociological Methods and Research, 12*(2), 119–142.

Schuman, H. and Presser, S. (1981). *Questions and Answers in Attitude Surveys*. New York: Academic Press.

Steiger, D. M., and Conroy, B. (in press), IVR: interactive voice response. In E. D. de Leeuw, J. J. Hox and D. A. Dillman, (eds) (in press), *International Handbook of Survey Methodology*. Mahwah, NJ: Lawrence Erlbaum Ass.

Sudman, S., and Bradburn, N. M. (1974). *Response Effects in Surveys*. Chicago: Aldine.

Tourangeau, R., Couper, M. P., and Steiger, D. M. (2003). Humanizing self-administered surveys: experiments on social presence in Web and IVR surveys. *Computers in Human Behavior, 19*, 1–24.

Tourangeau, R., and Smith, T. W. (1996). Asking sensitive questions: the impact of data collection mode, question format, and questions context. *Public Opinion Quarterly, 60*(2), 275–304.

Tourangeau, R., Steiger, D. M., and Wilson, D. (2002). Self-administered questions by telephone: evaluating interactive voice response. *Public Opinion Quarterly, 66*(2), 265–278.

Turner, C. F., Ku, L., Rogers, S. M., Lindberg, L. D., Pleck, J. H., and Sonenstein, F. L. (1998 May 8). Adolescent sexual behavior, drug use and violence: increased reporting with computer survey technology. *Science, 280,* 867–873.

van Bastelaer, R. A., Kerssemakers, F., and Sikkel, D. (1987). A test of the Netherlands' Continuous Labor Force Survey with hand-held computers: interviewer behavior and data quality. In J. H. W. Driessen, W. J. Keller, R. J. Mokken, E. H. J. Vrancken, W. F. M. de Vries, and J. W. W. A. Wit (Eds.), *Automation in Survey Processing* (pp. 37–54). Voorburg/Heerlen: Netherlands Central Bureau of Statistics (CBS Select 4).

CHAPTER 4

Mobile Web Surveys: A Preliminary Discussion of Methodological Implications

Marek Fuchs
Universität Kassel, Kassel, Hessen, Germany

4.1 INTRODUCTION

In recent years mobile devices have increasingly been used to access the World Wide Web. With new hardware solutions (smart phones, mobile digital assistants, Blackberries) and software protocols emerging (GPRS, HSDPA, UMTS), the mobile Internet will become standard in the near future. The pros and cons of one possible implementation of the mobile Internet for self-administered survey data collection are discussed in this chapter. In this implementation, respondents receive a survey invitation on their mobile phones either by short text message, by multimedia message, or by a live interviewer call. Potential respondents are invited to access the Internet using their mobile devices and log on to the Web survey (by clicking on a URL sent in a short text message). Respondents then fill in the standard HTML questionnaire using an auxiliary input device provided by the mobile application (12 button keypad, thumb wheel, or pointing device).

This methodology allows survey researcher to make use of the advantages of random respondent selection procedures (random digit dialing within the mobile telephone networks), which is a clear advantage compared to current Web surveys that rely mostly on random access panels being recruited by telephone or face-to-face screening or on less than optimal opt-in access panels. At the same time, noncontact should be reduced since respondents are reached using their mobile phones or other mobile devices. Also, nonresponse should be reduced because respondents can fill

Envisioning the Survey Interview of the Future, Edited by Frederick G. Conrad and Michael F. Schober
Copyright © 2008 John Wiley & Sons, Inc.

in the questionnaire at a time that is convenient for them. Less positive, mobile Web surveys may be especially prone to coverage error, because certain subgroups of the target population do not have access to the mobile Internet or are at least underrepresented. This may result in samples and estimates that are biased because of certain groups lacking in the sample. Also, mobile phone surveys might be especially prone to measurement error, which describes errors associated with the responses obtained from the survey participants because the survey was conducted using a mobile device. Possible measurement errors are described in a later section of this text, however, just to give an example, one can imagine that text input as a response to an open question is shorter and less complete than a handwritten response on a paper questionnaire or text input using a QWERTY keypad.

Since mobile Web surveys have not yet been adopted as a standard survey mode, the discussion of the advantages and disadvantages of this methodology cannot be assessed based on experimental data or experiences from field tests. Accordingly, this chapter aims to discuss the pros and cons of this survey mode using common wisdom from the survey methodology literature. Since the design discussed in this chapter combines features of self-administered and interviewer-administered surveys, the discussion of this new survey mode will draw on a body of literature that covers both self-administered and interviewer-administered surveys. Since the survey design discussed in this chapter is just one possible application of the mobile Internet to survey data collection, it should be stated clearly that other useful applications might emerge that offer even more advantages in order to reduce survey error and to enable valid survey estimates.

4.2 PROBLEMS WITH TRADITIONAL SURVEY MODES

4.2.1 Noncontact and Nonresponse Error

Currently, face-to-face surveys as well as telephone surveys are designed and conducted in a highly professional fashion; however, at the same time they face a decline in certain aspects of data quality. Besides a significant heterogeneity across countries, topics, and survey organizations (Engel et al., 2004), it is safe to assume that contact rates and cooperation rates for face-to-face surveys as well as for landline telephone surveys deteriorate gradually over time. In addition, random digit dialing (RDD) telephone surveys in the general population using landline phone numbers have to deal with an increasing coverage problem. Several recent studies have demonstrated these trends.

In a meta-analysis related to survey nonresponse, deLeeuw and deHeer (2001) collected information regarding large-scale ongoing governmental face-to-face surveys in 16 countries in Europe and the Americas. Based on a detailed assessment of the yearly waves, deLeeuw and deHeer (2001) were able to measure change in response rates, noncontact rates, and refusal rates. They found a significant increase of noncontact rates (+0.3 percentage points per year) as well as an increase of refusal rates (also +0.3 percentage points per year) across all 16 countries. Although these percentages are small, they may distort survey estimates, whose precision can be critical in policy and other important social decisions. In addition to the average increase of noncontact and refusals, they demonstrated differences across countries in terms of the size of this increase. For some countries the increase is much steeper than for others: for example, Denmark and Belgium show noncontact and refusal rates above average while in Slovenia and Australia the respective rates are considerably lower. Thus, survey nonresponse in face-to-face interviews is an international problem, which is more severe in some countries than others.

Similar yet more troubling trends are found in telephone surveys: analyzing data from a large scale ongoing Gallup RDD telephone survey over the time period October 1997 through September 2003, Tortora (2004) comes to the conclusion that contact rates and response rates are declining over time. Contact rates show a drop of about 16 percentage points while cooperation rates have declined about 11 percentage points, substantially higher than what was observed for face-to-face interviews. Also, Tortora (2004) documents a tremendous increase in the use of answering machines in the general population, which might be used for screening purposes and thus have a potentially negative impact on response rates.

Similarly, Curtin and colleagues (2005) report a significant decline in response rates in the Survey of Consumer Attitudes (SCA), a telephone survey conducted at the University of Michigan's Survey Research Center, Ann Arbor, Michigan since 1979. The findings suggest that SCA response rates have been marked by three distinct periods: a gradual decline from 1979 to 1989, a plateau from 1989 to 1996, followed by an even sharper decline after 1996 (Curtin et al., 2005). For the last period the decline averages 1.5 percentage points per year. Also, noncontacts have increased dramatically: since the late 1990s noncontacts have accounted for almost as much nonresponse as refusals (Curtin et al., 2005). Recent technological advances have introduced further problems reaching people when using landline telephones. Besides call screening, answering machines, caller ID, and call blocking, many households have adopted a rather strict and aversive communicational pattern that provides little room for surveys using landline telephones.

Based on a meta-analysis, Engel and Schnabel (2004) report only a moderate decrease of response rates for Germany in the 1990s compared to the 1970s and 1980s. Nevertheless, Engel and Schnabel (2004) point out that additional efforts during field work might have compensated for declining contact rates and cooperation rates, and thus, a simple comparison of contact rates over time might be misleading.

Even though these results are based on just a few selected surveys in selected countries, findings suggest a general trend toward declining response rates in RDD landline telephone surveys as well as in face-to-face interview surveys. Given the anecdotal reports from field operations regarding the increased efforts in contacting respondents, the documented development in response rates probably underestimates the real changes over time. Also, one should note the increase in terms of the cost per case as a consequence of the additional efforts. Whether or not declining response rates have increased nonresponse bias is not yet proved. Recent studies by Keeter and colleagues (2000), Curtin and colleagues (2005), and Merkle and Edelman (2002) suggest a rather small impact of nonresponse rates on survey results. Based on an extensive analysis of about 30 nonresponse studies, Groves (2006) comes to the conclusion that the response rate of a survey alone is not a sufficient predictor of the magnitude of nonresponse bias. Nevertheless, a decline of response rates increases the probability of nonresponse bias for those variables of interest that show a strong correlation to the likelihood to respond.

4.2.2 Coverage Error

In addition to the nonresponse problems discussed previously, a considerable proportion of all telephone households is about to leave the traditional landline telephone network by either adopting a mobile phone as their main communication device ("cord cutters" and "mobile-only," see Kuusela et al., 2007) or by switching to Voice-over-IP, which may for some providers include a completely new system of telephone numbers. This leads—at least for the traditional landline telephone surveys—to a considerable increase of coverage error.

For several industrialized countries, a drop in landline penetration rates is noticeable. Part of this trend is caused by portions of the younger telephone customers in Europe who choose to rely exclusively on wireless devices—at least during their years as young adults they never subscribe to a landline telephone service. Whether they do so once they settle down, live with a partner, or raise children is yet to be determined. Other subgroups among the previous landline customers have decided to abandon landline service; in part this may be motivated by a serious personal crisis (e.g., divorce) or by economic challenges (e.g., unemployment). In addition, in some less developed countries with geographically scattered populations, people have adopted wireless communication devices without ever having access to landline technology. As a consequence, a drop in landline penetration rates has occurred in several European countries since the late 1990s (Fuchs, 2006).

So far, estimates for the mobile-only population are still less reliable compared to the estimates regarding landline telephone usage. However, an increasing proportion of the general population has abandoned the landline telephone system as their main communication device—or they will do so in the near future. The actual proportion of mobile-only households/individuals varies from country to country; in Finland it has reached some 54% (TNS opinion & social, 2007). Also, there are important regional differences and sociodemographic differences in each country. In Italy, for example, the mobile-only population is far bigger on the Italian islands and in the scattered and less economic prosperous regions in the south (Kuusela et al., 2007). Also, it is known that younger people tend to belong to the mobile-only population to a larger degree (Kuusela and Notkola, 1999). In Germany, the proportion of mobile-only is much smaller (Fuchs, 2002a,b); however, the younger population in particular has adopted the mobile phone as their main communication device, which in turn will lead to a higher proportion of mobile-only in the future. Declining cost and affordable calling plans will contribute to this trend as well. As a consequence of this trend and fueled by the need for faster networks, Deutsche Telekom has announced it will shut down their traditional landline telephone network and transfer it to an IP-based data network by 2012.

Another technological innovation challenges the traditional landline telephone as a survey mode: Voice-over-IP (Steeh and Piekarski, 2006). So far, the proportion of Voice-over-IP (VoIP) users is still fairly small. Also, some providers still use derivates of the traditional telephone numbering system (some combination of area code + local subnet code + number). However, it is unknown whether new numbering systems for voice-over-IP, which are already in effect, will take over: for example, with the introduction of the new extended IP numbering system (Internet Protocol version 6, IPv6) in the near future it might well be that all telephone devices are addressed using their IP number. As an alternative, the ENUM numbering system—which allows portability of traditional telephone numbers and domain name server entries (a domain name server translates a domain name into an IP address, e.g., the domain www.umich.edu relates to—among others—the IP 141.211.144.186) might be used. Both systems—and other numbering systems still to emerge—might not be compatible with established RDD routines and survey researchers will be challenged with the need to differentiate telephone devices from other technical equipment using IP network addresses.

Given the noncontact rates and refusal rates of face-to-face surveys and telephone surveys and considering the increasing noncoverage error of landline telephone sampling frames, Web surveys have been considered to be a relatively cost-efficient means of data collection (Couper, 2000; Dillman, 2000; Vehovar et al., 2005). Given the fact that e-mail addresses are not listed for the general public and cannot be generated at random, random sampling is rarely possible. As a consequence, large-scale opt-in panels have emerged, providing researchers with huge populations of possible respondents who are—in principle—willing to participate in surveys (Göritz, 2007). Even though some panel providers undertake efforts to recruit panel members offline (Enright, 2006), for example, by RDD telephone surveys, many available panels consist mostly of self-selected members, which is problematic because their members may be quite different from randomly selected respondents and thus may distort survey estimates. Also, because Internet access still is more prominent in the middle and upper socioeconomic groups of the population, the membership of online access panels would not be a good representation for the general population even if all Internet users would have the same probability of becoming a member in one of the available access panels. In order to compensate for the biases associated with online access panels, some scholars promote propensity score weighting (Schonlau et al., 2003, 2004), which is a tool to weight respondents of a sample drawn from an online access panel according to the known biases. Given theses disadvantages of online access panels, Web surveys are mainly restricted to list-assisted studies in closed populations, where lists of all members of the target are available including their e-mail addresses. This holds at least until random procedures to generate working e-mail addresses become available—which is currently out of sight. Also, because Web surveys rely on self-administration, response rates are usually lower compared to fate-to-face and telephone surveys (Vehovar et al., 2002).

Given the sampling problems in online surveys and considering the coverage problems of landline telephone surveys as well as the nonresponse problems of both face-to-face and landline telephone surveys, researchers are looking to develop new survey modes that incorporate the benefits of random selection procedures, cost-efficient self-administration, and the use of interviewers motivating and convincing future respondents in the introductory phase of a survey interview. In recent years the mobile Internet has been identified as a mode that will allow such a survey design. The adoption of RDD sampling procedures for mobile phone numbers is underway (Buskirk and Callegaro, 2007; Häder and Gabler, 2006; Vehovar et al., 2005), and thus random respondent selection should be achievable. Because mobile phones offer verbal and visual communication channels at the same time, a survey mode using mobile devices could benefit from the advantages of interviewer-administration in terms of motivating and convincing reluctant respondents to participate. Finally, since self-administration is an essential characteristic of mobile Web surveys, this survey mode might benefit from lower survey costs per case. The feasibility of such a survey mode becomes plausible based on some recent technological developments.

4.3 TECHNOLOGICAL ADVANCEMENTS

Many advanced mobile phones and other mobile devices such as organizers and mobile digital assistants (MDAs = PDAs with mobile capabilities) offer Web browsers and Internet access using either traditional GPRS (or derivates thereof) or modern UMTS protocols. Several technological advancements and their implications for survey data collection have to be considered when using such hardware and software protocols for mobile Web surveys.

4.3.1 Integration of Diverse Hardware Components

Devices such as digital cameras, PDAs, mobile phones, MP3 players, radios, and mobile gaming devices are increasingly integrated into a single hand-held product. While some mobile devices specialize in some functionality—such as mobile telephone communication and camera function—in order to reduce the size and weight of individual components, others integrate almost every function of the previously separated hardware: for example, smart phones (including the iPhone) combine a mobile phone, organizer functionality, digital camera, MP3 player, and gaming device. Among those capabilities, mobile Web access is just one additional feature provided by most modern smart phones and MDAs.

Even though actual use of mobile Web access is still not widespread (Telephia, 2006), a substantial increase in mobile Web usage is likely. Many companies have started providing services especially designed for mobile Web devices (news networks, transportation providers, online auctioning, etc.) and mobile network providers offer plans and contracts that include mobile Web usage. Given the fast adoption of mobile technologies by customers and the resulting increase in penetration rates in recent years, it is safe to assume that mobile Web access will soon cover most of the population using mobile phones today.

4.3.2   New Network Technology

So far mobile Web usage suffers from narrow bandwidth and thus from slow download speed. However, with the emerging new mobile networks offering up to 2 megabits per second download speed and with smart phones that automatically choose the fastest network available (GSM, GPRS, UMTS, wireless LAN) entering the market, fast mobile Web access will soon be available, at least in metropolitan areas.

Given these technological advancements, mobile Web access becomes a realistic option for a large proportion of the mobile phone population. In the early days, WAP 1.0 technology allowed mobile devices to access WAP (low level HTML) pages designed specifically for mobile devices. More advanced pages including full HTML and plug-ins could not be visited due to the limited memory capabilities of mobile phones that prevented users from running browsers on their mobile devices. Also, network restrictions limited the downstream bandwidth. Although the introduction of WAP 1.2 in 1999 brought some improvements, the general limitations of WAP technology could not be overcome. The first smart phones allowing full access to the Internet including some plug-ins entered European and U.S. markets in 2005. Since then, most mobile phone producers offer medium-end and high-end devices with Web browsers installed. Already in 1999 i-mode started in Japan as one of the first mobile services offering mobile access to special pages. Currently, more than 50 million customers in Asia, Europe, and the United States make use of this technology, which relies on a reduced HTML standard and requires special mobile phones designed to display such content.

So far, the mobile Web is perceived to be slow and thus mobile Web usage is still in its infancy. Todays, there are only about 35 million users in the United States, resulting in a rather small fraction of the total mobile users; in Europe, the proportion is larger, although the penetration and representation of the general public is unknown. Also, it should be noted that the low bandwidth of the available networks still limits the speed and ease of using the mobile Internet. The situation today is very similar to the one a decade ago, when the landline Internet was still about to become a standard network in Europe and the United States. Given the momentum of technological advancements as well as the business ideas emerging, it is safe to assume that mobile Internet will soon become as popular and convenient as traditional landline Internet access.

Thus, survey researchers should be able to make use of the mobile telephone numbering system to generate random samples of potential respondents; contact them using live interviewer calls, short text messages, or multimedia messages; send them a URL to a Web survey; and ask them to complete a self-administered survey using their mobile devices. This technology is already being used by some market research companies in Hong Kong (Bacon-Shone and Lau, 2006), the United States (e.g., "Zoom"), and Europe (e.g., "Globalpark," see also Shermach, 2005; Townsend, 2005). However, the methodological implications of this method are mostly unknown. The remaining part of this chapter discusses this method in order to evaluate its feasibility for surveys in the general public and special populations.

4.4   DEMONSTRATION APPLICATION

In 2005 and 2006 a demonstration application for the mobile Web survey (see Figs. 4.1 and 4.2) was developed.

Figure 4.1 shows two possible invitations delivered to the respondent: on the left there is a traditional short text message indicating the topic of the study and the URL to the survey [see Neubarth et al., (2005) and Steeh et al., (forthcoming) for a test of short text messages as invitations to traditional Web surveys]. Alternatively (see example on the right side in Fig. 4.1), one could use multimedia messages (a short video clip sent to the respondent's mobile phone) showing a human interviewer or an animated agent (avatar) that reads the survey invitation to the respondent. Figure 4.2 shows a screen shot from a test questionnaire. Since almost every recent mobile Web device offers a color screen, questionnaires can adopt many features from traditional Web surveys. Also, a variation for different question types can be used, for example, single response closed-ended questions using radio buttons and multiple response questions using check boxes. However, as the example indicates, the small screen limits question complexity (see also Bates and Ramsey, 2004; Tjostheim, 2005). No matrix questions using multiple horizontally oriented items with Likert-type response options are suitable. Also, questions making use of pull-down menus and open questions are less than optimal.

FIGURE 4.1   Left: Invitation by short text message (with URL of survey). Right: Invitation by multimedia message (video clip).

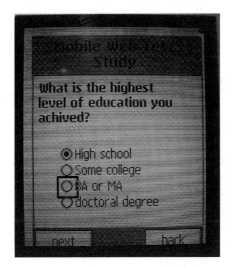

FIGURE 4.2 Question from example study as seen on mobile phone screen. (Source: www.event-evaluation.de\Studien\Mobile05\SurveyKit.php, accessible either by regular Web browser or mobile Web browser.)

Although this demonstration indicates the feasibility of such a survey design from a technical perspective, the methodological implications are still to be tested. So far, only a few publications are available; they all emphasize the promise of this approach in order to overcome noncontact and coverage problems. At the same time, they indicate the need for further methodological research before fully adopting this methodology for survey research. Also, making use of audiovisual elements in the initial phase of a survey interview can be seen as advantageous compared to simple text-based invitations, because the invitation is more likely being perceived as a request from a human being. Thus, a refusal is associated with an additional burden compared to refusing to a text-based request that relies on machine-like communicative pattern (short text message). This is crucial: when the traditional Web survey was developed in the mid/late-1990s, many survey researchers were irritated with the methodological standards of the early adopters of Web surveys for data collection. It took at least five to eight years of methodological research to assess Web surveys in terms of data quality using the well-developed concept of total survey error (see Couper, Chapter 3 in this volume; Groves, 1989; Groves et al., 2004) before researchers interested in substantive questions felt comfortable using Web survey methodology for their own studies.

With the mobile Web developing at such a fast pace, it would be optimal to avoid the situation for mobile Web surveys. Currently, some market research companies as well as some early adopters in governmental and academic research already make use of this methodology for transportation research, satisfaction research, quality of life research, and other fields (Bates and Ramsey, 2004; Tjostheim, 2005). Unfortunately, a systematic evaluation, of mobile Web surveys in terms of the survey error concept is not yet in sight. In order to promote and guide such evaluations, we will discuss methodological implications of the mobile Web surveys using standard methodological terminology.

4.5 SOURCES OF ERROR IN MOBILE WEB SURVEYS

The key advantages of the mobile Web survey result from the fact that it allows random sampling based on the mobile phone numbers, combined with the cost-efficient features of a self-administered Web survey and the contactability of mobile-only and hard-to-reach populations. At the same time, serious drawbacks in terms of noncoverage and measurement error are to be expected. In the following section, we assess the pros and cons of the mobile Web survey according to components of the total survey error (Groves et al., 2004). In this context, it is important to recognize that the mobile Web survey combines elements from traditional telephone surveys (using landline or mobile telephone numbers) as well as from traditional Web surveys. While the traditional telephone survey is interviewer-administered, the traditional Web survey relies on self-completion; while the traditional telephone survey allows random sampling, the traditional Web survey is often based on self-selected samples. Thus, a one-dimensional comparison is neither appropriate nor possible. We assess the methodological implications of this methodology in terms of the following components of total survey error.

4.5.1 Coverage Error

So far, the penetration rate of smart phones is rather low. Also, the technical ability of the general public to operate a mobile Web device is even smaller. Thus, whether people are actually able to take part in a mobile Web survey depends not only on the penetration rate of certain mobile devices but also on their ability to operate the mobile Internet and on the payment plan or mobile phone contract that includes mobile Web access at a reasonable price. Thus, we have to accept that complete coverage of the general public cannot be achieved in the near future. However, in order to assess mobile Web surveys, we have to compare the likely coverage error with that of other survey modes. Given the declining landline telephone penetration and the rising mobile-only population, the drawbacks of the mobile Web survey in terms of coverage error might soon diminish, not only because a growing population will adopt this technology but also because landline telephone surveys will suffer from increasing coverage problems in the future.

In contrast to fixed line Web surveys, the coverage error of mobile Web surveys will most likely be more pronounced. Internet penetration has reached a considerable level even among those who traditionally were not considered early adopters. The current coverage bias given the state of mobile Internet penetration rate is similar to the bias that was observable in the early days of the fixed line Internet. However, it depends on the future of broadband Internet access whether the fixed line Internet will experience a serious decline in penetration rates. As for now, fixed line Internet access seems to have approached a ceiling level in many European countries. Also, it is not known whether fixed line Internet access will remain on high levels once landline telephone technology usage drops. Given the fact that Internet access is increasingly provided by other than just the landline telephone network companies (e.g., by cable TV providers and electric companies), one could imagine home-based Internet access without landline telephone service.

Given the current state of mobile Internet access and considering the status of the other survey modes discussed so far, mobile Web surveys will most likely not be suitable as an exclusive survey mode for studies in the general population. However, in a mixed mode study it might well function as the survey mode of choice for certain subpopulations (e.g., the highly mobile and the mobile-only populations). This might especially be true for longitudinal studies, where respondents expect to be contacted (e.g., after a hospital release, in a time budget study, or when it comes to a traveler survey). Also, studies among special populations with a known high rate of mobile Internet access (e.g., the young or business people) or studies among registered users of mobile Web services would be natural choices for the mobile Web survey.

4.5.2 Sampling Error and Field Work

As with traditional random digit dialing (RDD) samples, mobile phone samples do not ask for area probability sampling or other kinds of heavily clustered designs. Even though cost-effective RDD sampling (Mitofsky, 1970; Waksberg, 1978) involves clustering to a certain degree, RDD samples usually are less prone to such design effects than face-to-face area probability samples. The same is true for RDD samples among mobile phone numbers; thus, from a sampling point of view, mobile Web surveys do not suffer from additional sampling error. However, one has to take into account the various payment plans across different countries. While in Europe a call to a mobile device is paid for by the caller only, in the United States the called party is charged for the connection as well. This may reduce the willingness of selected respondents to take part in such surveys. Also, it should be noted that cell phone numbers are not allowed in autodialed RDD samples, which puts an extra burden on interviewers who have to dial each number manually.

In addition, generally speaking, mobile phones are considered to be personal devices. As a consequence, no within-household selection procedure is necessary [see Binson et al., (2000) for an overview of such techniques]. Thus, the design effect in a random sample of mobile phone numbers in the general population should be smaller than with the sampling procedures typically used in other survey modes.

However, based on pilot studies in the mobile phone population (Fuchs, 2002a,b), we expect less efficient field work when contacting smart phone users. In part, this is caused by a smaller hit rate of working numbers when contacting RDD mobile phone numbers. The hit rate is roughly defined as the proportion of working numbers of all numbers generated using a RDD procedure. As this will be improved in the course of the future extensions of traditional RDD telephone surveys with mobile numbers [see Häder and Gabler, (2006) and Brick et al., (2006) for studies with dual frames], we can hope for developments that allow cost-efficient sampling without too many call attempts and calls to nonworking numbers.

Unfortunately for survey researchers, mobile phone numbers do not reveal any regional or local information. This prevents us from the benefits of regional stratification and geographic oversampling (if needed). In order to design cost-efficient samples that give maximum precision for a given amount of resources, researchers make use of stratification. State, county of residence, or municipality are widely used stratification criteria. Because mobile phone numbers have special prefixes (in Europe) that do not indicate the residence of its owner, regional stratification is not possible. For the same reason, it is not possible to apply disproportional stratification, which applies a higher selection probability for certain regions or municipalities. However, at least in some European countries, the prefix of mobile phone numbers contains some information regarding the type of mobile phone (prepaid or contract) and—given the fact that customers to most mobile phone service providers are regionally clustered— some proxy information regarding the region that the owner of the mobile phone lives in.

4.5.3   Nonresponse Error

Currently, noncontact and refusals are the main sources for nonresponse error in traditional face-to-face and telephone surveys. Given the mobile character of the smart phone, it is hoped that mobile-only populations and hard-to-reach populations can be contacted using mobile Web surveys. Also, respondents are not expected to complete the mobile Web surveys at the time of the receipt of the invitation. Rather, they can react to the inquiry at a time that is convenient for them. As a consequence, we propose that refusal might be less pronounced compared to traditional landline telephone surveys or mobile phone surveys. In contrast, short text messages (like i-mode messages in Japan) have a limited lifetime. Assuming that a mobile device is switched off for more than just a few days, it is unclear for how long the message remains available because mobile service providers have very diverse policies regarding undeliverable messages. Also, it is well known from Web surveys that invitation messages sent by e-mail can easily be deleted by potential respondents without considering the invitation to the survey. It is unknown if respondents in the general public invited without prior notice by multimedia message or short text message will be any more likely to respond to a mobile Web survey. If not, the advantage of a reduced noncontact rate might well be outweighed by the potential increase of the refusal rates. If so, overall nonresponse might be substantially reduced.

The mobile Web survey offers various ways to send out survey invitations: short text messages, video clips, live interviewer calls. Since pure text messages suffer from low salience, one could speculate regarding lower nonresponse rates for video invitations or live interviewer calls, or live interviewers using video telephony in order to invite respondents. Prerecorded video clips or live visual cues are far more intense signals that cannot be ignored so easily. Also, the social presence (Krysan and Couper, 2006) introduced by the visible interviewer might increase the willingness of respondents to participate because it gives the invitation a warmth that is lacking in other media such as text; however, it might drive away potential respondents who prefer self-administration.

While mobile phone surveys suffer from the unknown effects of the respondent being at a remote location or in a situation where others can overhear the conversation, mobile Web surveys are self-administered and do not require any voice input. For example, respondents who are commuting on a train might not be willing to complete a mobile phone survey because fellow commuters might overhear their side of the conversation. However, they might well be willing to complete a mobile Web survey while sitting in a train together with others or during a stop-over at an airport because the text input eliminates overhearing as a deterrent. So far, we can only speculate about which topics are suitable for a mobile Web survey; we just do not know to what extent respondents perceive sensitive questions as burdensome if answered in public. Since a mobile Web survey is not necessarily completed at home (even though—depending on the timing of the invitation—many respondents might well be at home), not all topics are suitable for a mobile Web survey. A location outside the respondent's home might not be perceived as an appropriate setting for any sensitive topic. Even though Smith (1997) and Pollner and Adams (1997) report only minor effects on data quality of third person being present during a face-to-face interview, answering sensitive questions in public might be uncomfortable for respondents.

Mobile Web surveys might require even shorter field periods than fixed Web surveys. Based on available findings, the field phase of ordinary Web surveys indicate a rather short field duration; after three days about 60–70% of all returning cases tend to be completed. Assuming a more intense usage pattern of mobile devices compared to Internet access, one might speculate that field work in mobile Web surveys might speed up even more. However, given the experiences from traditional Web surveys, one might expect many reminders would be necessary to boost response rates. Since reminders are sent by short text message or multimedia message, they are not free of charge (compared to e-mail reminders). Also, in order to reduce nonresponse, one has to consider incentives to motivate respondents to take the survey. Since mailing addresses are usually not known in advance, only electronic incentives are reasonable, for example, cash payments into the respondent's mobile phone account or vouchers for Internet mail order shops.

4.5.4   Measurement Error

Finally, it should be noted that the screen size of mobile phones and other mobile devices is still very much limited. So far, VGA standards are not met—not even by the displays of high-end devices. Thus, the complexity of the questionnaire and every single question is fairly limited—even if compared to an interactive Web survey that displays one question at a time. The length of the question wording as well as the number of response options needs to be restricted. Otherwise respondents need to scroll horizontally and vertically, which should have a negative impact on data quality.

Also, one can speculate that respondents participating in a mobile Web survey might be more heavily distracted from the actual survey question and the survey as a whole because of demanding navigational tasks. This effect was assumed to be responsible for various mode effects in traditional Web surveys (Fuchs, 2002c, 2003). In a mobile Web survey, this effect should be even more pronounced since it presumably requires more attention to details. Also, longer system response times and disturbances in the surroundings (if away from home when answering questions) might increase measurement error as well (item nonresponse, response order effect, question order effects). Whether the use of progress indicators might reduce some of these errors—similar to Web surveys (Couper et al., 2001)—is not yet clear.

One should also consider the impact of available input devices. Currently, mobile devices offer a variety of means of text input: some devices offer no text input at all, just the 12 button number keypad. Other devices offer rudimentary text input using multitap, which allows text input using the 12 button phone keypad, where each button represents 3–4 characters. This requires up to 4 pushes for each alphanumeric character. More advanced mobile devices include T9/Fasttap, which builds words based on combinations of buttons pushed; using this technique requires 1 key pressed per character with some additional keys in case of ambiguous combinations. Larger mobile devices offer a micro-QWERTY keypad, where the two halves of the QWERTY keypad are overlapping, or pointing device/stick for soft keypads displayed on the screen. Even though methodological studies (Deutskens et al., 2006) have demonstrated that keyboard input to open questions is not less rich or shorter than handwritten answers to open questions in paper and pencil surveys, one has to be careful in extending this finding to multitap or T9 text input on a mobile device. Whether this input device leads to abbreviations or whether it produces less detailed responses is not yet clear.

4.6 CONCLUSION

The picture regarding the various components of survey error is not yet clear. Neither have we a compelling understanding of the feasibility of the mobile Web survey and the sources of error involved, nor is an overall evaluation of the survey mode as a whole possible at this time. However, the ideas discussed here might work as a preliminary schedule for experiments regarding the methodological implications of the mobile Internet for survey data collection.

The survey mode discussed here is just one possible application of the mobile Internet to surveys. So far, the design discussed makes use of traditional self-administration for data collection: the questions are posed text based and the responses are collected by clicks on radio buttons and check boxes or—less feasible—by open alphanumeric input. Given the advancements in the field of mobile videotelephony, one could also imagine live interviewer-administered interviews using mobile devices. Interviewers could read all questions in the questionnaire or at least a certain portion of them before respondents are switched to an IVR survey system. Also, mixed mode designs might contribute to data quality: for example, during the course of a traditional telephone interview using mobile phones the interviewer could turn over the survey using a mobile Web application to the respondent who would fill in a CASI portion with sensitive questions. Afterwards the interviewer-administered survey could continue.

Because mobile hand-held devices offer the possibility to access the mobile Internet and to place a phone call at the same time, one could also think of respondent-initiated phone calls in order to clarify question meaning or to get help with nonstandard responses. Respondents who encounter problems while responding to the questionnaire could call the survey organization for support.

Human interviewers do not deliver completely standardized survey questions. Depending on mood, daily conditions, and experience, within-interviewer variation in behavior can be observed. Also physiological, psychological, and social characteristics of interviewers as well as language characteristics contribute to between-interviewer variation, which in turn increases variance and potentially produces biases. With the introduction of computer animated agents (talking heads, or more fully embodied avatars), researchers could increase the standardization of survey question administration. Since every question is administered identically, each respondent receives the same treatment. Even though it is unclear whether this really increases data quality—it might well be that a respondent-specific interviewer behavior is a far better way to standardized the survey experience from a respondent's point of view—one could think of agents becoming companions (assistants, technical advisors) to respondents working on a mobile Web questionnaire. When necessary, such an agent could more completely take on the role of an interviewer.

The self-administered mobile Web survey provides online access to paradata regarding certain aspects of respondent behavior (e.g., response time, change of already selected response categories). This information can potentially be used to develop respondent models that could inform survey systems regarding the motivational status of the respondent or whether or not the respondent needs further information in order to understand and answer a particular question. Thus, the available paradata could be used to trigger system-initiated motivational cues, definitions, or clarification of concepts (Conrad et al., 2007).

So far, the mobile Web survey has not been tested in a setting that allows for a methodological evaluation. As a consequence, much of what has been said is based on speculation and reasoning regarding similar survey modes. This chapter is meant to be an invitation to discuss the likely pros and cons of this mode and at the same time it is a call to start field-experimental evaluations or lab tests in order to assess the implications of this mode. Thus, once market, academic, and governmental researchers begin to adopt mobile Web surveys, survey methodologists and communication technologists must be ready and able to provide advice regarding the proper use of this technique.

REFERENCES

Bacon-Shone, J., and Lau, L. (2006). Mobile vs. fixed-line surveys in Hong Kong. Paper presented at the Second International Conference on Telephone Survey Methodology, Miami, FL, USA.

Bates, I., and Ramsey, B. (2004). Developing a mobile transportation survey system. Manuscript.

Binson, D., Canchola, J., and Catania, J. (2000). Random selection in a national telephone survey: a comparison of the Kish, next-birthday, and last-birthday methods. *Journal of Official Statistics, 16*, 53–59.

Brick, M., Dipko, S., Presser, S., Tucker, C., and Yaun, Y. (2006). Nonresponse bias in a dual frame sample of cell and landline numbers. *Public Opinion Quarterly, 70* (special issue), 780–793.

Buskirk, T., and Callegaro, M. (2007). Cellphone sampling. Encyclopedia of Survey Research (Chapter to appear in 2004).

Conrad, F., Schober, M., and Coiner, T. (2007). Bringing features of human dialogue to web surveys. *Applied Cognitive Psychology, 21*, 1–23.

Couper, M. (2000). Web surveys. A review of issues and approaches. *Public Opinion Quarterly, 64*, 464–494.

Couper, M., Traugott, M., and Lamais, M. (2001). Web survey design and administration. *Public Opinion Quarterly, 65*, 230–253.

Curtin, R., Presser, S., and Singer, E. (2005). Changes in telephone survey nonresponse over the past quarter century. *Public Opinion Quarterly, 69*, 87–98.

deLeeuw, E., and deHeer, W. (2001). *Trends in Household Survey Nonresponse: A Longitudinal and International Comparison.* New York: John Wiley & Sons.

Deutskens, E., de Ruyter, K., and Wetzels, M. (2006). An assessment of equivalence between online and mail surveys in service research. *Journal of Service Research, 8*(4), 346–355.

Dillman, D. (2000). Mail and internet surveys. *The Tailored Design Method*, 2nd ed. Hoboken, NJ: John Wiley & Sons.

Engel, U., Pötschke, M., Schnabel, C., and Simonson, J. (2004). *Nonresponse und Stichprobenqualität. Ausschöpfung in Umfragen der Markt- und Sozialforschung.* Frauhfurt a. M.: Avbeitskreis deutscher Markt- und Sozial forschuings-institute.

Engel, U., and Schnabel, C. (2004). *ADM-Projekt: Möglichkeiten der Erhöhung des Ausschöpfungsgrades in Umfragen der Markt- und Sozialforschung.* Bremen: University of Bremen.

Enright, A. (2006). Make the connection: recruit for online surveys offline as well. *Marketing News, 40*(6), 21–22.

Fuchs, M. (2000). Screen design and question order in a CAI instrument - Results from a usability field experiment. *Survey Methodology, 26*(2), 199–207.

Fuchs, M. (2002a). Kann man Umfragen per Handy durchführen? Ausschöpfung, Interview-Dauer und Item-Nonresponse im Vergleich mit einer Festnetz - Stichprobe. *Planung & Analyse, 729(2)*, 57–63.

Fuchs, M. (2002b). Eine CATI-Umfrage unter Handy-Nutzern. Methodische Erfahrungen aus einem Vergleich mit einer Festnetz-Stichprobe. In S. Gabler and S. Häder, (Eds.), *Methodische Probleme bei Telefonstichprobenziehung und–realisierung* (pp. 121–137). Münster: Waxmann.

Fuchs, M. (2002c). Comparing Internet and paper and pencil surveys. Question order effects and response order effects as indicators for differences of data collection techniques. *Egyptian Journal of Public Opinion Research, 3*(1) 21–43.

Fuchs, M. (2003). Kognitive Prozesse und Antwortverhalten in einer Internet-Befragung. *Österreichische Zeitschrift für Soziologie, 28*(4), 19–45.

Fuchs, M. (2006). Non-response and measurement error in mobile phone surveys. Paper presented at the Second International Conference on Telephone Survey Methodology, Miami, FL, USA.

Göritz, A. (2007). Using online panels in psychological research. In A. N. Joinson, K. Y. A. McKenna, T. Postwes and U. D. Reips (Eds). *The Oxford Handbook of Internet Psychology* (pp. 473–485). Oxford, UK: Oxford University Press.

Groves, R. (1989). *Survey Errors and Survey Costs*. New York: John Wiley & Sons.

Groves, R. (2006). Nonresponse rates and nonresponse bias in household surveys. *Public Opinion Quarterly, 70*, (special issue), 646–675.

Groves, R., Fowler, F., Couper, M., Lepkowski, J., Singer, E., and Tourangeau, R. (2004). *Survey Methodology*. Hoboken, NJ: John Wiley & Sons.

Häder, S., and Gabler, S. (2006). Neue Entwicklungen bei der Ziehung von Telefonstichproben in Deutschland. In F. Faulbaum and C. Wolf (Eds.), *Stichprobenqualität in Bevölkerungsumfragen*. (pp. 11–18). Bonn: IZ-Sozialwissenschaften.

Keeter, S., Miller, C., Kohut, A., Groves, R., and Presser, S. (2000). Consequences of reducing nonreponse in a national telephone survey. *Public Opinion Quarterly, 64*, 125–148.

Krysan, M., and Couper, M. (2006). Race of interviewer effects: What happens on the web? *International Journal of Internet Science, 1*(1), 17–28.

Kuusela, V., and Notkola, V. (1999). Survey quality and mobile phones. Paper presented at the International Conference on Survey Nonresponse, October 28–31, 1999.

Kuusela, V., Callegaro, M., and Vehovar, V. (2007). Influence of mobile telephones on telephone surveys. In J. M., Lepkowski, et al. (Eds.), *Advances in Telephone Survey Methodology*. (forthcoming) New York: John Wiley & Sons.

Merkle, D., and Edelman, M. (2002). Nonresponse in exit polls: a comprehensive analysis. In R. Groves, D. Dillman, J. Eltridge, and R. Little, (Eds.), *Survey Nonresponse* (pp. 243–258). Hoboken, NJ: John Wiley & Sons.

Mitofsky, W. (1970). Sampling of telephone households. CBS News memorandum.

Neubarth, W., Bosnjak, M., Bandilla, W., Couper, M., and Kaczmirek, L. (2005). Prenotification in online access panel surveys: e-mail versus mobile text messaging (SMS). Paper presented at the Consumer Personality & Research Conference 2005, Dubrovnic.

Pollner, M., and Adams, R. E. (1997). The effect of spouse presence on appraisals of emotional support and household strain. *Public Opinion Quarterly, 61*, 615–626.

Schonlau, M., Zapert, K., Payne Simon, L., Sanstad, K., Marcus, S., Adams, J., Spranca, M., Kan, H., Turner, R., and Berry, S., (2003). A comparison between responses from a propensity-weighted web survey and an identical RDD survey. *Social Science Computer Review, 21*(1), 128–138.

Schonlau, M., van Soest, A., Kapteyn, A., Couper, M., and Winter, J. (2004). Attempting to adjust for selection bias in Web surveys with propensity scores: the case of the Health and Retirement Survey (HRS). In *Proceedings of American Statistical Association, Section on Survey Research Methods*. Alexandria, VA: American Statistical Association.

Shermach, K. (2005). On-the-go polls. *Sales & Marketing Management, 157(6)*, 20.

Smith, T. (1997). The impact of the presence of others on a respondent's answers to questions. *International Journal of Public Opinion Research 9(1)*, 33–47.

Steeh, C., and Piekarski, L. (2006). Accommodating new technologies: the rejuvenation of telephone surveys. Presentation on the occasion of the Second International Conference on Telephone Survey Methodology, Miami, FL, USA.

Steeh, C., Buskirk, T., and Callegaro, M. (2007). Using text messages in U.S. mobile phone surveys. In *Field Methods. 19(1)*, 59–75.

Telephia (2006). Mobile Internet population jumps to 34.6 million, with email, weather and sports websites securing the highest reach. Press release, San Francisco, August 14, 2006.

Tjostheim, I. (2005). Mobile self-interviewing. An opportunity for location based marked research. Are privacy concerns a showstopper? Paper presented at a workshop of the Association for Survey Computing.

TNS Opinion & Social. (2007). E-communications household survey. Special Eurobarometer 174/66.3. Report to the European Commission. Brussels: European Commission.

Tortora, R. D. 2004. Response trends in a national random digit dial survey. *Metodološki zvezki*, 1, 21–32.

Townsend, L. (2005). The status of wireless survey solutions: the emerging "power of the thumb." *Journal of Interactive Advertising, 6*, 1, 52–58.

Vehovar, V. (2003). Mobile phone surveys. Paper presented at Gallup Europe: Measuring the Emotional Economy of Europe.

Vehovar, V., Batageli, Z., Loraz M., Katja, Z, M. (2002). Nonresponse in web surveys. In R. Groves, D. Dillman, J. Eltridge, and R. Little (Eds), *Survey Nonresponse* (pp. 229–242). Hoboken, NJ: John Wiley & Sons.

Vehovar, V., Dolnicar, V., Lozar Manfreda, K. (2005). Internet survey methods. In S. Best and B. Radcliff, (Eds.), *Polling America. An Encyclopaedia of Public Opinion.* (pp. 368–374). Westport, London: Greenwood.

Waksberg, J. (1978). Sampling methods for random digit dialling. *Journal of the American Statistical Association, 73*, 40–46.

CHAPTER 5

Video-Mediated Interactions and Surveys

Anne H. Anderson
University of Dundee, Dundee, Scotland, UK

5.1 INTRODUCTION

As the survey world has adopted new communication technologies for interviewing, from the telephone to the computer to the Internet (see Couper, Chapter 3 in this volume), there have also been concerns about the impact of each new technology on the quality and representativeness of the data collected. This chapter explores some of the potential advantages and disadvantages of another new communication technology for survey researchers: video conferencing. Video conferencing, which made its debut with picture phones at the New York World's Fair in 1964, resisted larger scale adoption until fairly recently; video conferencing in nonsurvey settings has now become far more commonplace, with corporations connecting remote offices via dedicated video lines, and home users casually using desktop video-conferencing software for personal purposes. Video-mediated interviews, in which the interviewer and respondent can see (at least a part of) each other via a video window, have thus become technologically feasible and affordable in an unprecedented way.

Video interviews offer several potential advantages for surveys, but there are also several gaps in our knowledge about the likely impact. The jury is thus still out on whether, when, and how survey researchers ought to adopt video. Nonetheless, a growing body of research on effects of video in other domains can inform survey researchers as they consider adopting today's technologies, as well as future technologies that make use of video.

Many rather overhyped claims have been made for video technologies, with strong unsubstantiated statements about "the death of distance," "every picture is worth a

Envisioning the Survey Interview of the Future, Edited by Frederick G. Conrad and Michael F. Schober
Copyright © 2008 John Wiley & Sons, Inc.

thousand words," and "70% of communication is nonverbal." In this chapter some of the empirical evidence of how the availability of visual signals, whether "live" or transmitted over communication networks, affects the way people carry out particular collaborative tasks will be presented. The aim is to identify where video technology could add value for survey researchers in terms of recruitment of respondents, completion rates, data quality, or cost. Many of the studies reported here offer comparisons of video conferencing with face-to-face conversations, to explore whether the interaction that takes place is equivalent on some key dimensions. In other studies video technology is evaluated in comparison with telephone interactions, to assess the benefits of the additional visual information. Only a few studies allow comparison of face-to-face, video conference, and telephone conversations.

5.2   VIDEO-CONFERENCING SYSTEMS

Video-conferencing technology takes several forms, from dedicated video-conference rooms in organizations, to picture phones, to desktop video conferencing via personal computers. Some of these variants are illustrated in Fig. 5.1.

These systems provide different experiences, from rather restricted, grainy images in the corner of the PC screen, which show jerky movements and have delayed sound transmission, through systems that provide full screen images that update as frequently as TV pictures, to specially designed rooms with large-scale displays. Figure 5.1 (top left) illustrates a dedicated video-conference suite as used in many business organizations. Figure 5.1 (center left) shows a typical desktop computer version of video conferencing, with other on-screen shared applications, which might be, for example, a selection of survey information or questions. Figure 5.1 (center right) illustrates one form of video phone. Figure 5.1 (bottom) shows another form of desktop video conferencing, where images of the other party fill the screen.

At the highest end (e.g., Hollywood mogul Jeffrey Katzenberg's custom deluxe video-conference system, Halo, which is being sold to many top firms at a cost of $550,000, as reported in *The Economist*, 14 Dec. 2005), systems with multiple cameras and excellent sound systems, and very high speed network connections, really do provide the feeling of "being there" for their users. More typically, most systems, operating without this kind of DreamWorks support, do not yet provide an experience equivalent to face-to-face interaction. This does not mean, however, that video technology could not offer benefits to the survey world. With the increasing availability of high bandwidth Internet connectivity and low cost video cameras, the cost comparisons with telephone interviewing over the next few years may make video conferencing an attractive alternative.

There are two key comparisons that should be considered in evaluating the potential of video conferencing, apart from cost. One concerns comparing video interviews with face-to-face interviews in terms of recruitment, completion, and data quality. The second comparison is whether video interviews offer advantages over traditional telephone interviews on these same criteria.

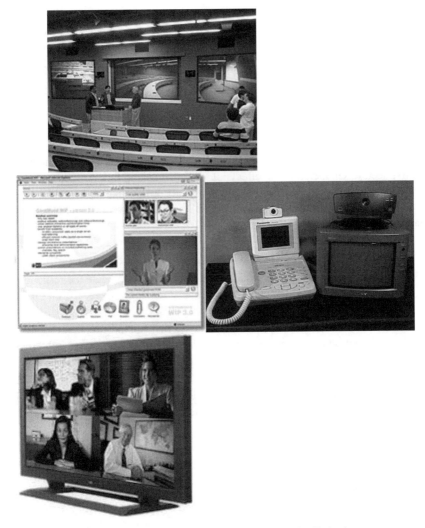

FIGURE 5.1 A Variety of Forms of Video Conferencing Technology.

In this chapter the potential impacts of the use of video-conference technology on surveys are considered. As yet there appears to be no existing research literature that directly explores the use of this form of technology on surveys. This chapter therefore draws on studies of the impacts of video communication in other domains, such as selection interviews, medical consultations, and problem-solving collaborations, where these highlight potentially applicable insights. Key dimensions are the likely impacts on the communication processes during an interview, the social interaction between interviewer and interviewee, and the satisfaction of both parties with the experience. In addition, a selection of salient features of surveys will be considered to attempt to identify if the introduction of video technology would affect these known phenomena. The key questions will be to highlight areas where video technology could add value, as well as to outline any potential pitfalls for surveys if this technology was deployed.

5.3   WHAT DO VISUAL CUES ADD TO INTERACTIONS? FACE-TO-FACE VERSUS TELEPHONE INTERVIEWS

If we want to explore the potential effects of video conferencing on survey interviews, we first need to characterize the effects of visual signals in face-to-face interviews, of which a survey interview is a special case. Face-to-face survey interviews compared to telephone or postal self-administered interviews have the following general characteristics:

1. Face-to-face interviews tend to have higher participation and completion rates.
2. Face-to-face interviews provide visual cues that can indicate the respondent's cognitive state (e.g., listening to the question, trying to recall relevant past experience, planning or formulating a response).
3. Face-to-face interviews provide visual cues that can indicate the respondent's emotional state (e.g., relaxed, bored, confused, frustrated, angry, embarrassed, upset).

If we take these features in turn, we can investigate the possible underlying mechanism and the likely impact on surveys. We can also consider the extent to which video conferencing has the same impact.

In face-to-face interaction, as we access the additional cognitive and affective information, we tend to feel more engaged or connected than we do in a telephone interaction. This greater sense of "social presence" (Short et al., 1976) may lead to greater participation and completion rates in face-to-face surveys. The lack of these visual channels in telephone interviews, and hence the lack of sense of engagement, seems to lead to respondents being less satisfied, and to decrease the quality of responses. For example, Holbrook, Green, and Krosnick (2003) analyzed three large-scale U.S. surveys, comparing face-to-face and telephone interviews. They found that telephone interviewees were less satisfied with the length of the interview than those interviewed face-to-face, despite the fact the telephone interviews were shorter. When the interviewers rated the telephone participants they assessed them as less cooperative and engaged in the interview than those interviewed face-to-face. Holbrook and colleagues note that the considerable dissatisfaction felt by telephone respondents is likely to make them less likely to participate in future surveys.

More telephone respondents were also reported as "satisficing," that is, giving answers without much thought. This kind of response is attributed to participants' lack of cognitive effort during the interview and is considered to be more of a problem in longer surveys or in the later stages of interviews. Holbrook et al. (2003) report that one issue is the lack of control over participants' other activities during telephone interviews, so respondents may be multitasking, for example, watching TV while the interview is taking place, and so may be paying less attention to the interview questions or their responses. This may have caused one of the findings of this study, that telephone respondents more often gave no opinions or undifferentiated ratings in responses to questions. Less well educated respondents showed the telephone satisficing responses most strongly.

Face-to-face interviews have also been shown to have advantages in terms of the likely honesty of the responses. In survey interviews, as in many other kinds of social interaction, people do not always present their true views, but rather offer opinions that seem socially acceptable. One of the larger scale investigations of this phenomenon (Holbrook et al., 2003) examined the impact of telephone and face-to-face interviews on socially desirable, biased responses. From analysis of three major U.S. surveys from three different decades, the finding was that telephone interviews produced more socially desirable responses. The view of Holbrook et al. (2003) is that the greater rapport and participant engagement that emerge in face-to-face interviews lead to more genuine responses.

Although "rapport" seems an intuitive concept, researchers have derived ways of assessing rapport and the impact of communication context. Face-to-face communication has been shown to offer advantages over telephone interactions. For example, in one lab-based study (Drolet and Morris, 2000) pairs of unacquainted participants communicated by telephone or face-to-face before playing a "conflict game." Rapport measures were derived from the judges' assessments of the video record of an interaction, and the similarity of the participants' nonverbal behavior, namely, similarity between the participants' facial, postural, and gestural movements. The finding was that face-to-face contact fosters rapport, which then helps people cooperate, while telephone interactions were less effective at establishing rapport.

5.3.1 Advantages of Face-to-Face Communication

The visual signals in face-to-face interactions also have a range of impacts on how people communicate. In face-to-face conversations, when we interact with a listener whom we can see, the visual cues we exchange can have a number of effects on the communication process. These effects have been documented on the process of turn-taking in conversations, where visual cues lead us to interrupt one another less often. We also tend to exchange lots of short turns of speaking, that is, the conversation is often highly interactive. Evidence for this has typically come from laboratory-based studies using problem-solving tasks, where a problem is displayed on-screen and the participants have to discuss this material to reach a solution. The studies described here use a map task; each participant has a similar but not identical schematic map shown on his/her screen and one participant's map shows a route that he/she has to describe so that the partner can duplicate this route on his/her map.

In one set of map task interactions (Boyle et al., 1994), participants who communicated face-to-face accomplished their goal more efficiently, as some of the relevant information was carried in the visual channel. So dialogues when speakers and listeners were visible to one another contained significantly fewer words and turns of talk. There was also a significant reduction in the number of interruptions.

Doherty-Sneddon, Anderson, O'Malley, Langton, Garrod, and Bruce (1997) extended this analysis to consider the impact of visual cues on how speakers achieved a shared understanding during problem-solving interactions. They found that the "grounding" (Clark and Marshall, 1981) process, that is, conversational partners' acknoweledgment that they understand each other well enough for the current task, altered when speakers could see one another: speakers less frequently asked listeners if they understood, and listeners in turn tend to ask fewer clarification questions. Analysis of the nonverbal behavior suggested that gazes were used to check understanding when speakers and listeners could see one another. When they could not use this visual channel, more of the checking had to be done verbally, and so the dialogues were longer.

The availability of visual signals may even alter the way we articulate individual words. Anderson, Bard, Sotillo, Doherty-Sneddon, and Newlands (1997b) found that when a listener looks directly at us, we speak more carefully: the individual words are pronounced more clearly. Most of the time in a face-to-face conversation, however, we are looking at something other than the speaker and thus the average word spoken in such a conversation is spoken *less* carefully than the average word from a conversation where we cannot see the listener.

5.3.2 Disadvantages of Face-to-Face Interaction

Is more information always best? Or are there situations where lower social presence technologies, such as postal surveys or telephone interviews, offer advantages? Telephone and self-administered surveys may offer respondents greater privacy, which might be desirable particularly when they are responding to questions on sensitive or personal issues. But the relationship between interview context and privacy is not always clear-cut or predictable; depending on a respondent's domestic circumstances, face-to-face interviews in the respondent's home may provide a greater sense of privacy than a telephone interview (the respondent may be able to judge the likelihood of responses being overheard), or less privacy.

Linked to privacy is the greater sense of anonymity felt by respondents in contexts other than direct face-to-face interviews. We might predict that socially desirable biased responses would be most prevalent where social presence is high, for example, in face-to-face interviews in the home, and less frequent where social presence is reduced, as in telephone interviews. The evidence from comparisons of face-to-face and telephone interviews, however, is rather mixed; as we saw previously, a large-scale longitudinal comparison found less apparent bias in face-to-face interviews. This may be due to a trade-off between the impacts of social presence and rapport. Rapport may be greater in face-to-face interactions and hence lead to greater honesty when compared with the increased anonymity of less rich communication media.

For some groups of respondents or for some very sensitive topics, face-to-face interviews may not be optimal. Ghanem, Hutton, Zenilman, Zimba, and Erbling (2005) compared face-to-face interviews with audio-CASI (the respondent answers a self-administered questionnaire hearing the questions through headphones on a laptop computer) on sexual behavior and illicit drug taking. They found that women and respondents under 25, but not men or older participants, produced less socially desirable responses in audio-CASI interviews concerning sexual activities. In contrast, no differences were found for reported drug use. These varied findings underline the fact that the literature shows somewhat inconclusive patterns of social desirability and interview medium effects.

5.4   VIDEO CONFERENCING VERSUS TELEPHONE INTERACTION

Here we wish to explore the extent to which video conferencing offers advantages over telephone interviews, in terms of social presence, rapport, satisfaction, and communication efficiency.

5.4.1   Social Presence

Face-to-face interactions are considered to be high in social presence due to the abundance of verbal and visual cues (gaze, gesture, facial expressions) that are available, while telephone or e-mail are regarded as providing less social presence. As Clark and Brennan (1991) outlined, different communication media can be distinguished in terms of their characteristics and how these will affect the grounding process, or how communicators achieve a common understanding. In these terms video conferences offer (1) visibility—participants can see one another; (2) audibility—participants can hear one another; (3) co-temporality—participants interact at the same time; and (4) sequentiality—the communicative messages that participants exchange follow one another in a known temporal order. Video conferences should provide more of a sense of presence than telephone conversations, which lack visibility, or e-mail, which lacks visibility, audibility, and co-temporality. The extent to which any particular video conference actually provides an equivalent sense of social presence to face-to-face interactions will depend on the quality of the system. Most current video-conference systems provide restricted views, and this and various other technical limitations mean they are likely to provide rather less social presence.

Visual cues can affect both the content of communication and social perceptions, as demonstrated in studies (Stephenson et al., 1976) of negotiations conducted via video or audio conference. Participants with differing views on labor relations role-played negotiations between management and unions. Where visual cues were lacking, the discussions were more impersonal and task-focused, and there was more overall disagreement and more negative expressions about opponents. Other social advantages of video conferences have been reported in similar studies, including more positive views of collaborators and more feeling of social engagement with them (Reid, 1977; Rutter, 1987; Short et al., 1976). The social dimensions of interactions seem to be more influenced by the increased social presence engendered by visual signals (see also Fussell et al., Chapter 11 in this volume).

There are a number of studies that show that video-conference technology can offer benefits in terms of the users' sense of social presence with a remote collaborator. Tang and Isaacs (1993) conducted workplace studies of groups of co-workers, located at three locations in the United States, who were using desktop video conferencing as part of their ongoing collaborations. People using video conferences felt that they were better able to manage gaps and pauses than when communicating by telephone. Tang, Isaacs, and Rua (1994) studied a multidisciplinary team of co-workers, located in three different buildings, who had to collaborate to achieve their work goals. Video users were enthusiastic about the technology and could use the visual signals from facial expressions and body postures to help interpret people's reactions when difficult issues were being discussed.

In one study in my lab (Anderson et al., 1996), we explored the use of video conferencing to support long distance international collaboration. In this study, students in Scotland had to communicate with The Netherlands via video or audio conferencing. The participants had to take the role of a customer who wishes to plan a holiday trip across the United States, by seeking advice and information from the travel agent in The Netherlands. In questionnaires completed after the interactions, we found that participants reported a significantly greater sense of social presence in the video conference condition, with 92% stating that they were "very aware" or "fairly aware" of the agent versus only 50% selecting these options after audio interactions. This kind of response to video conferencing may have impacts on survey participants' willingness to complete a survey interview, or to engage in the mental effort required to respond accurately in comparison to telephone interviews.

So video has been shown to offer advantages in terms of users' perceptions and in their greater sense of social presence with their interlocutor. But there are also domains where audio communication is quite adequate and video does not seem to add value. Video technology has been compared to audio-only communication in a number of early studies, in terms of problem-solving task outcomes. For simple problem-solving tasks, the addition of video does not usually influence the quality of output or time to completion (Chapanis et al., 1972, 1977; Short et al., 1976). Reid (1977) reviewed the literature and concluded that for such cognitive tasks, conducted in laboratories, speech was sufficient. The quality of the video is not likely to be a critical factor in this respect, as these studies showed that audio-only interactions produced as good outcomes as face-to-face interactions.

5.4.2   Potential Disadvantages of Video

Are there situations where telephone interactions are not only as good as video but may offer some advantages? Privacy is one area where the video-conference interaction might cause problems that would not occur in a telephone survey. In telemedicine, for example, concerns over privacy have been discussed as one possible reason for the slow take-up of telemedical facilities in the United States, despite substantial federal funding. In examining the use of medical consultation by video conference, Paul, Pearlson, and McDaniel (1999) found that even when appropriate protocols have been created to ensure that privacy is respected, problems can nonetheless arise. For example, privacy can be violated when an IT support person needs to be present or when other staff enter the video-conference facility during consultations (the equipment can sometimes be placed for technological reasons rather than for privacy). It may be possible to address such issues in survey interviews by following protocols, which, for example, determine that if technical difficulties arise requiring the presence of an IT support person, that an interview about a sensitive topic is suspended. Similarly, the interviewee should be made aware if there is a possibility of others at the interviewer's site overhearing/seeing the interview, and assurances about professionalism, anonymity, and so on should be given.

5.5 VIDEO CONFERENCES VERSUS FACE-TO-FACE INTERACTIONS

5.5.1 Communication

Research on video conferencing has shown that the technology can have a range of impacts on the communication process. Logically, one would expect the effects of video conferencing should be the same as the effects of visual signals in face-to-face communication. One would also expect that the technical characteristics of the systems used should determine the nature of the effects. One of the most widely replicated impacts of the use of video-conferencing technology is on the process of turn-taking. Systems that, due to bandwidth constraints, incorporate transmission delays, so that, for example, the video images and the sound signals are synchronized but both arrive with a $1/2$ second delay, cause considerable difficulties in conversations. In one of the studies we conducted (Anderson et al., 1997a), we compared participants' abilities to communicate and collaborate to complete a shared problem-solving task (the map task) using various technologies. One of the technologies we studied was a videophone system, which had a $1/2$ second delay in the transmission. This was compared with the same system with the video images disabled but audio still delayed, and with normal telephone interaction. We found that both delayed conditions led to much greater difficulty in achieving smooth turn-taking. Participants kept interrupting one another, with over 50% of turns containing an interruption. This compared unfavorably with normal telephone conversations without transmission delays, where we found only 15% of turns interrupted. This difficulty in turn-taking interfered with the participants' ability to collaborate to get their task completed successfully; task performance was 36% poorer in both conditions with delay. The addition of video images gave no benefits if this involved incorporating a delay, and in fact led to much more difficult communication.

Video conferences can offer advantages over telephone calls, but they are not usually able to provide as much of a sense of social presence as face-to-face interactions. Even this apparent shortcoming can offer some advantages. The human face is a physiologically arousing stimulus, and speakers can find the face of their listener distracting. So when speakers are planning a complex utterance or answering a difficult question, they will look away, and if they are encouraged to keep looking at the listener they will often become disfluent (Beattie, 1981). The extent of gaze aversion has been shown to relate to the cognitive difficulty of answering questions (Glenberg et al., 1998) or to the unreliability of an answer to a survey question (Conrad, Schober, and Dijkstra, 2004).

The evidence on the impact of faces mediated by technologies is mixed. Ehrlichman (1981) in a lab study found that looking at faces presented over video did not interfere with participants' speech planning when responding to factual questions being posed by a video interviewer. Research in our lab (Mullin et al., 2001) has shown that participants do not look as often at their interlocutor in a small desktop video conference window as they do in face-to-face interactions, even if the video image refreshes very frequently at 25 frames per second. In these studies the participants were collaborating to complete shared tasks; in one task the participants collaborated to design a tourist poster, and in another task one participant guided the other across a map presented on their computer screens. The relevant task materials were presented on part of the computer screen, with the video image also on screen (similar to the center left illustration in Fig. 5.1).

In contrast, in a lab prototype video system, where we had no network or other practical constraints, we were able to produce video-conference images that were presented as life-size images and the system was configured to support direct eye contact over the video link. When we explored performance using this system on a problem-solving task (the map task described previously), we found that participants looked at one another far *more* often than they did in equivalent face-to-face interactions. Perhaps as a result, participants may have become distracted. In this video context we found that significantly more needed to be said to complete a shared task than in audio-only interactions (Doherty-Sneddon et al., 1997).

Even this atypically realistic video setup does not necessarily produce an equivalent experience to face-to-face interactions—and, surprisingly, in some circumstances this might be helpful. Doherty-Sneddon and McAuley (2000) used this video-conference system, with life-size images and direct eye contact, to conduct interviews with children. The children (aged 6 and 10) took part in a staged event concerning lost property. They were later asked to recall the event and then were asked a series of open-ended, closed, and misleading questions about the incident, either in a face-to-face interview with an adult or via the high quality video system. The video and face-to-face interviews were compared. In the video interviews the children produced less incorrect information in their free recall and the younger children were far more resistant to misleading questions from an adult interviewer [e.g., "The blue car you came in today is really comfortable, isn't it?" (when the car was red)]. Doherty-Sneddon and McAuley suggest that the reduced social presence of the video interview lessens the younger children's sense of intimidation when being questioned by an adult and makes them more confident in reporting their recollections. In surveys where there are concerns over power/status effects and hence the ability of some groups of respondents to present their own views, video-conference technology may be advantageous. Data quality would be improved if, as Doherty-Sneddon and McAuley suggest, video lessens the feeling of intimidation felt by lower status participants in face-to-face interviews and hence the need to respond as the participant feels the interviewer expects.

5.5.2   Social Desirability Bias

From the existing literature it is quite hard to predict the impact of the use of video-conferencing technology on social desirability responses, but on balance video conferencing may offer some advantages over telephone interviews in eliciting genuine responses. Video conferencing could be used by interviewers to establish good rapport with interviewees, for example, by the use of visual signals such as nods or facial expressions, which encourage responses or show appropriate emotional responses to replies, so that the interviewee is encouraged to complete the survey and to share his/her true experiences/views. For very sensitive topics, however, it may be that a recorded video of the interviewer asking the questions with touch-screen responses may be the best mix for balancing rapport and privacy to maximize the accuracy of responses.

One particular aspect of the social desirability bias is the impact of the race of the interviewer on the responses given. It would be useful to know the likely impact of video technology on this "race-of-interviewer effect." In the traditional survey literature, there are several reports of significant race-of-interviewer effects in face-to-face interviews. Hatchett and Schuman (1976) investigated the responses of white respondents to African-American and white survey interviewers. They found that respondents gave more "liberal" responses when the interviewer was African-American, just as in earlier research on the responses of African-American survey interviewees responding to white survey interviewers. The interpretation given is that in face-to-face interviews, usually conducted in the respondents' homes, the respondents adjust what they present of their views to "get through the racial interaction with minimal tension." These phenomena have been replicated in more recent studies where African-American respondents report more favorable attitudes to whites if the interviewer is white (Anderson et al., 1988). White interviewees being interviewed by African-American interviewers show similar effects (Finkel et al., 1991).

Might the reduced social presence offered by video interviews allow respondents to share their true views more freely, as they are less concerned about the interviewer's reactions to what they say? The Doherty-Sneddon and McAuley (2000) findings that children are less likely to acquiesce with adults' questions in video interviews might suggest this. There is little available relevant data but one study by Couper (2002) offers a less optimistic view. Couper (2002) reports that race-of-interviewer effects occur in video-CASI interviews (Krysan and Couper, 2001), that is, when a video is shown on a laptop computer screen showing an interviewer reading survey questions for the respondent to answer.

In contrast, Kroek and Magnuson (1997) found that the reduced social presence of video offered some advantages for cross-race interviews. In a study of the use of video conferences in job selection, where the interviewers were predominantly white, African-American job candidates were significantly more favorable in their responses to video interviews than white or Hispanic groups. In their questionnaire responses after their interviews, African-American applicants also reported higher ratings for the perceived conversational 'flow' during the interview. Kroek and Magnuson speculate that the lack of direct eye contact, which is usually perceived as a disadvantage of video communication, may be beneficial if candidates feel vulnerable or uncomfortable, for example, when taking part in a cross-race interview with significant personal consequences such as job selection. Note, however, that Kroek and Magnuson are cautious about this finding, as the number of African-American applicants was relatively small.

For survey responses that directly address ethnicity issues, the moving images of the video interview may provide sufficient reminders of the interviewer's ethnicity, and hence perceived reactions, to make interviewees adjust their responses to avoid causing offense. For such topics telephone interviews may elicit more accurate responses. Video may offer advantages for general surveys if some respondents feel more relaxed being questioned by an interviewer from a different ethnic background than they would in a face-to-face setting.

5.5.3 Rapport

Many accounts of interviewing describe the importance of "rapport" between interviewer and interviewee. There are, however, a few problems with the concept, even in traditional face-to-face interviews, not only because there are relatively few empirical studies of the impact of rapport despite its apparent importance, but also because there is little consensus on the definition of rapport or its impact on interviews. In the existing studies on the influence of rapport on interview data, there are conflicting patterns of results (Goudy and Potter, 1976). Hill and Hall (1963) and Weiss (1968) both reported that greater rapport between interviewer and interviewee led to *less* valid responses. In contrast, other studies (Williams, 1968) reported the more expected pattern, namely, that greater rapport leads to more honest answers. After conducting their own research on rapport, Goudy and Potter concluded that given the lack of consensus on how to define and operationalize rapport, "further empirical studies may be useless." As we saw earlier, some progress has nonetheless been made in ways to operationalize and investigate rapport in comparing face-to-face and telephone interactions (Drolet and Morris, 2000).

Many articles about the use of video conferencing similarly stress the importance of establishing rapport, whether in distance education, telemedicine, or business (Dooley et al., 1999; Manning et al., 2000; Simpson et al., 2005), although again this is usually not operationally defined and there are few empirical studies of this aspect of video conferencing.

So although we might assume that rapport might be harder to achieve given the relative unfamiliarity of the medium and the impoverished visual cues compared to face-to-face communication, this may not always be true and even if it is it may offer advantages for some individuals in "sensitive" contexts.

In the domain of telepsychiatry/telecounseling, the issue of rapport is of particular concern, and the potential impact of technology on this has been a concern. In a single case study tracking the reactions of a psychiatrist and client over 10 sessions,

Ghosh, McLaren, and Watson (1997) report a reduction in the client's rated bond with the therapist in video-conferencing counseling compared to face-to-face support. The therapist quickly became used to the new technology and reported that he felt entirely comfortable by the third session, but the client did not show this degree of comfort or progress. The authors attribute much of this difference to the particular client's difficulty and conclude that effective therapy took place despite the client's perceptions that there were some difficulties around discussing sensitive issues in the video sessions. The role asymmetry between therapist and client/patient bears at least surface similarity to the different roles played by a survey interviewer and respondent. It is therefore possible that survey respondents will report lower levels of rapport with video than face-to-face interviewers, while video interviewers will report generally positive feelings about the communication in video interviews.

Miller and Gibson (2004) conducted detailed qualitative content analysis of recorded video-conference interviews in the supervision of 26 trainee psychologists, and they found some effects of the technology on perceived power and status. Trainee psychologists interacting with their supervisors via video conference felt less equal than those in traditional face-to-face supervision meetings. But the effects on emotional rapport were more mixed. Some trainees felt less willing to reveal emotional topics in video sessions compared to face-to-face and may have felt less engaged with their supervisor. However, some trainees felt more able to discuss emotional issues in video meetings and some preferred to use telephone rather than video sessions for emotional content.

One of the few empirical studies of the impact of different video technologies on rapport (Manning et al., 2000) investigated if self-ratings of rapport between clients and therapists were lower after video therapy sessions compared to face-to-face. The prediction was that the signal delay of some forms of video conference would be particularly detrimental to establishing rapport, and so the study compared three levels of video (zero delay, 300 ms, and 1000 ms delays) with face-to-face interactions. No significant differences were found in the self-reported rapport ratings of male clients. Female clients, in contrast, reported significantly lower rapport ratings in face-to-face counseling sessions. Manning et al. (2000) conclude that delay was not the issue but that the female clients felt more comfortable with their male counselors in the more distant video-mediated interactions.

Another recent study is also generally positive. Simpson, Bell, Knox, and Mitchell (2005) used video-conferencing technology to support six patients with eating disorders who lived in remote areas. One of their key concerns was whether the technology could be used to establish trust and rapport between patient and therapist. They found the technology was effective for all the clients in establishing a rewarding therapeutic relationship with their therapists. Indeed, three clients preferred the technology to traditional face-to-face interactions as they felt they had more control and could always turn off the monitors. Two clients expressed no preference and only one would have preferred traditional face-to-face contact. The authors speculate that this technology might offer particular benefits for "shame-based difficulties." We might consider this analogous to some of the personal issues around "antisocial behaviors" that some surveys need to probe.

From these studies it seems that, in general, a reasonable degree of rapport can be established during video conferences. While the rapport may not be as strong as that in a traditional face-to-face therapy session, this may not always be a disadvantage if sensitive issues are being addressed. The level of rapport that can be achieved via a video link will plausibly be adequate for most surveys, which are less emotionally demanding than telepsychiatry, at least if interviewers become used to the technology and make some effort particularly at the start of video sessions to make the interviewee feel welcome and relaxed.

5.5.4   More and Less Structured Interviews

In the previous sections the impact of video technology on various unconscious aspects of social behavior have been considered. But how do participants rate the usefulness or effectiveness of video-conference interviews? Does the nature or structure of the interview affect these judgments? How does video compare to face-to-face interactions in this respect? One domain where video technology has been tested in this way, and which has some similarities with surveys, is in selection and recruitment.

Chapman and Rowe (2002) investigated the effect of communication medium on genuine job applicants' perceptions of the attractiveness of organizations, interviewer friendliness, and performance. In the study, half the applicants were interviewed over a reasonably high quality video link (Video frame rate of 12–14 frames per second) and half by traditional face-to-face interviews. Although the applicants rated the interviewer's performance less favorably in video interviews, there was also an interaction of medium and interview structure. Less structured interviews led to more favorable perceptions of the organization when the interviews were conducted face-to-face. In contrast, more structured interviews led to *more* favorable perceptions of the organization when conducted via video. Chapman and Rowe speculate that the structured interview with its formal question–answer format may be particularly suitable for video, as it may help overcome the difficulties of rapid exchanges of turns of speaking, characteristic of more informal interactions (see Section 5.5.1). Because survey or market research interviews tend to be similar to the more structured interviews in Chapman and Rowe's study, survey participants might be more willing to participate in future surveys conducted by an organization using video conferencing, given that it increases favorable perceptions of the organization.

Another factor is relevant to structure in interviews. In an earlier study, 75% of interviewers reported that they changed their interview script when using video-conference technology (Chapman and Rowe, 2002). The likely causes include technological restrictions of many video systems, which can lead to changes in the surface structure of conversations, such as increasing the length of turns of speech, prolonging a turn of speech rather than allowing another speaker to take a turn of talk, and trying to reduce the number of interruptions. Depending on the nature of the changes to the script, effects on survey data quality are likely to be detrimental although some sorts of changes to the script can have positive effects on response accuracy (Conrad and Schober, 2000). This suggests that survey designers should particularly attend to effects of video mediation on implementing standardized interviewing in desirable ways.

5.5.5 Caveats

There are a few important caveats that have to be kept in mind when we consider the implications for surveys of the findings in the research literature on video conferencing. First, most of the studies of communication in video conferences have involved unscripted problem-solving interactions of various kinds. These are rather different from survey interviews in a number of ways. In particular, they are usually highly interactive; speakers and listeners swap these two roles frequently and rapidly, unlike in most survey interviews, where the interviewer is in control of the interaction throughout and is following a script.

Second, the content of the problem-solving interactions is often concerned with the present physical environment, and so both parties have access to the facts. In contrast, survey questions tend to concern concepts that are not immediately visible to both parties, such as the respondent's past experiences, activities, or attitudes, and interviewers do not have access to what an answer ought to be. So the nature of the topics, the kinds of misinterpretations, and the potential solutions for resolving misinterpretations differ in certain fundamental ways. For example, in a map task, participants may struggle to know where to go if directed "Can you see the woods?" when their map shows two woods. Here the participants are unlikely to have difficulty interpreting the terms used, but only in identifying the intended referent. Unlike in most surveys, there is no need to recall personal experiences to answer the questions, and no judgment to be made about how individual circumstances fit the question.

In contrast, the kinds of communication problems encountered in surveys can involve knowing how one's personal circumstances map onto terms as defined by survey designers. As described by Schober and Conrad (1997), respondents' circumstances may not map onto survey definitions in a straightforward way, leading to a different sort of clarification dialogue than one sees in a map task:

I: How many hours do you usually work?
R: Does "usually" mean on average?

As they report (see also Conrad and Schober, 2000), nearly every survey question includes terms that some respondents will have trouble knowing how to answer because of their particular circumstances and conceptualizations [see Suessbrick et al. (2000) on the surprising variability in interpreting what counts as having smoked 100 cigarettes in one's entire life]. Other response problems in surveys can result from trouble remembering, trouble formulating answers, and other aspects of the entire response process, none of which are at issue in a map task. So it remains to be seen what the effects of video conferencing will be in different kinds of surveys, and the effects may differ for surveys about nonsensitive facts and behaviors versus opinions and attitudes versus sensitive behaviors.

The other caveat concerning the implications of the existing research literature on video conferencing is the demographics of the participants in previous research. Most studies have involved pairs of university students or groups of business people. The survey research community has to consider a much wider cross section of the population. As yet we know very little about how this wider community will react to video-conference interviewing, and the landscape is changing quickly as more and more people use video conferencing for business and personal purposes.

5.6   VIDEO CONFERENCE VERSUS VIDEO CONFERENCE: TECHNOLOGICAL FEATURES

So far we have made general comparisons of video conferences with telephone or face-to-face interactions. Video-conference systems, however, differ greatly in their technical characteristics. As illustrated in Fig. 5.1, the size of the images displayed and their location, whether as a small or larger image on a desktop computer or as larger display in a specialist video-conference room, can differ substantially. Even within today's desktop displays the type of video image varies. One dimension on which systems differ is the quality of the signal transmission (image resolution) and the frequency with which images are updated (frame rate). People can detect these differences in quality and respond to them. Designers of video systems often choose to maximize spatial resolution and color (i.e., image qualities) rather than frame rate when there are bandwidth restrictions on the network. This may not be the best design decision for effective video interviews.

Low frame rate video has been reported to cause physiological stress responses (Wilson and Sasse, 2000). These quality differences can affect not only the way communication proceeds but also technology users' ability to process certain important nonverbal signals. Ehrlich, Schiano, and Sheridan (2000) found that the ability to detect people's emotions from video clips was improved in moving images even when these had poorer image resolution. Horn (2001) found that interviewers' ability to detect lies, in a simulated job interview, was significantly reduced when the video conference link had very low frame rate (updates only 3 times per second). Quality effects are not always straightforward, however. In a similar study that compared three levels of image resolution, combined with three levels of frame rate, Horn, Olson, and Karasik (2002) found that lie detection was significantly better in video interviews with poorer image resolution.

On balance, it seems that the technological feature that is most likely to be beneficial for survey interviews is adequate frame rate, as adequate frame rates (12–15 per second or more) seem to offer advantages for smooth, interactive communication, as well as providing better cues for the ability to process nonverbal emotional signals. The ability to respond to these kinds of nonverbal signals may be important in social dimensions of survey interviews such as the process of establishing rapport. Responding to such signals may also allow interviewers to maintain interviewees' engagement and thus to help ensure interview completion and data quality.

The relative importance of the quality of the video signal and the audio signal has been compared in several studies: good quality audio is critical. For example, a survey of telemedical implementations by Paul et al. (1999) explored the impact of the quality of the technology. The quality of the video images was not perceived to be a problem but the quality of the audio often was. Issues concerned the delay in transmission, which impeded the ability to conduct normal conversation. This study also highlights the mundane issues around successful implementation of video-conference technology. These include things such as the acoustics of the room chosen for the conference facility. One of the IT professionals involved in this trial was quoted as follows: "if you expect good acoustics from a hollow concrete box you've got another think coming."

In addition, there were problems with microphones. Such low tech aspects of the deployment of video conferencing should not be underestimated. In several years of experience of using this technology in my lab, we found that huge improvements were often made by very simple things like ensuring adequate lighting, providing higher quality microphones and speakers, and taking time to ensure these were properly adjusted before video conferences.

So one clear recommendation for survey researchers is that if the video technology that is being considered would involve audio transmission delays this is likely to cause real difficulties in interviews. As Houtkoop-Steenstra (2002) illustrates, the complex structure of many survey questions already provides many challenges to interviewers and respondents, particularly the latter as they try to use "normal" conversational behaviors to take turns of talk. Houtkoop-Steenstra illustrates how this can lead to misinterpretations, for example, when an interviewee responds before the clarification of the terms of the question have been read by the interviewer. If the benefits of video-conference technology are to be achieved in survey interviews, it will be very important that the turn-taking process does not become even more difficult because of poor quality video conferencing with signal delay.

The rate at which a given video-conference system updates the images transmitted (frame rate) has also been shown to influence the communication process. Even high frame rate video-conference systems, with images updating 25 times per second, can interfere with turn-taking. In a recently published paper, we report on a comparison of high frame rate VMC and face-to-face interaction (Anderson, 2006). The VMC participants achieved just as good task outcomes as those in face-to-face interactions, and their dialogues were very similar in length and content. The video conversations, however, contained significantly more interruptions than the face-to-face interactions (an average of 16% versus 8% of turns contained an interruption). With this standard of video conferencing it seems that, although there are effects on turn-taking, these are not so distracting that they have a serious impact on collaboration—although they may cause some strain to the participants.

Howarth and Anderson (2007) compared high frame rate video-conferencing and face-to-face problem-solving interactions. The study again showed that tasks can be completed just as effectively via video-conference interactions, and that the dialogues in the two contexts are similar in length. Here the particular interest, however, was the way new topics of conversation were grounded in the two settings. We explored the impact of the technology on how interactive participants were when they introduced new topics in the conversation, which concerned giving route directions. We compared the extent to which participants used interactive sequences at these points in their dialogues. That is, we noted the number of times speakers asked questions about their listener's knowledge BEFORE they gave instructions. From earlier research, (Anderson and Boyle, 1994) we knew that such preparatory questions are very effective for establishing common ground (share discourse context) because they elicit responses from listeners which highlight any differences in knowledge or perspective, without the listener's having to interrupt or initiate a clarification. This is illustrated in the extract below in the way the landmark "burnt forest" is introduced.

Speaker A: "OK?"

Speaker B: "Right."

Speaker A: "Right starting at the top left isn't it?"

Speaker B: "Right."

Speaker A: "Right—if you come down sort of south–south east. You got a burnt forest?"

Speaker B: "Aye."

Speaker A: "Now come down round the left of the burnt forest . . . ."

We compared the frequency of these kinds of exchanges to less interactive introduction sequences, where the new landmark "camera shop" was introduced, embedded in the ongoing instruction:

Speaker A: "Right start g- go down ten centimeters down the left hand side of the camera shop."

Speaker B: "Mhm mhm."

Speaker A: "And round the bottom of the parked van . . . ."

In video conferences, the proportion of interactive introductions was significantly lower (0.31) than in face-to-face conversations (0.47). When participants conducted their interactions under time pressure, they also used significantly fewer interactive sequences in both face-to-face and video-conference sessions. Time pressure effects did not interact with the influence of technology. It seems that the novelty of the technology may cause some strain on participants and make them less interactive in their communications. This behavior in video-conference interactions seems somewhat analogous to the increased rates of less effortful responses, the so-called satisificing described by Krosnick (1991, 1999) in survey responding.

For survey researchers, it is worth remembering that novel communicative settings may cause strain—and to bear this in mind when matching surveys of different lengths and for different demographic groups, to means of delivery. It may be that very lengthy interviews may not be the most suitable for initial explorations with new technologies such as video conferencing. Respondents, notably those with less educational qualifications, have been found to make less effort to respond informatively, partic- ularly after some time in telephone interviews compared to face-to-face interviews. We might learn from this when considering the best way to deploy technologies such as video conferencing.

Humans are adaptable communicators, however, and it may be we learn to take communicating by video conference in our stride just as we have adapted to earlier communication technologies, from the telephone to the Internet to the cell phone.

We have some evidence of the ways in which video-conference users adapt their communicative style. Jackson, Anderson, McEwan, and Mullin (2000) explored how speakers in video conferences grounded new topics in their conversation. The study compared how this was done in two different video-conference settings, high quality (with frame rates of 25 frames per second) compared to low quality (with frame rates of 5 frames per second). We found that participants became more communicatively cautious that they had understood one another in the low frame rate sessions.

The task for participants was to design a tourist poster, and this involved describing photographs of locations. The lengths of the descriptions provided were significantly longer in the low frame rate condition when photographs were introduced, and on second and subsequent mentions. It seemed that given the quality of the visual feed- back, speakers could not be confident that their listener understood them. They very quickly and unconsciously adapted to this constraint by providing more information to ensure they were achieving common ground.

The way participants rapidly adapt to the constraints of new communication tech- nologies and their limitations can be important. In one study we conducted on the impact of video-conferencing technology on communication, we found that the way participants perceived the system and its limitations was important (Sanford et al., 2004). We compared the way participants communicated when using two different low frame rate (5 frames per second) video-conference systems. In one system the audio communication was "open channel": both participants could speak at any time. This of course is analogous to face-to-face conversation and is generally regarded as a desirable feature of a video-conference system. In the second, "click to speak" system, only one participant could speak at any time and (s)he had to click an icon on screen to open the audio channel.

We analyzed the communication process and task outcomes of these two settings, and compared the results to those obtained in face-to-face interactions. Contrary to expectations, dialogues from the open channel video conferences led to significantly poorer task outcomes. Participants in these interactions interrupted one another a good deal more often (17% of turns with interruptions) than those in the 'click to speak' video conferences (only 4% of turns contained interruptions). The apparently less sophisticated system where speakers had to "click to speak" seemed to lead partici- pants to adjust their communicative style. In these conferences, speakers took longer to complete the task, they said more and they asked one another more clarification questions. It seemed they had adjusted to the restrictions of the system.

In contrast, in open channel video conferences, participants communicated much more as they would in face-to-face interactions. The quality of the video-conference system, with its very modest frame rate, did not provide good enough visual information to support this style of interaction and the overall outcome of the exchanges deteriorated.

So when bandwidth and technology are limited, it is best to make these limitations clear to participants, who may well be able to adapt and overcome these shortcomings.

In sum, video-conferencing technology clearly can have a number of noticeable effects on communication. The most ubiquitous of these is probably the impact on turn-taking, with speakers generally more likely to interrupt one another during a video conference. The extent of the problem and the impact of the increased level of interruptions depend on the technical characteristics of the system and the nature of the interaction. Provided a reasonable quality level is achieved, with good audio with no delay and frame rates of over 12 refreshes per second, the technology is likely to support effective interactions and to be suitable for either standardized or more collaborative interviewing. Given that Conrad and Schober (2000) have shown distinct advantages of the latter for survey data quality, I propose that any introduction of technology must be able to support this more conversational style of interaction. The performance of any given video-conference technology must be assessed against this requirement, but the aforementioned standards are potential quality benchmarks. The technology is also likely to cause speakers to make subtle changes to the way they communicate, noticeably on the way they attempt to ensure they achieve a shared understanding with their listener. These differences, although of research interest, do not seem likely to have major impacts on the quality of survey data.

5.7   CONCLUSIONS

Video-conference technology is becoming widely available and potentially offers another mechanism to recruit and retain participants for surveys. As has been emerging throughout the discussion in this chapter, video clearly adds information when people are interacting with one another. The question for survey interviews is: When does this extra information help and when does it hurt? Does increased awareness of the interviewer relative to a telephone interaction lead to greater rapport and higher completion rates or does it increase social presence and reduce candour in reporting sensitive behaviors? How are video interviews different from face-to-face interviews; that is, what is the impact of physical copresence or its absence when the interviewer can be seen? What exactly are the cues passed in face-to-face interviews that might not be passed via video, or conversely, are there some cues that are more salient in video than face-to-face interviews?

Video interviews are likely to appeal more to some demographic groups than others, at least initially. It may be that those sections of the population who are more technologically experienced, such as younger members of the population and young men in particular, may be attracted to surveys conducted via this technology. This may offer some advantages in recruitment. In the United Kingdom there is considerable concern over the absence of certain groups, notably young men, in surveys and national census; a video technology solution might be a suitable additional route to engaging such hard-to-reach groups.

Some older or less technologically experienced groups may not be sufficiently confident or proficient to use such technology unsupported, which suggests that the notion of video interview centers may have to be considered. The privacy dimensions, both positive and negative, of such an approach should be carefully considered. The video interview center option does offer the opportunity to ensure that everything is done to maximize the quality of the video interaction, including the low tech issues highlighted earlier. The growing availability of this technology in medical centers, schools, colleges, and businesses means that for different types of surveys this may well be a viable option.

As has been discussed throughout this chapter, video-conferencing technology is in fact a whole range of possible technologies, and the suitability of adopting a particular technological solution for a particular survey needs careful consideration. The research evidence suggests that reasonable quality video conferencing is likely to offer increased sense of social presence over telephone or e-mail. This in turn is likely to lead to greater rapport and engagement, which is likely to increase survey data quality.

This highlights the overarching conclusion for this chapter. Video-conference technology potentially offers several advantages for surveys, *but* research on its detailed impacts on real survey interviews is urgently needed. Pilot studies of this kind should be undertaken before any widespread adoption is planned.

REFERENCES

Anderson, A. H. (2006). Achieving understanding in face-to-face and video-mediated multi-party interactions. *Discourse Processes, 41*(3), 251–287.

Anderson, A. H. and Boyle, E. (1994). Forms of introduction in dialogues: their discourse contexts and communicative consequences. *Language and Cognitive Processes, 9,* 101–122.

Anderson, A. H., Newlands, A., Mullin, J., Fleming, A. M., Doherty-Sneddon, G., and van der Velden, J. (1996). Impact of video-mediated communication on simulated service encounters. *Interacting with Computers, 8*(2), 193–206.

Anderson, A. H., O'Malley, C., Doherty-Sneddon, G., Langton, S., Newlands, A., Mullin, J., Fleming, A. M., and van der Velden, J. (1997a). The impact of VMC on collaborative problem-solving: an analysis of task performance, communication process and user satisfaction. In K. Finn, A. Sellen, and S. Wilbur (Eds.), *Video-Mediated Communication* (pp. 133–155). Mahwah, NJ: Lawrence Erlbaum.

Anderson, A. H., Bard, E., Sotillo, C., Doherty-Sneddon, G., and Newlands, A. (1997b). The effects of face-to-face communication on the intelligibility of speech. *Perception and Psychophysics, 59,* 580–592.

Anderson, B., Silver, B., and Abramson, P. (1988). The effects of the race of the interviewer on race-related attitudes to black respondents in SRC/CPS National Election Surveys. *Public Opinion Quarterly, 52,* 289–324.

Beattie, G. (1981). A further investigation of the cognitive interference hypothesis of gaze patterns during conversation. *British Journal of Social Psychology, 20,* 243–248.

Boyle, E., Anderson, A. H., and Newlands, A. (1994). The effects of eye-contact on dialogue and performance in a co-operative problem-solving task. *Language and Speech, 37*(1), 1–20.

Chapanis, A., Ochsman, R., Parrish, R., and Weeks, G. (1972). Studies in interactive communication: I. The effects of four communication modes on the behavior of teams during cooperative problem solving. *Human Factors, 14*, 487–509.

Chapanis, A., Ochsman, R., Parrish, R., and Weeks, G. (1977). Studies in interactive communication: II. The effects of four communication modes on the linguistic performance of teams during cooperative problem solving. *Human Factors, 19*, 487–509.

Chapman, D., and Rowe, P. (2002). The influence of video conference technology and interview structure on the recruiting structure of the employment interview: a field experiment. *International Journal of Selection and Assessment, 10*(3), 185–197.

Clark, H. H., and Brennan, S. (1991). Grounding in communication. In L. Resnick, J. Levice, and S. Teasley (Eds.), *Perspectives on Socially Shared Cognition* (pp.127–149). Washington DC: American Psychological Association.

Clark, H. H., and Marshall, C. (1981). Definite refernce and mutual knowledge. In A. K. Joshi, B. L. Webber, and I. Sag (Eds.), *Elements of Discourse Understanding* (pp. 10–62). Cambridge: Cambridge University Press.

Conrad, F., and Schober, M. (2000). Clarifying question meaning in a household telephone survey. *Public Opinion Quarterly, 64*, 1–28.

Conrad, F. G., Schober, M. F., and Dijkstra, W. (2004). Nonverbal cues of respondents' needs for clarification in survey interviews. In *Proceedings of the American Statistical Association, Section on Survey Research Methods.* Alexandria, VA: American Statistical Association.

Couper, M. (2002). New technologies and survey data collection: challenges and opportunities. Invited Address, International Conference on Improving Surveys, Copenhagen, August 2002.

Doherty-Sneddon, G., Anderson, A. H., O'Malley, C., Langton, S., Garrod, S., and Bruce, V. (1997). Face-to-face and video mediated communication: a comparison of dialogue structure and task performance. *Journal of Experimental Psychology: Applied, 3*, 1–21.

Doherty-Sneddon, G., and McAuley, S. (2000). Influence of video mediation on adult–child interviews for the use of the live link with child witnesses. *Applied Cognitive Psychology, 14*, 379–392.

Dooley, K., Van Laanen, P., and Fletcher, R. (1999). Food safety instructor training using distance education. *Journal of Extension, 37*, 3–10.

Drolet, A., and Morris, M. (2000). Rapport in conflict resolution: accounting for how nonverbal exchange fosters coordination on mutually beneficial settlements to mixed motive conflicts. *Journal of Experimental Social Psychology, 36*, 26–50.

Ehrilch, S., Schiano, D., and Sheridan, K. (2000). Communicating facial affect: it's not the realism it's the motion. *Extended Abstracts of CHI 2000* (pp. 251–252). Washington DC: ACM.

Ehrlichman, H. (1981). From gaze aversion to eye-movement suppression: an investigation of the cognitive interference explanation of gaze patterns during conversation. *British Journal of Social Psychology, 20*, 233–241.

Finkel, S., Guterbock, T., and Borg, M. (1991). Race of interviewer effects in a pre-election poll: Virginia 1989. *Public Opinion Quarterly, 55*, 313–330.

Ghanem, K., Hutton, H., Zenilman, J., Zimba, R., and Erbling, E. (2005). Audio-computer assisted self interview and face to face interview modes in assessing response bias among STD clinic patients. *Sexually Transmitted Infection, 81*, 421–425.

Ghosh, G., McLaren, P., and Watson, J. (1997). Evaluating the alliance in videolink teletherapy. *Journal of Telemedicine and Telecare, 2*, 33–35.

Glenberg, A., Shroeder, J., and Robertson, D. (1998). Averting gaze disengages the environment and facilitates remembering. *Memory and Cognition, 26*, 651–658.

Goudy, W., and Potter, H. (1976). Interview rapport: demise of a concept. *Public Opinion Quarterly, 39*, 529–543.

Hatchett, S., and Schuman, H. (1976). White respondents and race-of-interviewer effects. *Public Opinion Quarterly, 39*(4), 523–528.

Hill, R., and Hall, N. (1963). A note on rapport and the quality of interview data. *Southwestern Social Science Quarterly, 44*(3), 247–255.

Holbrook, A., Green, M., and Krosnick, J. (2003). Telephone versus face-to-face interviewing of national probability samples with long questionnaires. *Public Opinion Quarterly, 67*, 79–125.

Horn, D. (2001). Is seeing believing? Detecting deception in technologically mediated communication. *Extended Abstracts of CHI 2001* (pp. 297–298). Washington DC: ACM.

Horn, D., Olson., J., and Karasik, L. (2001). The effects of spatial and temporal video distortion on lie detection performance, *Extended Abstracts of CHI 2002* (pp. 714–715). Washington DC: ACM.

Houtkoop-Steenstra, H. (2002). Questioning turn format and turn-taking problems in standardized interviews. In D. N. Maynard, H. Houtkoop-Steenstra, N. C. Schaeffer and J. Van der Zouwen (Eds.). *Standardization and tacit knowledge. Interaction and practice in the survey interview.* (pp. 243–261). New York: John Wiley, & Sons, Inc.

Howarth, B., and Anderson, A. H. (2007). Introducing objects in spoken dialogue: the influence of conversational setting and cognitive load on the articulation and use of referring expressions. *Language and Cognitive Processes 22*(2), 272–296.

Jackson, M., Anderson, A. H., McEwan, R., and Mullin, J. (2000). Impact of video frame rate on communicative behavior in two and four party groups. In *Proceedings of Computer Supported Cooperative Work 2000* (pp. 11–20). Washington DC: ACM.

Kroek, G., and Magnusen, K. (1997). Employer and job candidate reactions to video conference job interviewing. *International Journal of Selection and Assessment, 5*(2), 137–142.

Krosnick, J. (1991). Response strategies for coping with the cognitive demands of attitude measurement in surveys. *Applied Cognitive Psychology, 5*, 213–236.

Krosnick, J. (1999). Survey research. *Annual Review of Psychology, 50*, 537–567.

Krysan, M., and Couper, M. (2001). Using virtual interviews to explore race of interviewer effects. Paper presented to the Annual Conference of the American Association for Public Opinion Research, Montreal, Canada, May 2001.

Manning, T., Goetz, E., and Street, R. (2000). Signal delay effects on rapport in telepsychiatry. *CyberPsychology and Behavior, 3*(2), 119–127.

Miller, R., and Gibson, A. (2004). Supervision by video conference with rural probationary psychologists. *CAL-Laborate*, June 2004, 22–28.

Mullin, J., Anderson, A. H., Smallwood, L., Jackson, M., and Katsavras, E. (2001). Eye-tracking explorations in multimedia communications. In A. Blandford, J. Vanderdonkt, and P. Gray (Eds.), *People and Computers XV* (pp. 376–382). London: Springer.

Paul, D., Pearlson, K., and McDaniel, R. (1999). Assessing technological barriers to telemedicine: technology-management implications. *IEEE Transactions on Engineering Management, 46*(3), 279–288.

Reid, A. (1977). Comparing the telephone to face-to-face interaction. In I. Pool (Ed.), *The Social Impact of the Telephone* (pp. 386–414). Cambridge, MA: MIT Press.

Rutter, M. (1987). *Communicating by Telephone*. Oxford: Pergamon Press.

Sanford, A., Anderson, A. H., and Mullin, J. (2004). Audio channel constraints in video-mediated communication. *Interacting with Computers, 16*(6), 1069–1094.

Schober, M. F., and Conrad, F. G. (1997). Does conversational interviewing reduce survey measurement error? *Public Opinion Quarterly, 61*, 576–602.

Short, J., Williams, E., and Christie, B. (1976). *The Social Psychology of Telecommunications*. London: John Wiley & Sons.

Simpson, S., Bell, L., Knox, J., and Mitchell, D. (2005). Therapy via video conferencing: a route to client empowerment. *Clinical Psychology and Psychotherapy, 12*, 156–165.

Stephenson, G., Ayling, K., and Rutter, D. (1976). The role of visual communication in social exchange. *British Journal of Social and Clinical Psychology, 15*, 113–120.

Suessbrick, A. L., Schober, M. F., and Conrad, F. G. (2000). Different respondents interpret ordinary questions quite differently. In *Proceedings of the American Statistical Association, Section on Survey Research Methods*. Alexandria, VA: American Statistical Association.

Tang, J., and Isaacs, E. (1993). Why do users like video? Studies of multimedia collaboration. *Computer Supported Co-operative Work, 1*(3), 163–196.

Tang. J., Isaacs, E., and Rua, M. (1994). Supporting distributed groups with a montage of lightweight interactions. In *Proceedings of Computer Supported Co-operative Work 1994* (pp. 23–34). Washington DC: ACM.

Weiss, C. H. (1968). The validity of welfare mothers' interview responses. *Public Opinion Quarterly, 32*(4), 622–633.

Williams, J. A. (1968). The interviewer role performance—further note on bias in information interview. *Public Opinion Quarterly, 32*(2), 287–294.

Wilson, G., and Sasse, M. A. (2000). Do users always know what's good for them? Utilising physiological responses to assess media quality. In S. McDonald, Y. Waern, and G. Cockton (Eds.), *People and Computers XIV* (pp. 327–339). London: Springer.

CHAPTER 6

The Speech IVR as a Survey Interviewing Methodology

Jonathan Bloom

SpeechCycle, Inc., New York, New York

6.1 INTRODUCTION

Speech recognition technology has made great strides in quality over the past decade, and usage has increased in step with this improvement. The technology has many applications, but it has made its most noticeable impact on the telephone, acting as a virtual agent for large businesses and government agencies. Most readers of this chapter have likely called and spoken to an interactive voice response (IVR) system that recognized their speech—or did not recognize it, as the case may be. These speech IVR systems have become commonplace, answering calls from individuals when they contact airlines, banks, telephone companies, their county's government, insurance companies, and even prisons.[1]

However, in the field of survey interviewing, speech IVR systems have not kept apace. Of the larger government surveys, speech IVR systems are being used today in surveys for the Current Employment Statistics (CES) program, but only for businesses that do not have access to touchtone. Not many other established surveys are using the technology. This is likely due to several factors, which we will discuss later. But one major reason worth noting up front is that the survey community is still quite busy validating the merits of touchtone IVR systems—that is, phone systems that respond exclusively to telephone keypad input. Until the costs and benefits of this more established technology are clarified, speech IVR systems will likely be seen as "around the corner."

---

[1] Maricopa County Prison's phone line even offers a spoken option for "self surrender."

Envisioning the Survey Interview of the Future, Edited by Frederick G. Conrad and Michael F. Schober
Copyright © 2008 John Wiley & Sons, Inc.

Once the research into touchtone IVR systems settles down, speech will certainly receive more attention—positive or negative. Survey researchers should be ready with an agenda for vetting the technology and ensuring that it meets certain criteria required for obtaining valid survey data. The primary goals of this chapter are to (1) lay out the inherent strengths and weaknesses of speech technology, (2) point out possible benefits and risks of applying the technology to survey interviewing, and (3) compare it to two closely related survey methodologies—touchtone IVR systems and computer-assisted telephone interviews (CATIs). Some of the points regarding speech technology in the survey realm will be backed by data, but others are merely the author systems estimation because the data do not yet exist.

One final note before we begin. Although researchers are generally quite enthusiastic, the fit between speech technology and survey interviewing is far from given at this point. We should not just be asking whether speech IVR systems can be used for the purpose of survey interviewing; we must also ask *why* we would use it in this way. As Mick Couper points out in chapter 3 in this volume, any new technology must be at the service of better data, better experience, and greater efficiency. If we do not feel that speech IVR systems hold this promise, then the survey community must look elsewhere for these improvements.

6.2 A BRIEF INTRODUCTION TO SPEECH RECOGNITION

Speech recognition technology involves the mapping of spoken language to text stored in computer memory.[2] The computer systems memory includes some number of words or phrases in text form, called the "grammar" or "vocabulary." The goal is to map the incoming acoustic signal from a speaker systems voice to the right word or phrase in the grammar.

Let's say a person calls a speech IVR system to take a survey and is asked for the state in which she resides. The caller says "California." The speaker systems acoustic signal is first picked up by the phone systems microphone and transmitted to a computer somewhere in another location. The acoustic signal then goes through "feature extraction," where it is analyzed for certain (quite abstract) qualities or parameters. Each incoming speech signal varies in its parameter values. The items in the computer's grammar (e.g., "Massachusetts," "Nevada," "Repeat that") also have parameter values associated with them. The values of the words in the grammar were obtained earlier by analyzing potentially thousands of recorded utterances of those words or phrases.[3] The values of the incoming signal ("California") are then compared to the values of items in the grammar. The computer returns an item from the

---

[2] This is often confused with its counterpart "text to speech" (TTS), in which text in the computer's memory is converted into speech by way of a synthetic voice. We will also briefly discuss TTS.

[3] This is quite a lot of overhead to consider for a small survey interview. However, most speech recognition software on the market today comes with grammars that have default "language models" built in. Accuracy will be only "good" at first, but will improve as more data—specific to the survey's calling population—begin to come in.

grammar that most closely matches the incoming signal and also returns a "confidence score" for the returned item. As the name suggests, the confidence score indicates the recognizer's confidence that it picked the right match out of the computer's grammar. Hopefully, the system picks "California" out of its grammar, the confidence score is high, and thus the system officially recognizes "California." The grammar changes with each question, looking quite different for a yes/no question than for a request for a zip code.

Many variables can compromise the confidence of the recognition, and they are more or less the same variables that can compromise human hearing. For example, the speaker may be located in a noisy environment such as a car or a house full of children and pets.

Also, people do not speak fluently. Speech production is rife with false starts, repairs, and filled pauses such as "um" and "uh." There is some evidence that people are more fluent when faced with a speech recognition system (Oviatt, 1995), but there is little that a person can do to avoid sneezing, coughing, and soon. Because speech IVR systems utilize telephones and phone networks to carry the speech signal to the recognizer, the signal may also be compromised by a bad connection. Speech recognition systems are trained on many different accents and dialects, but very strong accents may also be met with degraded accuracy.[4]

The confidence can also be compromised by the size of the grammar and the confusability of the items in the grammar. Accuracy on yes/no questions is usually quite high because there are only two words (and perhaps a handful of synonyms) that the system can recognize, and also because the words do not sound like one another. Accuracy is comparatively lower when the grammar includes, for example, all of the train stations in the United States served by Amtrak®. There are hundreds of items in the grammar and some of them are highly confusable (e.g., Newark Penn Station and New York Penn Station).

Poor recognition accuracy is obviously one of the biggest challenges facing speech IVR systems as a legitimate means of survey interviewing. One of the major benefits of the telephone as a medium for interviewing is supposed to be that people can easily respond to survey questions from their homes or on the road. The noise of an active home, a busy street, or a bad connection should not stand in the way much more than it does for human interviewers on recognition accuracy measures if speech recognition is to be a sufficient method of survey data collection. Studies suggest that human listeners score an order of magnitude better on recognition accuracy measures than they do on speech recognition systems (Lippmann, 1997, as cited in Moore, 2003; Meyer et al., 2006). Meyer et al. (2006) compared the recognition accuracy of computers and humans using a database of three-syllable nonsense words that varied in speaking rate, volume, intonation, and dialect. Computer and human participants were required to recognize the middle phoneme of each nonsense word. They found an average accuracy rating of 99.4% for human listeners and 74.0% for the speech

---

[4] Unlike speech recognition used for dictation software, speech IVR systems also cannot take advantage of linguistic context to help identify a word. Dictation software can look at preceding and subsequent words to narrow down the possible matches for a target word.

recognition system. Other studies also conclude that, whereas speech recognition technology varies quite a bit in accuracy depending on variables like noise levels, human recognition is more robust, staying consistently high even in the face of the aforementioned hurdles (Deshmukh et al., 1996).

This problem may seem daunting, but as we will see in the next section, speech technology can greatly improve its chances by exploiting many of the conversational strategies that human addressees use when faced with uncertainty about a speaker systems utterance.

6.3   SPEECH RECOGNITION APPLIED TO IVR SYSTEMS

So far, we have discussed speech IVR systems at the level of a single respondent input. A speech IVR system is a dialogue, and so it must not only be the listener, but must also take turns being the speaker. It must be capable of speech output. Output for speech IVR systems (and touchtone IVR systems) comes in two forms: recorded human speech or computer-generated speech, also known as text-to-speech (TTS). The effects of using TTS versus human recordings on survey data are still being debated (see Couper et al., 2004). But it should be noted that TTS improves at an impressive rate, becoming more understandable with each new release. Therefore, the results of the studies today could require updating fairly soon. TTS is reaching a turning point, after which the distinction between it and human recorded speech may become undetectable. Recently, developers of TTS started using a new strategy that strings together phonemes of actual human voices. This new "concatenative" approach to TTS has made a marked improvement in the quality of TTS when compared to traditional "formant" methods, in which all TTS audio is generated by the computer.

The speech IVR system cannot just recognize speech and generate speech. Speech IVR systems try to simulate human conversations, and human conversations are rule-governed activities (Clark, 1996; Sacks et al., 1974). In the case of a survey interview, the rules may be quite simple when the respondent can hear the question clearly and the respondent and the interviewer understand the question the same way, or at least both parties assume they do. Take, for example, this portion of an interview from a survey regarding tobacco use and opinions about tobacco use (Suessbrick, 2005):

> I: About how long has it been since you last smoked cigarettes every day.
> R: Um: s: uh since . I think February of ninety-two: I quit so: its like nine: . two three four five six seven eight nine, seven years.
> I: Alright.

The dialogue goes quite smoothly. The interviewer poses a question and the respondent, after some calculations, answers "seven." The interviewer then provides a backchannel, "Alright," to let the respondent know that his/her answer has been heard and recorded.

If a speech IVR system only needed to play out a question and recognize (a more constrained version of) the above answer, the job of handling a survey interview would be quite simple. However, consider the following question and answer session that takes place further along in the same interview.

I: About how long has it been since you COMPLETELY stopped smoking.

R: Since um nineteen ninety-two.

I: Okay, so how many . years would that be

R: That's um: . two three four five six seven eight nine, that's seven years, wait didn't you ask me that same question before?

I: No we asked you uh . this is w- we asked you how long you had smoked but now we're asking you about how long its been since you completely STOPPED smoking.

R: Okay.

I: Do you think that was *for*

R: *No no* I- it was another question where I answered seven years?

I: Oh you think maybe you misunderstood that ques*tion*?

R: *Yeah* uh I don't want to uh.

I: Well it's about how long has its been since you last . smoked cigarettes every day

R: Right, oh okay, got it

I: Okay so that was *uh*

R: *Yeah* that was still seven *years*

I: *Alright.*

R: Mm-hm.

I: Now again, how about how long has it been since you completely stopped smoking cigarettes . sort of,

R: Yeah. Seven years.

I: Alright.

R: And a couple of months.

During this interaction, it becomes clear that the respondent misunderstood an earlier question and the respondent and interviewer must now return to answer it again. The respondent may not have heard the question properly the first time, or heard it but did not understand it. Conceptual misalignments between interviewer and respondent do happen and are especially hard to resolve when traditional approaches to standardization (as defined by Fowler and Mangione, 1990) are followed (Conrad and Schober, 1999, 2000; Schober and Conrad, 1997; Schober et al., 2004). In one instance, Suessbrick, Schober, and Conrad (2000) found conceptual misalignments occurring almost as often as alignment.

Given that respondents may not hear a question or may not understand a question, or may think they understand a question when they do not, the rules that govern turn-taking in speech IVR systems must be flexible. The speech IVR systems of today cannot hope to handle the dialogue above, but most of them do have error handling

DialogModule™			ZIP Code
Entering from			
[This space usually contains names of other modules from which we enter the current module]			

Prompts		
Type	Name	Wording
Initial	sr6125i010	"Finally, what's the 5 digit ZIP code of the billing address for your card? <3 second pause> If your billing address is not in the US, say 'foreign card'"
Timeout 1	sr6125t010	"Sorry, I didn't hear you. For verification purposes, I need the 5 digit ZIP code of the billing address on your credit card. You can either say or enter the ZIP code."
Timeout 2	sr6125u010	"Sorry, I still didn't hear you. If you're unsure of what to do say help or press star. Otherwise, please enter your ZIP code on your keypad."
Retry 1	sr6125r010	"Sorry, I didn't understand. Please say your ZIP code again, one digit at a time or enter it on your keypad."
Retry 2	sr6125s010	"Sorry, I still didn't understand. If you're unsure of what to do say help or press star. Otherwise, please enter your ZIP code on your keypad."
Help	sr6125h010	"For security purposes, I can't submit your payment information without the 5 digit ZIP code of the address where you receive your credit card bill. You can either say the ZIP code or enter it on your keypad."

Option	Vocabulary	DTMF	Slot value*	Action	Confirm.
Digits	<digit_string>	5/9 Digit ZIP code	CCZipCode = defined	[This space would mention the next module to go to after the person says a zip code that we confidently recognize]	If necessary
Foreign Card	Foreign Card	<...>	ForeignCard = defined	[This space would mention the next module to go to after the person says "foreign card"]	If necessary

Confirmation Prompts			
Option	Name	Wording	Result
Foreign Card	sr6125c010	"… you have a foreign card …"	*"I think you said you have a foreign card, is that correct?"*
Digits	Default confirmation, as handled by DialogModule™		*"I think you said zero six eight five three, is that correct?"*

FIGURE 6.1 Amtrak module handling zip code collection. Spoken prompts are in bold.

in place that can address many common problems. Figure 6.1 shows the basic inner workings for one conversational turn in a speech IVR system that handles train travel. The prompts that are played out to the caller are in bold.

This module handles the play-out of one "initial" prompt that requests information from the caller. In this case it is "Finally, what systems the 5 digit ZIP code of the billing address for your card? < 3 second pause> If your billing address is not in the US, say 'foreign card'." In a survey interview, this would be where the actual survey question is asked; for example, "About how long has it been since you *completely* stopped smoking?"

The respondent may do many things in response to the initial prompt. One possibility is that he/she does nothing: the respondent does not speak. If the respondent does not know how to answer the question, this is a distinct possibility. In such cases, speech IVR systems have a fallback called a "timeout 1" prompt, which asks the caller the question again. One commonly used strategy in timeout prompts is to elaborate on the original question, trying to predict why the caller may not have answered. However, if this were a traditional survey interview in which clarification of a question is not permitted, the timeout prompt may need to simply repeat the initial prompt. Survey interview designers need to consider how to take advantage of these timeout prompts. The respondent may understand the question but remain silent because he/she doesn't know how to interact with the system. One possibility is that the timeout prompts do not change the question wording, but elaborate on how the respondent can be more successful in interacting with the speech IVR system. For example, in the module shown in Fig. 6.1, the system mentions touchtone input as an alternative modality to speech (it says "please say *or enter*"). In doing this, interviewers may be able to increase response rates while keeping question wording the same.

If the caller does not respond again, the module includes a "timeout 2" prompt, often asking the initial question again, but also mentioning either the agent option or a help option, which we will discuss in a moment. If the person does nothing a third time, most speech IVR systems transfer to a call center agent. If this is not a possibility with survey interviews, the system could tell the person to call back when they are ready or able to respond, or simply keep reusing timeouts 1 and 2 ad infinitum.

Even if the caller does respond, the response may be recognized with "low confidence" for some or all of the reasons mentioned earlier (noisy environment, thick accent, etc.). If the system does not recognize the caller, most speech IVR systems come with a built-in "retry" prompt that often looks similar to a timeout prompt. Often the only difference is that the timeout prompt starts with "Sorry I didn't hear you" and a retry prompt starts with "Sorry I didn't understand." Like timeouts, there are also "retry 2" prompts in case the person is understood with low confidence a second time.

There is also a context-sensitive help prompt included in this module, which is played out if the caller explicitly asks for help. The help prompt elaborates on the initial prompt. The retries and timeouts do this as well, but the help prompt is supposed to be even more detailed, the assumption being that if the caller explicitly asked for help, then the caller is willing to listen to a longer explanation.

Few people interacting with speech IVR systems ever explicitly ask for help and so these prompts are rarely heard by callers. That few people explicitly ask for help in speech IVR systems aligns with findings in other types of discourse that suggest people rarely ask for help (Graesser and McMahen, 1993). Nonetheless, for those people who do use it, a help prompt cannot just be a repeat of the initial prompt. For survey interviews, this can be a challenge depending on the survey designer's definition of standardization. As mentioned earlier, Fowler and Mangione (1990) established guidelines of standardization that are followed by many organizations. These guidelines are designed to minimize error by keeping question wording invariable. If the respondent is having trouble answering, the interviewer is only allowed to guide respondents with "neutral probes" like "Whatever it means to you." A speech IVR system could be made to respond in such a manner if a respondent asked for help.

Some feel that standardization should not be applied to wording, but to meaning, so that the question is understood the same way by the interviewer and all respondents (Suchman and Jordan, 1990). More open-ended collaboration of question meaning during survey interviews can lead to higher response accuracy (Conrad and Schober, 2000). If one takes a more collaborative approach, then the help prompt in a speech IVR system—as well as timeout and retry prompts—can be put to better use. It may be able to provide more information that answers many of the respondents' questions. Of course, there is evidence that allowing for elaboration on question meaning can make (simulated) speech IVR interviews longer (Bloom, 1999), which may be a cost issue. Also, providing a help prompt is not the same as real-time human–human collaboration. Speech IVR system designers must figure out in advance what elements of a survey question might confuse respondents, and put that information in the help prompts. But there is no guarantee that the designers will predict accurately.

In addition to initial, timeout, retry, and help prompts, these modules also include "confirmation prompts." In certain cases, the caller may be recognized by the system with "medium confidence." The confidence was not high enough to simply move on to the next question, but also not low enough to retry. Like human addressees in such situations, the system responds by saying something like "I think you said the zip code was oh six eight five three, is that right?"[5] These confirmation prompts are used pretty commonly, except for yes/no questions, because that can lead to serious confusion ("I think you said 'no.' Is that right?" How does one respond to that if the answer is "no" and then how does the speech IVR system interpret that "no" answer?). Confirmation prompts are unique to speech IVR systems; they have no equivalent in touchtone IVR systems.

One could look at such confirmation prompts in the context of a survey interview and think of them as the equivalent of Fowler and Mangione systems "neutral probes." However, whether neutral probes are actually neutral is up for debate. There is evidence from human–human interviews that such probes could be used to lead respondents to a specific answer (Schober and Conrad, 2002). Subtle cues in the interviewer's presentation of the probe can be picked up by respondents and interpreted as hints that the answer being provided is incorrect. In the context of a speech IVR system, we cannot be sure if confirmation prompts would be interpreted as leading the respondent one way or another. People may assume computers do not understand or employ such conversational knowledge. If that turns out to be the case, then there is no concern (although assuming less capability on behalf of the IVR system may have its downsides too, e.g., users feeling the need to segment each word when speaking). However, we need to be mindful of such possibilities when considering speech IVR systems for the purpose of survey interviewing.

Finally, in speech IVR modules there are usually one or more active "global commands." Global commands are commands that are available throughout the interaction. For example, "repeat that" is often available in all conversational turns, regardless of context. Where the speech IVR system could use retry prompts when confidence was low, people speaking to speech IVR systems can use "repeat that" when *their* confidence is low. "Repeat that" is a powerful resource, given the serial and ephemeral nature of speech (as opposed to parallel and persistent interfaces like the Web).

So far, we have taken a high-level tour of speech IVR systems. We have seen that speech IVR systems are comprised of speech recognition input, TTS or human recorded output, both combined with the aforementioned conversational strategies. In the process we have discussed many strengths and risks of this technology when

---

[5] Designers figure out the medium confidence thresholds by trial and error. They set the distinction between high, medium, and low confidence somewhat arbitrarily at first, and then listen to caller responses to the confirmation prompts. They then set the thresholds so that approximately 50% of the responses to confirmations are "yes" and 50% are "no."

considering it as a method of gathering survey data. Most of these risks and benefits were inherent in the technology. In the next section, we will take a more comparative approach, itemizing the pros and cons of speech IVR systems when compared to touchtone IVR systems and to CATIs.

6.4 SPEECH IVR SYSTEMS COMPARED TO CATIs

A large body of research exists comparing surveys administered using touchtone IVR systems to those administered using CATIs. Far less research has been done to compare speech IVR systems to CATIs. However, when comparing speech IVR systems to CATIs, the benefits of speech IVR systems may overlap with those of touchtone IVR systems, given that both are telephony-based self-administered methods of data collection.

So far, the data suggests that touchtone IVR systems do offer some benefits over CATIs. Most importantly, interviews using touchtone IVR systems cost less than CATIs (Miller-Steiger, 2006). Although speech IVR systems do cost more to build and maintain than touchtone IVR systems (because grammars need to be written and improved over time), the cost is still less than hiring and training survey interviewers (Clayton and Winter, 1992).

In addition, respondents are more likely to provide socially undesirable responses with a touchtone IVR system than they are with human interviewers, for example, when answering questions about tobacco use (Currivan et al., 2004) and sexual activity (Tourangeau and Smith, 1998; Villarroel et al., 2006). At first blush, it appears that "turning down" the social presence of the interviewer makes respondents more willing to admit to socially undesirable behaviors. But the relation between social presence and socially desirable answers may be less a matter of degree and more a binary distinction—"human" versus "not human." Couper, Singer, and Tourangeau (2004) found that touchtone IVR systems lead to more socially undesirable admissions than using human interviewers. But no difference was found between automated systems that varied in level of "humanity." More specifically, they found no difference in socially undesirable responding when employing IVR systems using recorded human speech versus computer-generated speech (also known as text-to-speech, or TTS). This suggests that if people know the "speaker" is not a live human then the similarity of its voice to that of a live human voice does not matter.

Clifford Nass' lab at Stanford used a speech IVR system to conduct a similar study, although the human interviewer condition was removed so that a computer produced all output (Nass et al., 2003). They did find an effect of the IVR systems "humanity," but they found that disclosures *dropped* when they used TTS versus recorded human speech. It is hard to tell what accounts for the different results, given that there are many differences in the methodologies of the two studies. However, one interesting possibility is that the use of speech recognition in the latter study—as opposed to touchtone—impacted the direction of the effect. Is it possible that once we are confined to just comparing automated systems (and no longer comparing them to human interviewers) that what counts is not level of the output systems "humanity" but rather the internal *consistency* of humanity across the input and output modalities? Perhaps TTS aligns better with touchtone input and human-recorded speech aligns better with speech recognition. When the inputs and outputs of the technology are paired in terms of their proximity to human conversation, respondents may be more willing to open up to the system.

The point is, touchtone IVR systems seem to increase respondent willingness to give answers that are socially undesirable, and hopefully, this benefit can be generalized to speech IVR systems as well. The question arises whether the addition of speech input (as opposed to touchtone input) might lessen the effect, given that respondents are interacting in a way that is more like what they do with a human interviewer. Because respondents are engaged in a simulated conversation, will this turn on social presence? In addition, people may be less likely to divulge sensitive information with a speech IVR system because they are vocalizing. If anyone else is nearby, the respondent may feel uncomfortable speaking an answer. We have found that touchtone entry in our speech IVR systems (we usually offer both) is highest when callers are asked for sensitive information, like credit card numbers or account passwords.

Another benefit of speech IVR systems over human interviewers (touchtone IVR systems share this benefit as well) is the added control the survey designers have over standardization. With speech IVR systems, all participants hear the exact same questions presented the exact same way. The variation in presentation across respondents is removed. In addition, as we have already discussed, survey designers can control the amount of elaboration that the system provides when misunderstandings occur. A survey designer can follow the guidance of Fowler and Mangione, or take a more "collaborative" approach (Schober and Conrad, 1997, 2002) and try to clarify question meaning during fallback prompts. The system can offer whatever amount of elaboration the designers are comfortable providing—up to the limits of what the dialogue design can handle. This last point is important. If a survey designer wished to create a survey interview that was highly interactive, CATIs may be preferred over speech IVR systems. The vast majority of speech IVR systems are "system-initiated," meaning that the system guides the direction of the conversation. For example, the first words spoken in the conversation are uttered by the speech IVR system, not the respondent. From there, the IVR system sets the agenda, leaving no room for the respondent to answer questions out of order, or ask his own questions (e.g., "How long will this take?"). However, it is possible, in principle, to design a speech IVR system with "mixed initiative" (system and user/respondent control the dialogue); but this would require substantially more advanced dialogue management than what is used in current systems. (For a discussion of the costs and benefits of mixed initiative in interviews, see Schober et al., 2004).

Speech IVR systems also have problems when compared to CATIs. Most troublesome is the lower accuracy of speech IVR systems, which we have already mentioned. If a survey population includes many people who will have heavy accents, for example, then speech IVR systems are risky.

But speech technology need not be discarded outright for this reason. Accuracy issues vary depending on context. Speech IVR systems do quite well with recognition of yes/no responses and digit collection. Therefore, speech IVR systems can still be put to good use with specific kinds of data collection. For example, the Current Employment Statistics (CES) program conducted by the U.S. Bureau of Labor Statistics collects employment data from thousands of nonagricultural businesses each month. The nature of the data being collected is relatively simple from a speech recognition standpoint. It is comprised predominantly of digits, which is one of current speech recognition technology's stronger suits.[6]

In addition people learn to interact with speech systems upon repeated use. If a survey is going to be targeting a specific sample longitudinally, then speech IVR systems become more viable. Again, the CES stands out as an appropriate survey for speech technology. The same companies respond to the same questions over time. Observing speech recognition data on the CES in 1989, Clayton and Winter (1992) found that error due to recognition problems decreased after a respondent's first month of usage. Although recognition accuracy is a major concern when compared to other modalities, the problem will not be as severe if the technology is applied to the right type of survey: that is, one collects recognizable data such as "yes/no" and numeric responses from the native-speaking and clearly speaking respondents who use the system on a recurring basis.

Aside from accuracy, another likely shortcoming is actually found in research with touchtone IVR systems. They show higher breakoff rates than CATIs (Miller-Steiger, 2006). This may be because an automated system puts little if any social pressure on a respondent to continue. Apparently, the same "low social presence" that allows for more truthful responses also gives respondents more freedom to hang up. Therefore, touchtone IVR systems are preferable for surveys that are shorter in length and include shorter scales (Dillman et al., 2002, as cited in Miller-Steiger, 2006). These limitations are nontrivial, especially in light of the fact that telephone surveys of any kind, CATIs or otherwise, lend themselves to shorter scales and surveys (Dillman et al., 2002). With automated systems, they may need to be even shorter.

Clayton and Winter (1992) again found heartening results when speech was used for the CES. Looking at more than 1000 respondents, they found over the course of two years (1989 and 1990) that response rates for speech, touchtone, and CATIs were all comparable, with mail lagging far behind (the authors do not report statistical significance). The response rates for speech, touchtone, and CATIs were all higher than 90%. Speech never dropped below 85% in any given month. This is surprising, given the state of the art at the time of this research. They were using dated TTS technology and could only allow for digits spoken one at a time—for example, "four one two" as opposed to "four hundred and twelve."

Even if response rates are comparable to those of CATIs (and I am somewhat skeptical this will happen any time soon), respondents will most likely prefer talking to a human, at least about nonsensitive topics. In the design of commercial systems, we often hear customers mention that they appreciate speech IVR systems over humans

---

[6] Also, there is no reason that survey designers must choose *between* speech and touchtone. The two modalities are blended quite successfully in today systems commercial IVR systems (e.g., "If you're calling about a missed appointment, say 'yes' or press 1. Otherwise, say 'no' or press 2.").

because speech IVR systems do not mumble due to exhaustion or "cop an attitude." But these individuals are the exception rather than the rule. Anecdotally, we hear far more often that these systems are cold and unfeeling (and that they "pretend" to be warm).

6.5   SPEECH IVR SYSTEMS COMPARED TO TOUCHTONE IVR SYSTEMS

Speech IVR systems offer many benefits over touchtone. For one, people seem to prefer speech over touchtone. At Nuance, we reviewed 25 applications, including 289 individual speech modules, and 1.6 million caller responses. In the cases where prompting explicitly mentioned the availability of both modalities, approximately 70% of callers still used speech on the first try. This percentage did not decline on retries and timeouts that mentioned both modalities.

When Clayton and Winter (1992) asked respondents for their preference between the speech version and the touchtone version of the CES, 60% preferred speech over touchtone, while 32% preferred touchtone, and 10% did not respond. Again, those numbers are interesting considering the dated TTS and single digit input. Their respondents also experienced the speech IVR as shorter, even though the calls were on average 20 seconds longer due to additional instructions.

Another benefit of speech is that it facilitates the input of certain types of data. The CES involves numeric input and yes/no questions, so touchtone and speech are both viable options. What if participants had to provide an address? Entering addresses using touchtone is quite labor intensive. The same can be said for names, dates, and the selection of items from long lists (e.g., names of countries). If a survey includes data of this sort, then speech becomes a much more compelling interface. Borrowing an example from the commercial sector, Amtrak used a touchtone IVR system up until 2001. When callers had to specify the arrival and departure stations, they were required to enter the first three letters of both the arrival and departure city names. So if the caller was traveling to Boston South Station, they would need to enter 2-6-7. This requires them to hunt for the numbers on the phone systems keypad that correspond to the three letters. That alone makes the interface more difficult than speech. But also consider that the caller still needs to identify which station in Boston selecting from a menu (there are three stations served by Amtrak in the Boston area). Since switching to speech, the Amtrak IVR system now asks for the station name instead of the city, for example, "Boston South Station." It would not have made sense to ask for the station instead of the city name with touchtone, because the first three letters are not enough to disambiguate between the three (Boston South Station, Boston North Station, and Boston Back Bay Station). By using speech, we have turned two steps into one ("city + station" becomes "station"), and made that single step less labor intensive (removed hunting for keys on the keypad).

Because humans are designed to communicate via speech (Pinker, 1994; Pinker and Bloom, 1990), the prompts in a speech IVR system can also be shorter than those in a touchtone IVR system. No explicit directions are required. For example, with a speech IVR system, a yes/no question like, "In the past year have you seen a medical doctor?" can be asked without any extra wording (the question comes from the Current Population Survey). With a touchtone IVR system, additional verbiage is required to explain the options: "In the past year have you seen a medical doctor? Press 1 for 'yes' or 2 for 'no'." Of course, the caveat here is that humans are not necessarily designed to communicate using speech *to interact with a computer*. With computers, common ground can be harder to establish (see Brennan, 1991, 1998). What words does the system know? What does it remember from the last conversational turn? From the last call? Because of this, directions may be required when the interaction runs into problems (i.e., in timeout and retry prompts).

Speech IVR systems also offer better ergonomics. When a respondent is using a phone that has the touchtone keypad on the handset, she does not need to take the phone away from her head in order to respond (Mingay, 2000). This is the case for some landline phones as well. For longer surveys, differences like this could have a considerable impact on call duration and also on the respondent systems opinion of the survey.

There is no "confidence level" associated with touchtone input. It is an "all or nothing" modality, where a selection was made or it was not. The certainty of the touchtone IVR system could be considered a benefit, but there are negative aspects associated with that certainty. Groves (2005) suggested that the best data that survey researchers could hope for would, among other things, include informative paradata. With speech IVR systems, the messier input can also be seen as rich paradata that can be logged and analyzed. We can hear the respondent systems level of anger or disinterest when he answers a question. This information could also be used to train the recognizer, so that if it later picks up "anger" in a response, it could digress from the survey path and assure the caller that, for example, the survey is almost complete. With touchtone entry, there is no way to tell how angrily the respondent pressed a key.

This is a double-edged sword. The rich paradata provided by speech IVR systems is also their primary problem. Just like the comparison between speech IVR systems and CATIs, response accuracy is again an issue when compared to touchtone IVR systems. Assuming that the respondent understands the question properly and assuming that she clearly enunciates her answer in a quiet environment, the recognizer may recognize the right response with high confidence, but it might also correctly recognize the right response with low confidence, or incorrectly recognize the wrong response with high confidence. And we cannot even assume that the respondent systems speech and environment are ideal. Some respondents may respond eloquently, but with a thick accent.[7] The caller may have side conversations (e.g., "Honey, did we buy any furniture in the past year?"). They may be calling from a city apartment with open windows on a major truck route, or they may simply sniff. All of these possibilities

---

[7] Respondents who speak English as their second or third language may also prefer touchtone, possibly even over a human interviewer, because touchtone only makes demands on their English comprehension abilities, not on their production abilities (assuming for a moment that this is a survey in an English-speaking country).

add up to longer calls and the risk of data error. Speech recognition technology is improving constantly. Recognizers are getting better at distinguishing signal from noise and accurately interpreting the signal, but this does not mean they are close to reaching the accuracy of touchtone.

Also, there are language limitations. Speech recognizers exist for many different languages at the time of this writing (e.g., U.S. English U.K. English, Australian English, Spanish, French, German, Portuguese, Hebrew, Japanese, Mandarin Chinese), but the list is obviously not exhaustive.

Another smaller downside to speech IVR systems is the memory overhead required during menu contexts. The caller must listen to the options in the menu and try to rehearse the wording of the most appropriate options while listening for potentially better options. With touchtone, the caller can externalize his memory by simply placing a finger on a key while listening to the rest of the options.

Finally, there is the question of social presence. At this point it is hard to say what effect speech IVR systems will have on social presence when compared to touchtone IVR systems and CATIs. Speech IVR systems could strike a happy medium between CATIs and touchtone systems. Speech IVR systems more closely simulate a human conversation than do touchtone IVR systems, and because of this, social presence theoretically increases. The effect may be that speech IVR systems offer the best of both worlds, keeping truthful responses to sensitive questions high due to its automation, while keeping respondent breakoffs low because of its similarity to human conversation. This is, of course, the hope. Medium social presence could also backfire, leading to the worst of both worlds. Speech IVR systems could lead to a drop in truthful responses to sensitive questions because they simulate a human conversation and unacceptable breakoff rates because of its automation. These are empirical questions waiting to be answered.

6.6 CONCLUSION

Speech technology comes with many potential benefits and shortcomings. Based on those discussed in this chapter, we can start to identify the kind of situations in which speech IVR systems would make sense as an alternative to CATIs or touchtone IVR systems. I envision the following optimal scenario for speech:

> A researcher for a pharmaceutical company is curious about the regularity with which people use its products. She wants to get weekly information over a period of one year. She decides to set up a longitudinal survey including a sample of several hundred individuals. Ninety percent speak English as their first language. The researcher has not been given the budget to hire interviewers for an extended period of time, but is concerned about the low response rates of mailed-out questionnaires. The data are sensitive, so the designer feels it would probably be best to have the respondents interact with an automated system. The company has given its blessing for her to work with the IT department and use its phone network. She would consider a Web survey, but the population being surveyed includes many busy individuals who do not have time to answer these questions while at work or home, and

would prefer to answer the questions while "on the go." The answers to the questions are mostly simple binary yes/no questions ("Are you currently taking your medication?") and some short numbers (e.g., a four-digit security code). There is also a question that requires the entry of an item from a long list (one or more medication names). The survey is short...only about ten questions.

This is a very specific scenario, but it is optimal for speech. One could still justify speech if a few of these situational factors were different. For example, a company might want to convey a "cutting edge" image and have little interest in the fact that touchtone could easily be used for all of its survey questions. The point is that speech IVR systems might be appropriate for a range of data collection situations, perhaps more so for some than others, but the survey designer needs to carefully consider the pros and cons of the particular technology in the particular measurement context.

We have not yet touched upon the potential of speech IVR systems. Touchtone technology has matured and does not seem to be improving at this point. Although some of the technologies available to human interviewers are making CATIs more efficient, the interviewers themselves are not necessarily getting any better or worse. Speech technology is relatively new, and the accuracy of the recognizers, the clarity of TTS voices, and the intricacy of the dialogue designs are steadily improving. Innovations are continually changing the landscape.

For example, new architectures for dialogue management are allowing speech IVR systems to approach a "mixed initiative design" in which respondents can answer more than one question at a time, and move to other questions in the dialogue. For example, if the IVR system asks the caller "What systems the departure station?" the caller can respond "New York Penn Station . . . oh . . . and I have a Triple A discount." Because this technology is relatively new, development time for any given application is still considerably longer than it is for applications using traditional "system-initiated designs." However, like any other technology, cost and effort can be expected to decrease over time. For certain surveys, where the order of responses is not crucial, this technology may provide benefits.

Another technology that may be helpful is called "hotword" (McGlashan et al., 2002). In certain modules, the recognizer can be set to only listen for a specific item and ignore anything else that might be heard. The benefit here is that the system can sit quietly while the caller does other things without concern about interruption by background noise. For example, if a respondent needs to provide a social security number, cannot recall the number, and needs to get their social security card, the IVR system can wait while the person gets it. It might say "Okay, I'll wait while you get that. When you're ready, just say 'I'm back'." Then the system will only be listening for "I'm back" and any other noise will be recognized and rejected. If a dog barks in that time, or if the person is walking with the phone and breathing heavily, the system will be less likely to make a false positive and proceed to the next question.

More subtle improvements are also happening, in the form of improved recognition algorithms. With each new release of a recognizer, the developers have figured out a way to make it *slightly* more accurate. These slight improvements are quite noticeable when one looks over decades. Clayton and Winter were working with technology in 1992 whose use today would be unthinkable.

The point is, even if one is skeptical of speech technology as useful for survey interviews, the landscape is constantly changing and it may be more viable with every passing year. It is similar to the old saying about the state of Florida: "If you don't like the weather, wait five minutes."

REFERENCES

Bloom, J. E. (1999). Linguistic markers of respondent uncertainty during computer-administered survey interviews. Doctoral dissertation, New School University.

Brennan, S. E. (1991). Conversation with and through computers. *User Modeling and User-Adapted Interaction, 1*, 67–86.

Brennan, S. E. (1998). The grounding problem in conversation with and through computers. In S. R. Fussell and R. J. Kreuz (Eds.), *Social and Cognitive Psychological Approaches to Interpersonal Communication* (pp. 201–225). Hillsdale, NJ: Lawrence Erlbaum.

Clark, H. H. (1996). *Using Language*. Cambridge, MA: Cambridge University Press.

Clayton, R. L., and Winter, D. L. S. (1992). Speech data entry: results of a test of voice recognition for survey data collection. *Journal of Official Statistics, 8*(3), 377–388.

Conrad, F. G., and Schober, M. F. (1999). A conversational approach to text-based computer-administered questionnaires. In *Proceedings of the 3rd International Conference on Survey and Statistical Computing* (pp. 91–101). Chesham, UK.

Conrad, F. G., and Schober, M. F. (2000). Clarifying question meaning in a household telephone survey. *Public Opinion Quarterly, 64*, 1–28.

Couper, M. P., Singer, E., and Tourangeau, R. (2004). Does voice matter? An interactive voice response (IVR) experiment. *Journal of Official Statistics, 20*(3), 1–20.

Currivan, D., Nyman, A. L., Turner, C. F., and Biener, L. (2004). Does telephone audio computer-assisted survey interviewing improve the accuracy of prevalence estimates of youth smoking? Evidence from the UMass Tobacco Study. *Public Opinion Quarterly, 68*, 542–564.

Deshmukh, N., Duncan, R. J., Ganapathiraju, A., and Picone, J. (1996). Benchmarking human performance for continuous speech recognition. In *Proceedings of the ICSLP-1996* (pp. 2486–2489). Philadelphia, PA.

Dillman, D. A., Phelps G.,Tortora R., Swift K., Kohrell J., and Berck. J., (2002). Response rate and measurement differences in mixed mode surveys using mail, telephone, interactive voice response and the internet. *Draft paper*, retrieved 1, August 2006, from http://survey.sesrc.wsu.edu/dillman/papers/Mixed%20Mode%20ppr%20_with%20Gallup_%20POQ.pdf.

Fowler, F. J., and Mangione, T. W. (1990). *Standardized Survey Interviewing: Minimizing Interviewer-Related Error.* Newbury Park, CA: SAGE Publications.

Graesser, A. C., and McMahen, C. L. (1993). Anomalous information triggers questions when adults solve quantitative problems and comprehend stories. *Journal of Educational Psychology, 85*(1), 136–151.

Groves, R. M. (2005). Dimensions of surveys in the future. Paper presented at the University of Michigan Workshop, Designing the Survey Interview of the Future, Ann Arbor, MI.

Lippmann, R. P. (1997). Speech recognition by machines and humans. *Speech Communication, 22*(1), 1–16.

McGlashan, S., Burnett, D., Danielsen, P., Ferrans, J., Hunt, A., Karam, G., Ladd, D., et al. (2002). Voice Extensible Markup Language (VoiceXML) Version 2.0. Retrieved 19 February, 2007 from http://www.w3.org/TR/2002/WD-voicexml20-20020424/.

Meyer, B., Wesker, T., Brand, T., Mertins, A., and Kollmeier, B. (2006). A human–machine comparison in speech recognition based on a logatome corpus. Paper presented at the Speech Recognition and Intrinsic Variation Workshop (SRIV2006), Toulouse, France. Retrieved 1 August, 2006, from ISCA Archive http://www.isca-speech.org/archive/sriv2006.

Miller-Steiger, D. (2006). Interactive voice response (IVR) and sources of survey error. Paper presented at Telephone Survey Methodology II Conference. Miami, FL.

Mingay, D. M. (2000). Is telephone audio computer-assisted self-interviewing (T-ACASI) a method whose time has come? In *Proceedings of the American Statistical Association Section on Survey Research Methods*, (pp. 1062–1067). Alexandria, VA: American Statistical Association.

Moore, R. K. (2003). A comparison of the data requirements of automatic speech recognition systems and human listeners. In *Proceedings of Eurospeech 2003* (pp. 2582–2584).

Nass, C., Robles, E., Bienenstock, H., Treinen, M., and Heenan, C. (2003). Voice-based disclosure systems: effects of modality, gender of prompt, and gender of user. *International Journal of Speech Technology, 6*(2), 113–121.

Oviatt S. (1995). Predicting spoken disfluencies during human–computer interaction. *Computer Speech Language, 9*(1), 19–36.

Pinker, S. (1994). *The Language Instinct*. New York: HarperCollins.

Pinker, S., and Bloom, P. (1990). Natural languages and natural selection. *Behavioral and Brain Sciences, 13*, 707–784.

Sacks, H., Schegloff, E., and Jefferson, G. (1974). A simplest systematics for the organization of turn-taking in conversation. *Language, 50*, 696–735.

Schober, M. F., and Conrad, F. G. (1997). Does conversational interviewing improve survey data quality beyond the laboratory? In *Proceedings of the American Statistical Association, Section on Survey Research Methods* (pp. 910–915). Alexandria, VA: American Statistical Association.

Schober, M. F., and Conrad, F. G. (2002). A collaborative view of standardized survey interviews. In D. Maynard, H. Houtkoop-Steenstra, N. C. Schaeffer, and J. Van der Zouwen (Eds.), *Standardization and Tacit Knowledge: Interaction and Practice in the Survey Interview* (pp. 67–94). Hoboken, NJ: John Wiley & Sons.

Schober, M. F., Conrad, F. G., and Fricker, S. S. (2004). Misunderstanding standardized language in research interviews. *Applied Cognitive Psychology, 18*, 169–188.

Suchman, L., and Jordan, B. (1990). Interactional troubles in face-to-face survey interviews. *Journal of the American Statistical Association, 85*(409), 232–241.

Suessbrick, A. (2005). Coordinating conceptual misalignment in discourse and the limits of clarification. Doctoral dissertation, New School University. *Dissertation Abstracts International, 66B* (01), 589.

Suessbrick, A., Schober, M. F., and Conrad, F. G. (2000). Different respondents interpret ordinary questions quite differently. In *Proceedings of the American Statistical Association, Section on Survey Methods Research* (pp. 907–912). Alexandria, VA: American Statistical Association.

Tourangeau, R., and Smith, T. W. (1998). Collecting sensitive information with different modes of data collection. In M. P. Couper, R. P. Baker, J. Bethlehem, J. Martin, W. L. Nicholls II, and J. M. O'Reilly (Eds.), *Computer Assisted Survey Information Collection* (pp. 431– 453). Hoboken, NJ: John Wiley & Sons.

Villarroel, M. A., Turner, C. F., Eggleston, E. E., Al-Tayyib, A. A., Rogers, S. M., Roman, A. M., et al. (2006). Same-gender sex in the USA: impact of T-ACASI on prevalence estimates. *Public Opinion Quarterly, 70*(2), 166–196.

CHAPTER 7

Automating the Survey Interview with Dynamic Multimodal Interfaces

Michael Johnston

AT&T Labs Research, Florham Park, New Jersey

7.1   CHALLENGES FOR THE SURVEY INTERVIEW OF THE FUTURE

Survey interviews play a critical role in many aspects of government and business. Fundamental policy decisions, including funding levels for critical spending programs, are informed in large part by quantitative survey data collected primarily by telephone and in-person interviews. These surveys provide one of the few tangible indicators of the success or failure of specific economic and social policies. In addition to their use in government, survey interview techniques in market research play a critical role in evaluating products and services and in making business planning decisions.

However, rapid social and technological changes in the nature of human communication threaten the viability of current survey techniques and the availability of critical measures that inform economic and social policy. As pointed out by Fuchs (Chapter 4 in this volume), the replacement of landline telephones with mobile telephones and Voice-over-IP (VoIP) Internet telephony poses significant challenges to the effectiveness of traditional sampling techniques for selection of telephone survey respondents. Another critical technological factor regarding VoIP is that the telephone area code no longer provides a reliable indication of geographical location. A VoIP phone can in principle be connected anywhere on the Internet and the area code can remain the same. The announcement that the German telephone company Deutsche Telecom will close down its publicly switched telephone network by 2012 provides a compelling indicator of the scale and urgency of the problem. Social factors, such as the rapid adoption of "Do Not Call" lists, also play an important role because members of the public often do not distinguish between telemarketing and legitimate scientific surveys.

In addition to issues regarding the viability of landline telephony as a "channel" for accessing respondents, survey interviews also face significant challenges in terms of cost. The size of survey samples and response quality are limited by the cost of training and employing survey interviewers. The creation of new technologies enabling effective automated surveys or semi-automated systems that improve the productivity of face-to-face interviewers will be critical in reducing these costs.

With our increasingly linguistically diverse population, another challenge for survey interview methodology is to ensure that results are not skewed by factors such as the native language of the respondent. It is important for human interviewers to avoid misclassification of responses caused by language and cultural differences. This is also an important consideration for automated systems utilizing speech recognition. Non-native speakers typically experience lower speech recognition performance. Potentially this can be overcome by employing a multimodal interface that provides the user with alternate input methods such as handwritten input or a structured graphical user interface. Going beyond this, through incorporation of multilingual aspects to the interface and perhaps the use of machine translation, it should be feasible to develop multimodal survey interfaces that adapt dynamically to the native language of the speaker.

Similarly, it is important that survey results be able to incorporate the responses of respondents with physical or mental disabilities, the elderly, and the young. Multimodal interfaces can also play a role here as an assistive technology. The availability of different modalities and adaptivity in the dialogue strategy employed by the system can allow users to employ the modes best suited to their capabilities and preferences.

In this chapter, we explore the possibility that new technologies such as natural spoken dialogue systems and multimodal interfaces, combining conversational and graphical interaction, could provide effective methodologies for automated and semi-automated administration of survey interviews and help to overcome these challenges. We compare the relative benefits and shortfalls of these new techniques to existing automated and non-automated techniques such as human interviewers and Web surveys. We also explore the potential use of multimodal interfaces as an aid to survey interviewers.

It is important to distinguish between approaches in which the same system can be delivered in different independent modalities (e.g., both over the telephone and on the Web), which we will refer to as *multichannel*, and truly *multimodal* systems, which enable the user within a single session or even dialogue turn to use a combination of different input modalities. The term *recognition* refers to the process of automatically recognizing user input. For example, *speech recognition* is the process of automatically determining the transcription of a user's speech signal. We use the terms *mode* and *modality* interchangeably to refer to specific input methods such as speech, pen, or touch.

7.2   CURRENT SURVEY INTERVIEW METHODOLOGIES

The central goal of survey research is to provide accurate answers to the questions posed. For a survey technique to be effective, it must enable a sufficient number of representative respondents to take the survey and complete it in full; it must provide accurate responses, while controlling the time and training needed. In this section, we discuss the benefits and shortfalls of three different modes of collecting survey data: traditional human interviews, current Web surveys, and computer-assisted survey interviews (e.g., video CASI).

7.2.1   Human Interviewer

The vast majority of public opinion polls and factual surveys reported today are based on telephone and face-to-face interviews conducted by human interviewers. This approach actively engages the respondent (user) in a way that mailed surveys or current Web surveys cannot. Also, compared to automated approaches there is less potential for recognition errors. Human ability to recognize and understand what the user says far exceeds the current limits of speech recognition and understanding technology. One important disadvantage of the approach is the risk of interviewer-related error (Fowler and Mangione, 1990); that is, any systematic effect of particular interviewers on survey responses. Exactly those adaptive perceptual abilities that make humans more effective at recognizing and understanding language and conducting effective dialogue can lead to variance between interviewers and interviewer-related error in survey results. For example, if interviewers pose questions in different ways and use different strategies for clarification and follow-up, this can result in different patterns of answers for different interviewers, i.e. interviewer related error. Human speakers dynamically tailor and adapt their choice of language in everything from vocabulary, to complexity of syntax, to overall dialogue flow in response to feedback from the hearer, and not all speakers will adapt in exactly the same ways. One approach to overcoming this problem is to employ standardized survey interview techniques (Fowler and Mangione, 1990), which attempt to regularize the way in which interviews are conducted. Guidelines are set limiting the extent to which interviewers should provide feedback and clarification of concepts. However, Schober and Conrad (2002) argue convincingly that, given the collaborative nature of human communication, even with standardized interviewing techniques such as neutral probing, interviewers can't avoid influencing response. They argue for a more collaborative conversational approach to survey interviews and demonstrate experimentally that allowing interviewers to use a full range of approaches for grounding concepts improves response accuracy compared to standardized interview techniques. (See also Suchman and Jordan, 1990).

Another problem with human interviews, at least those conducted over the phone, is that there is no way to provide persistent or parallel presentation of information. This may lead to additional cognitive load on the respondent and could slow down the interview process. Another disadvantage is the synchrony of the approach. The respondent has to answer the survey when the interviewer calls and there is no way for the respondent to answer in an asynchronous fashion—that is, when time permits. Perhaps the most critical limiting factor on human interview techniques is the cost of training and employing professional survey interviewers, which in almost all cases must limit the amount of data that can be collected. Automated approaches, in contrast, while in some cases more costly to set up initially, potentially can be operated continually with lower cost.

7.2.2 Web Surveys: Graphical User Interface

Interactive Web surveys, in which the user is presented with a graphical user interface, enable users to complete a survey by filling out a series of interactive forms using a mouse and keyboard as input modalities. Figure 7.1 provides a mock-up example of a simple Webpage for completion of part of the Consumer Price Index Survey from the Bureau of Labor Statistics.

This approach has several important benefits. Compared to spoken interaction, a graphical user interface can present more information. Multiple options can be presented *simultaneously* and information is *persistent*. While there is cost in implementing and hosting a Web survey, the infrastructure and personnel required are far less than for telephone interview. Web surveys also have the benefit that they can be answered asynchronously; that is, there is no need to answer the survey when it arrives. Compared to spoken interaction, there is less opportunity for recognition error or mishearing of responses.

FIGURE 7.1   Mock Web form for Bureau of Labor Statistics Consumer Price Index Survey.

However, there are also important problems and challenges in the use of Web user interfaces for survey tasks. As pointed out by Schober et al. (2003a), while in most Web applications users are *obtaining* information (e.g., flight times, stock information, news), in surveys users *provide* information. As a result users are less likely to seek clarification of concepts and there is potential for reduced response accuracy (Schober et al., 2003b). Drawing on their experiments with human interviewers, Schober and Conrad argue that survey interviews should at the very least provide easy access to clarification of key concepts, and better still they should actually engage in ordinary conversational interaction with the user (Schober and Conrad, 1997).

Another potential shortfall of purely graphical interaction is that it lacks many of the cues available in spoken conversational interaction that provide indicators of possible differences of understanding between the participants, conceptual misalignment (Schober, 1998); in this case the participants are the survey designer and the respondent. Ehlen (2005) examines a range of cues in spoken survey responses with a speech interface including speech timing, prosody, and lexical cues and their connection to accuracy of responses. One critical finding (Ehlen et al., 2005, in press) is the utility of a so-called Goldilocks range for response time (not too fast, not too slow), which provides a strong indicator of conceptual alignment. When respondents answer outside this range it indicates increased risk of misunderstanding of the question. One problem with using traditional graphical interfaces for surveys is that they lack these real-time aspects of conversation. It remains to be seen whether there is some utility to click timing, but given the nature of graphical interaction it cannot be relied on in the same way as speech timing. See Conrad, Schober, and Coiner (2007, in press) for a discussion of modeling click timing as an indicator of comprehension difficulty and the need for clarification. One issue is that, in a graphical interface, users may change responses multiple times before a final answer is exposed to the application. One possible avenue of exploration is to design Web interface technology that goes beyond the typical Web interaction paradigm and tracks client side user behavior in more detail, enabling collection of paradata (Couper and Lyberg, 2005), data regarding how respondents complete survey questions, such as how often the user changes a radio button selection, in addition to the user's final response. Clearly, this poses significant problems in terms of privacy; see Marx (Chapter 13 in this volume) and Konstan et al. (Chapter 12 in this volume). Another possible problem with graphical interfaces is that they lack the engagement of conversational interaction, which could potentially discourage participation.

7.2.3 Computer-Assisted Survey Interviews

A third approach to survey interviews, CASI (computer-assisted survey interviewing), utilizes both a human interviewer and a computer system, typically a laptop[1] brought to the respondent's residence. In some cases the computer is used to provide a mechanism for respondents to type in rather than speak sensitive responses (Tourangeau and

---

[1] Nusser et al. (2002) describe prototype systems using hand-helds and wearable computers designed to assist in data collection in the field.

Smith, 1998). In one technique, video-CASI (Krysan and Couper, 2003), digital video is played as a stimulus on a laptop. In a related approach, Tourangeau, Couper, and Steiger (2003) utilize Web survey screens with still photos. An important benefit of adding a computer to the human interview process is that information can be presented in a parallel and persistent fashion to the respondent. Video-CASI offers the possibility of enabling multiple interviewers to engage the user with an identical human conversational stimulus. The main drawback of the approach, however, is its cost in terms of personnel and time. The multimodal interface approach to survey interviews introduced in this chapter can be seen as a natural progression from video-CASI, in which the range of interactions is extended and cost of administration may be reduced if multimodal survey interfaces can be self-administered, eliminating or reducing interviewer costs. Furthermore, once built, multimodal interfaces can be more rapidly tailored to specific tasks, while video-CASI re-quires rerecording of video segments in order to change the survey.

7.3 SPOKEN DIALOGUE SYSTEMS

Interactive voice response (IVR) systems capable of answering telephone calls and providing callers with various kinds of information or routing their calls have existed for decades. However, it is only recently that the speech recognition, language understanding, and dialogue management technologies used in these systems have matured to the point where a spoken language dialogue system could effectively be used to conduct automated survey interviews in a conversational style.

Many deployed systems limit the user responses to touchtone inputs, "Press 1 for ...", which for anything other than a very simple set of survey questions would be tedious, complicated, and prone to errors from users confusing the options. The majority of deployed commercial dialogue systems supporting speech input (see Bloom, Chapter 6 in this volume) support system initiative dialogue in which the users' speech inputs are limited to short concise responses (Fig. 7.2). These systems lack the conversational competence needed to recognize and overcome conceptual misalignment and may not provide the high response accuracy needed for survey applications.

In recent years natural language systems supporting mixed initiative dialogue have been developed (Chu-Carroll and Carpenter, 1999; Gorin et al., 1997; Gupta et al., 2005) and have been deployed in commercial applications, where they provide call routing and customer information. Application areas include pharmaceutical, retail,

- System: *Which state?*

- User: *Michigan*

- System: *Which city?*

- User: *Ann Arbor*

FIGURE 7.2   System initiative dialogue.

and, telecom. They allow users to respond with unconstrained natural speech, and, based on salient phrases within the input and the current context, they can decide how to respond to or route a call. Critical in enabling the development of these systems are data-driven language understanding techniques, which utilize machine learning algorithms such as Boosting (Schapire, 1999) and Support Vector Machines (Vapnik, 1995) to classify user responses.

These algorithms have typically been applied to tasks such as routing a customer service call, but a similar approach could effectively be applied to classifying user responses to survey questions. The benefit would be that the user is free to respond as he wishes and does not have to learn a specific vocabulary in order to complete the survey. This is critical since respondents will often not have interacted with the system before and cannot be expected to learn how to use it. As in the Web survey case, they are providing information rather than obtaining it and so are unlikely to make a concerted effort to avoid recognition errors. Since these systems allow for more free form inputs, it will be possible to allow for a more conversational style of interaction. As a result, misalignment cues are available (Ehlen, 2005; Ehlen et al. (in press)), such as repeats, false starts, prosodic information, and pauses, and they can be used to identify potential conceptual misalignment and trigger clarification dialogue. One promising approach would be to apply problematic dialogue detection techniques (Walker et al., 2000) using Ehlen's misalignment cues as features. In problematic dialogue detection, a classifier is trained in order to predict, based on automatically derived features from an initial sequence of dialogue turns, whether a dialogue is likely to end poorly. Similarly, a statistical classifier could be trained based on annotated data to classify interactions with respect to their likelihood of conceptual misalignment based on automatically derived misalignment cues.

Although these new technologies for understanding allow users flexibility in their inputs, it is important to note that survey tasks are limited in ways that will make the task significantly easier than other automated service tasks. By their nature, the vocabulary required for answering individual survey questions is quite limited, and the range of possible dialogue moves at any point is quite constrained. In contrast, tasks such booking a flight or planning a route involve far more complex sequences of dialogue moves and extensive vocabulary. Another important consideration in the adoption of statistically trained methods for language understanding and dialogue, is that, over time, accurate estimates of system performance can be obtained and these can be used in predicting how much survey error could have been introduced by the automated survey methodology.

One significant, practical, though perhaps short-lived, advantage of this approach is the ubiquitous nature of the telephone as an access device. Most people are familiar with telephones and have access to a telephone of some kind, either landline, mobile, or IP phone.

While there is the potential now for sufficiently robust and accurate classification of user responses, problems still remain with speech-only systems in terms of output. There is no way to present information in parallel and users will have to listen to lengthy prompts, and since the information is not persistent, speech-only output can pose heavy memory demands on the user. For example, when interviewers orally present a list of response options, respondents show a preference for the last (i.e., most recently presented) options, presumably because this is more easily remembered than the options presented earlier (Schwarz et al., 1992).

Another potential problem with this approach is that in order to build an effective spoken dialogue system, especially for a diverse population of novice users, large amounts of training data need to be collected and annotated. These techniques may therefore be well suited to surveys that are frequently repeated. In the initial phases, the survey would have to be administered by a human interviewer and their responses recorded and tagged or the data would have to be collected using a Wizard of Oz technique (Dahlbäck et al., 1993; Fraser and Gilbert., 1991). Once sufficient data is collected, the automated system could be trained and deployed for the rest of the survey. One benefit of using data-driven response classification techniques is that they come with well understood accuracy measures. These potentially can be incorporated into estimates of survey error. If, over a large corpus of data, the classification error is 10%, then this can be used as a weighting factor in estimating the impact of the automated system on survey error.

7.4 MULTIMODAL INTERFACES

Multimodal interfaces are systems that support user input and/or system output over multiple different modes within a single interaction session with the user. Possible input modalities include speech, eye gaze, hand gesture, pen, and touch. Possible output modalities include synthetic speech, graphical displays, and gestures, either on screen using highlighting or the motions of an embodied conversational agent (Cassell, 2000) toward objects on the screen or to the surrounding environment. Some multimodal interfaces allow users to use either one mode or the other to convey their intention to the system. For example, they might either say "five bedrooms" (speech input modality) or push a button on the screen that is labeled "5 bedrooms" (GUI input modality), or write "5 brdm " with a pen on a touch-sensitive display (pen input modality). We refer to this as *simultaneous multimodality*. Other systems accept single commands, which combine multiple different modes. For example, the user might say "zoom in here" while simultaneously circling an area on a map display using a pen. We refer to this as *composite multimodality*. In order to support composite multimodality, some kind of multimodal integration or "fusion" component is needed (Allgayer et al., 1989; Johnston, 1998; Johnston and Bangalore, 2005; Johnston et al., 1997; Koons et al., 1993).

Systems capable of accepting combinations of spoken or typed input with pen or hand gestures have existed since at least the early 1980s. Carbonell (1970) described a geography tutor that accepted both typed input and mouse input. The "put-that-there" system (Bolt, 1980) was one of the first to support composite multimodality. It combined spoken input with gestures made at a map display using a 3D mouse. CUBRICON (Neal and Shapiro, 1991) provides spoken and mouse input to a map display. Multimodal interfaces have previously been applied to government applications. For example, the XTRA system (Allgayer et al., 1989) provided a multimodal interface for help with tax forms. Koons et al. (1993) present multimodal interfaces that support parallel input, from gestures, gaze, and speech. Cohen et al. (1997) describe Quick-Set, a mobile system that supports synchronous input by speech and pen gestures for controlling distributed interactive simulations. Cheyer and Julia (1995) present a map-based pen/voice interface for accessing information about hotels. Vo (1998) developed a toolkit for building speech/pen multimodal interfaces, which has been used to build a city guide application (QuickTour), a multimodal appointment scheduler (Jeanie-II), and a multimodal football simulator (Quarterback). The SmartKom project (Wahlster, 2002, 2004) addresses multimodal interaction using combinations of speech with pen or hand gesture, in mobile, home, and public kiosk settings. The SmartWeb project provides a mobile multimodal interface to the semantic web (Reithinger et al., 2005).

There is also a growing body of work on generation of output distributed over multiple modes. This typically involves generating text or speech in combination with static or dynamic graphical displays. For example, MAGIC (Dalal et al., 1996) combines speech generation with graphical displays to provide information on a patient's condition while WIP (Wahlster et al., 1993) and PPP (André et al., 1996) provide depictions of three-dimensional objects combined with text or speech. We refer the reader to André (2002) for a detailed overview of work in both multimodal input processing and multimodal output generation.

The critical motivating factor for multimodal interfaces is their naturalness. Human communication is multimodal, combining speech with pointing, physical gestures, and facial expression and so if automated systems are to communicate naturally with users they will need multimodal capabilities. Another critical factor is the flexibility offered by a multimodal interface. If multiple modes are available in an interface, it can enable interaction using whichever mode or combination of modes are most appropriate given the user's preferences, the task at hand, the characteristics of the device, and the nature of the environment. For example, specific kinds of content are best expressed in particular modes. In many cases, pen is more effective for indicating spatial location than speech. In the example in Fig. 7.3, the area specified by drawing

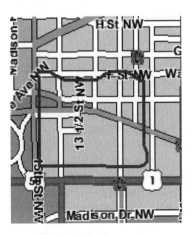

FIGURE 7.3   Area gesture.

a rectangle would be tedious if not impossible to specify precisely using speech. Note that this is not true of all spatial content; for example, in some cases it may be easier to say "along the bank of the river" in speech rather than trying to trace along the river bank with a pen.

Another important advantage of having multiple modes is that the user can switch modes to avoid errors or when noise or privacy are a concern (Oviatt and van Gent, 1996). For example, a user may want to use pen or touch to enter a password rather than speaking if the user can be overheard. We return to this issue in Section 7.5 with respect to responses to sensitive questions.

Multimodal interfaces are playing a critical role in the migration of human–computer interaction from the desktop onto mobile devices, such as personal digital assistants, tablet computers, and next-generation smartphones. The small screen real estate and lack of keyboard or mouse on these devices limits the complexity of the graphical user interface that can be provided and natural modalities such as speech and pen are needed.

There is also a growing body of empirical motivation for multimodal interfaces. Oviatt (1997) shows task performance and user preference advantages for multi-modal interaction compared to speech only or pen only. Oviatt (1999), Bangalore and Johnston (2000) and Johnston and Bangalore (2005) show that fusion of seman-tic content from different modes can compensate for errors in recognition of speech and/or gesture inputs.

In addition to mobile devices, multimodal interfaces are also useful for other keyboard-less settings such as wall-size displays (Corradini and Cohen, 2002) and public information kiosks (Johnston and Bangalore, 2004). With the ongoing voice enablement of the Web, multimodal interfaces can also play a role in enhancing the desktop browsing experience. Possible applications include spatial and location-based tasks such as providing city guides and travel information and intelligence and logistics tasks. They can also be used for mobile workers, including doctors, technicians, and law enforcement. More generally, in order to deal with complex tasks and information, user interfaces employ some kind of graphical visualization. In essence, visualization translates the parameters of complex data to spatial dimensions (along with color), and once the interface is spatial it is well suited to multimodal interaction.

To provide a concrete example of a multimodal system supporting composite multi-modality, we describe here the MATCH system (Johnston et al., 2001, 2002). MATCH stands for Multimodal Access To City Help. The system provides an interactive guide and navigation for information-rich urban environments such as New York City. The MATCH system provides entertainment and subway information for New York City and Washington, D. C. The user interacts with a graphical interface displaying restau-rant listings and a dynamic map showing locations and street information. The inputs can be speech, drawing on the display with a stylus, or synchronous multimodal com-binations of the two modes. The user can ask for the review, cuisine, phone number, address, or other information about restaurants and subway directions to locations. The system responds by generating multimodal presentations that combine synthetic speech output with dynamically changing graphics on a map display, such as zooming

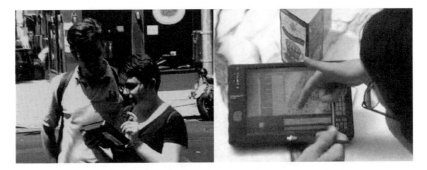

FIGURE 7.4   Multimodal interface on mobile tablet.

the map, displaying route segments, and highlighting graphical labels. There are both mobile (Fig. 7.4) and kiosk (Fig. 7.5) versions of the system and it can run either standalone or in a networked configuration. The kiosk version incorporates a life-like talking head.

For example, a user can request to see restaurants using the spoken command "show cheap Italian restaurants in Chelsea." The system will then zoom to the appropriate

FIGURE 7.5   MATCH system on public kiosk.

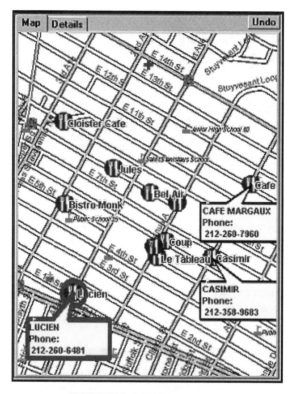

FIGURE 7.6   MATCH map display.

map location and show the locations of restaurants on the map (Fig. 7.6). Alternatively, the user could give the same command multimodally by circling an area on the map and saying "show cheap Italian restaurants in this neighborhood." If the immediate environment is too noisy or public, the same command can be given completely in pen as in Fig. 7.7, by circling an area and writing *cheap* and *Italian*.

Similarly, if the user says "phone numbers for these two restaurants" and circles two restaurants as in Fig. 7.8A, the system will draw a callout with the restaurant name

FIGURE 7.7   Unimodal pen command.

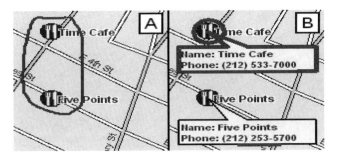

FIGURE 7.8 Information seeking commands.

and number and say, for example, "Time Cafe can be reached at 212-533-7000," for each restaurant in turn (Fig. 7.8B). If the immediate environment is too noisy or public, the same command can be given completely in pen by circling the restaurants and writing *phone* (Fig. 7.9).

The system also provides subway directions (Fig. 7.10). For example, if the user says "How do I get to this place?" and circles one of the restaurants displayed on the map, the system will ask "Where do you want to go from?" The user can then respond with speech; for example, "25th Street and 3rd Avenue," with pen by writing *25th St & 3rd Ave*, or multimodally, for example, "from here" (with a circle gesture indicating the location). Once the user has specified her starting point, the system then calculates the optimal subway route and generates a multimodal presentation coordinating graphical

FIGURE 7.9 Unimodal pen.

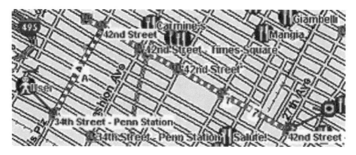

FIGURE 7.10 Subway route presentation.

presentation of each stage of the route with spoken instructions indicating the series of actions the user needs to take (Fig. 7.11).

The underlying technologies that enable multimodal interfaces include speech recognition, gesture and handwriting recognition, natural language understanding, dialogue, multimodal generation and presentation planning, and speech synthesis. Figure 7.11 shows how these various technologies work together in the MATCH system to enable multimodal interaction.

Users interact with the system through a multimodal user interface (MUI) which runs in a Web browser. The user's speech is collected and sent to a speech recognition server, which returns a transcription of what the user said (words). When the user draws on the map, the ink is captured and any objects potentially selected, such as currently displayed restaurants, are identified. The electronic ink is broken into a lattice of strokes and sent to both gesture and handwriting recognition components,

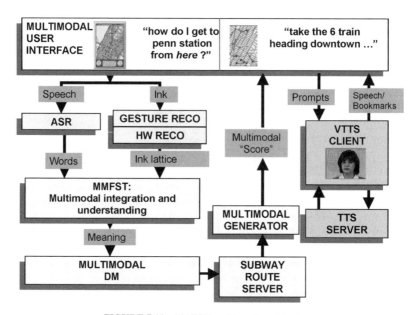

FIGURE 7.11 MATCH multimodal architecture.

which enrich this stroke lattice with possible classifications of strokes and stroke combinations. The gesture recognizer uses a variant of the template matching approach described in Rubine (1991). This recognizes symbolic gestures such as lines, areas, points, arrows, and so on. The stroke lattice is then converted into an ink lattice, which represents all of the possible interpretations of the user's ink as either symbolic gestures or handwritten word. The word and ink lattices are integrated and assigned a combined meaning representation by the multimodal integration and understanding component (MMFST) (Johnston and Bangalore, 2005; Johnston et al., 2002). This provides an output representation of the meaning, which is passed to a multimodal dialogue manager (MDM). If additional information or confirmation is required, the MDM enters into a short information gathering dialogue with the user. Once a command or query is complete, it is passed to the multimodal generation component (MMGEN), which builds a multimodal *score* indicating a coordinated sequence of graphical actions and spoken prompts. The MUI then coordinates presentation of graphical content with synthetic speech output.

Many of the early prototype multimodal user interfaces were standalone demonstration systems. However, new technologies such as Voice-over-IP (VoIP) are enabling the delivery of a multimodal user experience combining graphical interaction with spoken input and output over the Web. These advances are critical for delivery of multimodal interfaces to large numbers of users, something which will be critical if they are to be used for self-administered survey tasks.

7.5   MULTIMODAL INTERFACES FOR SURVEYS

Multimodal interfaces combining graphical interaction with conversational interaction have great potential as a methodology for the survey interview of the future. They overcome many of the limitations of unimodal interfaces by combining the benefits of both spoken and graphical interfaces. In the following sections, we examine these benefits in the context of both self-administered and computer-assisted survey interviews and also for "assistant" interfaces for survey interviewers through consideration of a number of example usage scenarios.

7.5.1   Multimodal Interfaces for Self-administrated and Computer-Assisted Surveys

Multimodal interfaces have a potential role to play both in enabling self-administration of a broader range of potentially more complex surveys and in providing more powerful and flexible tools for computer-assisted surveys in the CASI paradigm. The specific benefits they offer for both of these are:

- The ability to express content, both for system output and user input, in the most appropriate mode available.
- Combination of the engagement of conversational interaction with the parallel and persistent presentation of information characteristic of graphical interfaces.

- Enabling recovery from errors through mode switching.
- Facilitating inclusion of non-native respondents.
- Facilitating inclusion of physically challenged respondents.
- Enriching the survey experience by making it more entertaining.
- Enabling more rapid authoring and tailoring of survey content.

In order to illustrate these points, consider the example of multimodal self-administration of the housing survey used in the determination of the Consumer Price Index. Direct questions to the user may be most easily expressed in speech, for example, "How many bedrooms are there in your house?" while specific spatial content or diagrams are easier to express graphically. For example, rather than trying to provide in words a complex formal definition, such as what is required for a room to be classified as bedroom (e.g., size, having an exterior window), it may be easier to present diagrams or pictures of example room layouts. This feature may be of particular relevance for inclusion of respondents who are not native speakers of English, if they have difficulty with more complex spoken definitions in English. In this case, the system is using multimodal output, a combination of spoken prompts with graphical presentations. Such multimodal presentations can also be more tightly synchronized. For example, the system could provide a definition using speech synchronized with highlighting on a graphical display in order to engage the respondent and improve his/her comprehension. In addition to diagrams, the graphical aspect also enables parallel and persistent presentation of information, allowing for faster interaction by keeping the spoken prompts short and to the point, thus reducing cognitive load on the respondent. For example, multiple options for response can be presented as icons or buttons, which the user can quickly select among using a mouse or touchscreen. This avoids the need for long and unwieldy spoken clarification prompts, for example, "Did you mean 1 . . . 2. . . 3. . . .?" We expect the benefits found by Christian and Dillman for using tangible "show cards" in face-to-face interviews for response options to apply also to multimodal interfaces combining graphics and speech. Multimodality can also play an important role in making user input more flexible. It can alleviate recognition error problems (Oviatt and van Gent, 1996). Since users are not trapped in one mode, they can switch, for example, from speech to GUI. As in the case of certain CASI systems, multimodal interfaces can allow users to switch to a more private mode in order to enter sensitive information. For example, in a domestic violence survey, respondents could indicate where they were injured by a perpetrator using on-screen diagrams, and indicate how they were injured using handwriting, speech, or GUI buttons.

For many surveys it is critical that the results not be skewed by other factors such as the native language of the respondent. Non-native speakers may experience less accurate speech recognition from automated systems, depending on the data used to train the acoustic model, but this can be overcome in a multimodal interface by providing alternate input methods such as handwritten input or a structured graphical user interface. Going beyond this, through incorporation of multilingual aspects to the interface and perhaps the use of machine translation, it should be feasible to develop multimodal survey interfaces that adapt to the native language of the speaker. An important factor here is that if multimodality and multilinguality can be achieved within a single system, there are less likely to be the kind of "mode effects" you might expect if different systems were built for each mode or language, since there is a single underlying processing architecture and quite possibly a common dialogue or interaction controller for all of the different modes of interaction.

Compared to unimodal graphical Web surveys, multimodal systems are more conversational in nature and can engage the user more directly. For example, the system can engage in clarification dialogue with the user to overcome instances of conceptual misalignment. Misalignment cues, such as the Goldilocks range of response times, can be used to identify problems and trigger clarification dialogue. In contrast to in-person interviews, like Web surveys, a self-administered multimodal survey can be taken at the respondent's leisure and does not depend on the availability of a human interviewer. One possible approach is to deliver a link to the survey in e-mail, which opens a Webpage with multimodal content when the user clicks on a link.

Another potential benefit of a multimodal approach is to enrich the survey experience. As it becomes increasingly difficult to access respondents, new methods are needed to encourage participation. Multimodal interfaces combining graphics with speech may be more engaging and encourage participation. Support for this hypothesis can be found in work by André et al. (1999), who found that while animated interface agents did not necessarily improve user performance in information seeking tasks, users did enjoy the experience more, potentially decreasing stress and increasing motivation.

Multimodal interfaces can also play an assistive technology role in enabling access to surveys for respondents with physical or mental disabilities. The availability of different modes and adaptivity in the dialogue strategy can allow users to employ the modes best suited to their capabilities and preferences.

All of these benefits extend readily to the computer-assisted paradigm as well. One important advantage of using an interactive multimodal system compared to pre-recorded video, as in video-CASI, is that the survey material can be more rapidly authored and potentially tailored dynamically to the specific circumstances of the respondent. One interesting research and engineering challenge is to develop techniques that would enable survey researchers and interviewers to author the flow of their own interactive multimodal surveys at an abstract level without specialized knowledge of the underlying speech, graphics, and language processing technology.

For the computer-assisted case, the technology is already in place for deployment of multimodal survey interfaces. These kinds of interfaces already run standalone or in a networked configuration on the existing kinds of tablet computers and similar hardware already being used for CASI work. For self-administration, significant challenges remain depending on the delivery mechanism. Certain kinds of surveys could be administered through public information kiosks such as a customer satisfaction survey in a hotel or store lobby. Again for this case, there are less technological challenges, since the device can be selected, designed, and controlled by the organization administering the survey. The more problematic case is delivery of multimodal surveys over the Web, since in that case there is no control over the hardware and software used. Users' computer systems will have highly varied audio and graphical capabilities, and it is important that this not overly influence survey results. Also, in order to build robust spoken and multimodal dialogue systems, data needs to be collected and annotated for training. Today the telephone remains the ubiquitous access device for most of the world, but the rise of Internet telephony and wireless will continue to erode the base of landline users and listed numbers. New survey technologies need to be developed today so they can be used in a future where the majority of the population have broadband access to the Internet through desktop or mobile computers. In the case of mobile handsets, another issue concerns variability of the quality of the audio channel depending on the carrier and network.

Another potential use case for multimodal survey technologies is for in-situ evaluation of devices. For example, the survey could measure the subjective experience of driving a prototype car in a simulator, or in fact directly in a new vehicle, or user satisfaction with machinery such as copy machines or printers. The system can present spoken questions to the user and the respondent can refer to the relevant place on the dashboard or device using gaze (assuming eye tracking as input) and speaking his/her oral assessment of that feature (e.g., a numerical satisfaction score). Similarly, this kind of multimodal assessment combining pointing/gaze with voice could be used to measure reactions to live (or recorded) music for which a wave form or other graphical representation is displayed, allowing the respondent to point to a particular place in the music and then register an assessment.

7.5.2   Multimodal Interfaces for the Survey Interviewer

So far we have focused on multimodal interfaces where the respondent uses the system directly. Another promising application of multimodal interfaces is as a direct assistant to the interviewer. As in the self-administered case, there can be significant benefits to graphical presentation and input of information. Nusser et al. (2002) describe the use of wearable and hand-held computers by interviewers to display maps and geospatial information. This graphical interaction can be augmented with speech commands and annotations in order to simplify mobile entry of data and access to information. For example, consider the case of agricultural surveys, which can involve detailed spatial presentation and annotation of information such as the types of crops on map displays. The interviewer can indicate with a pen on a map display which fields he is asking questions about while reading standardized questions from a text display, possibly also displaying pictures of particular crops or livestock to make sure the farmer is thinking of the same item as the survey sponsors. In addition to this multimodal presentation of information, the interviewer could use integrated multimodal input in order to annotate the map. For example, the interviewer could point to a field on the display and say the type of crop, and speech recognition and multimodal integration and understanding techniques could be used to translate this directly into a database entry which the interviewer (and potentially also the farmer) can confirm. Another possible design avoids the use of speech recognition and instead directly associates audio sound recordings with spatial regions for later playback. In addition to or as an alternative to speech, handwriting recording or recognition could be supported, allowing the interviewer to specify crop locations using a combination of free-hand drawing on a map with handwritten annotations or commands. For example, the interviewer might say or write "millet planted 15 March 2007" and draw out a rectangle on a map or satellite image indicating the field in question. One can also envision a kind of collaborative multimodal interaction, where the interviewer asks the farmer to draw out the appropriate region on the map, then uses speech, handwriting, or GUI interaction to specify the crop, date of planting, and other information. In this case, one person's gesture is integrated with another person's spoken, handwritten, or GUI input.

Another compelling input modality for mobile entry of data is the position of the device itself, determined using a built-in global positioning system (GPS). This could be used to provide many additional benefits, including more accurate recording of respondent or interviewer location, tailoring of questionnaire to account for local factors (e.g., differences in vocabulary), and even to help the field interviewer in dynamically rescheduling appointments while he/she is mobile.

Another possible application of this kind of multimodal interaction for the interviewer is "Listing"—that is, creating a sample set of residences (or establishments) as part of an "area sample." This is usually done by the interviewer while driving or walking around the blocks that have been randomly selected from within a sampling unit from within a county, and so forth. The system could allow the interviewer to talk to the system while walking in order to make multimodal annotations. With the addition of a camera to the device, still images and video can be used as an additional modality. The user can photograph a residence to identify it if the address is not obvious from the street and cross-index this with GPS information and spoken annotations. The interviewer could even video record the approach to a residence in order to rapidly record literal instructions on how to find the unit. The system would incorporate the recognized speech or processed photograph or tagged video into the list of to-be-interviewed homes, businesses, and so on. This approach is a multimodal extension to the approach taken by Nusser and Fox (2002) using tablet devices and GPS. Of particular interest here is the potential for alignment and multimodal integration of content captured from different modes.

Another use case for this kind of mobile multimodal interface is for "Intercept Surveys" such as exit polls and consumer studies in malls. Mobility is critical in these settings since respondents have limited time and may be moving while the interview takes place. A tablet (instead of laptop) might be especially helpful to interviewers who need to conduct quick interviews with people who are busy doing something else. The interviewer can navigate by speaking to the system in order to avoid manual interaction (e.g., scrolling and dragging and clicking on a small display), particularly if conducting the interview while walking and while needing to maintain eye contact with the respondent. The interviewer can use a pen or touch to indicate the respondent's precinct or other geographic identifiers and could also play speech or video files of campaign ads or products to get reactions from the respondent. The interviewer could enter most answers by clicking, typing, or speaking. The interviewer could also present options (text and radio buttons or pictures) and then pass the tablet to the respondent for pen or touch selection of options to provide privacy in a public place, just as is currently done with CASI.

7.6   CONCLUSION

Table 7.1 presents an overview of the relative merits of human interview, CASI, Web survey, spoken dialogue systems, and multimodal dialogue systems for survey tasks. Schober, Conrad, and colleagues have demonstrated experimentally the benefits of collaborative conversational interaction for survey interviews compared to standard techniques. However, conversational interaction takes longer and in the case of automated telephone systems, conversational speech is harder to recognize and will be subject to more errors. Graphical Web surveys avoid recognition errors but do not engage the user and lack access to the conceptual misalignment cues (Ehlen, 2005), which can be used to detect understanding problems. Human–computer interfaces are rapidly evolving to support natural input modalities such as speech and pen and can now enable a coordinated multimodal user experience combining speech and pen with graphical user interfaces delivered over the Web and on mobile devices. These technologies can enable a new generation of automated surveys, which combine the engagement of spoken interaction with the robustness and speed of graphical interaction. This will allow for collaborative conversational interaction with less recognition errors, easier recovery from errors, and availability of cues for prediction of conceptual misalignment.

However, substantial research challenges remain in the application of these new technologies. It is critical that the community act now and study in more detail the impact of these technologies on survey tasks. Developing technology for this application area will take several years, and within that timeframe the viability of the existing paradigm of survey research will surely come into question. Investigation is needed to

TABLE 7.1   Comparison of Survey Technologies

Factor	Human	CASI	Web	Spoken	Multimodal
Conversational interaction	✓	✓	✗	✓	✓
Parallel presentation	✗	✓	✓	✗	✓
Persistent presentation	✗	✓	✓	✗	✓
Actively engages user	✓	✓	✗	✓	✓
Mode switching	✗	✓	✗	✗	✓
Avoid recognition errors	✓	✓	✓	✗	✓
Faster completion	✗	✗	✓	✗	✓
Take survey when you want	✗	✗	✓	✗	✓
Reduces cost	✗	✗	✓	✓	✓
No training data	✓	✓	✓	✗	✗

better understand which content should be conveyed in speech and which graphically for the survey tasks. Speech input can be encouraged in order to provide access to misalignment cues but could result in more errors in some cases. The balance between these factors needs to be better understood. Substantial challenges also remain in the authoring, development, and testing procedures for multimodal automated surveys. Tools are needed so that effective multimodal surveys can be implemented and administered by survey researchers without detailed technical experience in multimodal interface design. Another avenue to explore is the creation of semi-automated systems with a human in the loop where spoken and multimodal analysis is used as an aid to a human interviewer.

REFERENCES

Algayer, J., Hat-busch, K., Kobsa, A. Reddig, C., Reithinger, N., and Schmauks, D. (1989). XTRA: a natural language access system to expert systems. *International Journal of Man-Machine Studies, 31*(2), pp. 161–195.

André, E. (2002). Natural language in multimedia/multimodal systems. In R. Mitkov (Ed.) *Handbook of Computational Linguistics.* New York: Oxford University Press.

André, E., Müller, J., and Rist, T. (1996). The PPP persona: a multipurpose animated presentation agent. In *Advanced Visual Interfaces* (pp. 245–247). New York: ACM Press.

André, E., Rist, T., and Müller, J. (1999). Employing AI methods to control the behavior of animated interface agents. *Applied Artificial Intelligence. 13*, pp. 415–448.

Bangalore, S., and Johnston, M. (2000). Integrating multimodal language processing with speech recognition. In *Proceedings of International Conference on Spoken Language Processing.* Beijing, China.

Bolt, R. A. (1980). Put-that-there: voice and gesture at the graphics interface. In *Proceedings of the 7th Annual Conference on Computer Graphics and Interactive Techniques*, Seattle, pp. 262–270.

Carbonell, J. (1970). AI in CAI: an artificial intelligence approach to computer assisted instruction. *IEEE Transactions on Man-Machine Systems, 11*(4), pp. 190–202.

Cassell, J. (2000). More than just another pretty face: embodied conversational agents. *Communications of the ACM, 43*(4), pp. 70–78.

Cheyer, A., and Julia, L. (1995). Multimodal maps: an agent-based approach. In *Proceedings of the International Conference on Cooperative Multimodal Communication*, pp. 103–113.

Chu-Carroll, J., and Carpenter, B. (1999). Vector-based natural language call routing. *Computational Linguistics, 25*(3), pp. 361–387.

Cohen, P. R., Johnston, M., McGee, D., Oviatt, S. L., Pittman, J., Smith, I., Chen, L., and Clow, J. (1997). QuickSet: multimodal interaction for distributed applications. In *Proceedings of the Fifth ACM International Multimedia Conference.* New York: ACM Press.

Conrad, F. G., Schober, M. F., and Coiner, T. (2007). Bringing features of dialogue to web surveys. Applied Cognitive Psychology, in press.

Couper, M. P., and Lyberg, L. E. (2005). The use of paradata in survey research. In *Proceedings of the 54th Session of the International Statistical Institut*e, Sydney, Australia, April 2005.

Corradini, A., and Cohen, P. R. (2002). Multimodal speech–gesture interface for hands-free painting on a virtual paper using partial recurrent neural networks as gesture recognizer. In *Proceedings of the International Joint Conference on Artificial Neural Networks (IJCNN'02)*, Vol.III, May 12–17, Honolulu (HI, USA), pp. 2293–2298.

Dahlbäck, N., Jönsson, A., and Ahrenberg, L. (1993). Wizard of Oz-studies—why and how. In *Workshop on Intelligent User Interfaces*, Orlando, FL.

Dalal, M., Feiner, S., McKeown, K., Pan, S., Zhou, M. X., Höllerer, T., Shaw, J., Feng, Y., and Fromer, J. (1996). Negotiation for automated generation of temporal multimedia presentations. *ACM Multimedia*, pp. 55–64.

Ehlen, P. (2005). The dynamic role of some conversational cues in the process of referential alignment. Ph.D. Thesis. The New School University.

Ehlen, P., Schober, M. F., and Conrad, F. G. (2005). Modeling speech disfluency to predict conceptual misalignment in speech survey interfaces. In *Proceedings of the Symposium on Dialogue Modeling and Generation,* 15th Annual Meeting of the Society for Text and Discourse, Vrije Universiteit, Amsterdam, July 2005.

Ehlen, P., Schober, M. F., and Conrad, F. G. (in press). Modeling speech disfluency to predict conceptual misalignment in speech survey interfaces. *Discourse Processes.*

Fraser, N., and Gilbert, N. S. (1991). Simulating speech systems. *Computer Speech and Language, 5,* pp. 81–99.

Fowler, F. J., and Mangione, T. W. (1990). *Standardized Survey Intewiewing: Minimizing Interviewer-Related Error.* Newbury Park, CA: Sage Publications.

Gorin, A., Riccardi, G., and Wright, J. (1997). How may I help you? *Speech Communication, 23,* pp. 113–127.

Gupta, N., Tur, G., Hakkani-Tur, D., Bangalore, S., Riccardi, G., and Rahim, M. (2005). The AT & T spoken language understanding system. *IEEE Transactions on Speech and Audio, 99,* pp. 1–10.

Johnston, M. (1998). Unification-based multimodal parsing. In *Proceedings of the 17th International Conference on Computational Linguistics and 36th Annual Meeting of the Association for Computational Linguistics, (COLING-ACL 98)*, pp. 624–630.

Johnston, M., and Bangalore, S. (2004). MATCHKiosk: a multimodal interactive city guide. In *Proceedings of the ACL-2004 Interactive Posters/Demonstrations Session*, pp. 222–225.

Johnston, M., and Bangalore, S. (2005). Finite-state multimodal integration and understanding. *Journal of Natural Language Engineering, 11*(2), pp. 159–187.

Johnston, M., Bangalore, S., and Vasireddy, G. (2001). MATCH: Multimodal Access To City Help. In *Proceedings of ASRU 2001 Workshop.* Madonna di Campiglio, Italy.

Johnston, M., Bangalore, S., Vasireddy, G., Stent, A., Ehlen, P., Walker, M., Whittaker, S., and Maloor, P. (2002). MATCH: an architecture for multimodal dialogue systems. In *Proceedings of the 40th Annual Meeting of the Association for Computational Linguistics*, pp. 376–383.

Johnston, M., Cohen, P. R., McGee, D., Oviatt, S. L., Pittman, J., and Smith, I. (1997). Unification-based multimodal integration. In *Proceedings of the 35th Annual Meeting of the Association for Computational Linguistics*. Madrid, Spain.

Koons, D., Sparrell, C., and Thorisson, K. (1993). Integrating simultaneous input from speech, gaze and hand gestures. In M. Maybury (Ed.), *Intelligent Multimedia Interfaces* (pp. 257–276). Cambridge, MA: MIT Press.

Krysan, M., and Couper, M. P. (2003). Race in the live and the virtual interview: racial deference, social desirability, and activation effects in attitude surveys. *Social Psychology Quarterly, 66*(4), Special Issue: Race, Racism, and Discrimination, pp. 364–383.

Neal, J. G., and Shapiro, S. C. (1991). Intelligent multimedia interface technology. In J. W. Sullivan and S. W. Tyler (Eds.), *Intelligent User Interfaces* (pp. 45–68). New York: ACM Press Addison Wesley.

Nusser, S., Miller, L., Clarke, K., and Goodchild, M. (2002). Geospatial IT for mobile field data collection. *Communications of the ACM, 46*(1), pp. 63–64.

Nusser, S., and Fox, J. E. (2002). Using digital maps and GPS for planning and navigation in field surveys. Iowa State University, Department of Statistics Preprint 2002–09.

Oviatt, S. L. (1997). Multimodal interactive maps: designing for human performance, *Human–Computer Interaction*, pp. 93–129.

Oviatt, S. L., and van Gent, R. (1996). Error resolution during multimodal human-computer interaction. In *Proceedings of International Conference on Spoken Language Processing*, Vol. I, pp. 204–207.

Oviatt, S. L. (1999). Mutual disambiguation of recognition errors in a multimodal architecture. CHI 99. pp. 576–583.

Reithinger, N., Bergweiler, S., Engel, R., Herzog, G., Pfleger, N., Romanelli, M., and Sonntag, D. (2005). A look under the hood—design and development of the first SmartWeb system demonstrator. In *Proceedings of the Seventh International Conference on Multimodal Interfaces*, pp. 159–166.

Rubine, D. (1991). Specifying gestures by example. *Computer Graphics, 25*(4), pp. 329–337.

Schober, M. F. (1998). Conversational evidence for rethinking meaning. *Social Research* Special issue: "Conversation", *65*(3), pp. 511–534.

Schober, M. F., and Conrad, F. G. (1997). Does conversational interviewing reduce survey measurement error? *Public Opinion Quarterly, 61*, pp. 576–602. Reprinted in N.G. Fielding (Ed.), (2005). *Interviewing, Vol. 1 (SAGE Benchmarks in Social Research Methods Series)*. Thousand Oaks, CA: Sage Publications.

Schober, M. F., and Conrad, F. G. (2002). A collaborative view of standardized survey interviews. In D. W. Maynard, H. Houtkoup-Streenstra, N. C. Schaffer, and J. van der Zouwen (Eds.), *Standardization and Tacit Knowledge: Interaction and Practice in the Survey Interview*. Hoboken, NJ: John Wiley & Sons.

Schober, M. F., Conrad, F. G., Ehlen, P., and Fricker, S. S. (2003a). How Web surveys differ from other kinds of user interfaces. In *Proceedings of the American Statistical Association, Section on Survey Research Methods*. Alexandria, VA: American Statistical Association.

Schober, M. F., Conrad, F. G., Ehlen, P., Lind, L. H., and Coiner, T. F. (2003b). Initiative and clarification in Web-based surveys. In *Proceedings of 2003 AAAI Spring Symposium on Natural Language Generation in Spoken and Written Dialogue*. Technical Report SS-03-06, pp. 125–132. Menlo Park, CA: AAAI Press.

Schapire, R. E. (1999). A brief introduction to Boosting. *Proceedings of IJCAI*, pp. 1401–1406.

Schwarz, N., Hippler, H.-J., and Noelle-Neumann, E. (1992). A cognitive model of response-order effects in survey measurement. In N. Schwarz and S. Sudman (Eds.), *Context Effects in Social and Psychological Research* (pp. 187–201). New York: Springer-Verlag.

Suchman, L., and Jordan, B. (1990). International troubles in face-to-face survey interviews. *Journal of the American Statistical Association*, 85(409), pp. 232–253

Tourangeau, R., Couper, M. P., and Steiger, D. M. (2003). Humanizing self-administered surveys: experiments on social presence in Web and IVR surveys. *Computers in Human Behavior, 19*, pp. 1–24.

Tourangeau, R., and Smith, T. W. (1998). Collecting sensitive information with different modes of data collection. In M. P. Couper, R. P. Baker, J. Bethlehem, C. Z. F. Clark, J. Martin, W. L. Nicholls II, and J. M. O'Reilly (Eds.), *Computer Assisted Survey Information Collection* (pp. 431–453). Hoboken, NJ: John Wiley & Sons.

Vapnik, V. N. (1995). *The Nature of Statistical Learning Theory*. New York: Springer.

Vo, M. T. (1998). A framework and toolkit for the construction of multimodal learning interfaces. Ph.D. thesis, Carnegie Mellon University.

Wahlster, W. (2002). SmartKom: fusion and fission of speech, gestures, and facial expressions. In *Proceedings of the 1st International Workshop on Man–Machine Symbiotic Systems*, Kyoto, Japan, pp. 213–225.

Wahlster, W. (2004). *SmartKom: Foundations of Multimodal Dialogue Systems*. New York: Springer-Verlag.

Wahlster, W., André, E., Finkler, W., Profitlich, H. J., and Rist, T. (1993). Plan-based integration of natural language and graphics generation. *Artificial Intelligence, 63*, pp. 387–427.

Walker, M., Langkilde, I., Wright, J., Gorin, A., and Litman, D. (2000). Learning to predict problematic situations in a spoken dialogue system: experiments with "How May I Help You?" In *Proceedings of the 1st Meeting of the NAACL*. Seattle, Washington, USA, pp. 210–217.

CHAPTER 8

Is It Self-Administration If the Computer Gives You Encouraging Looks?

Justine Cassell and Peter Miller
Northwestern University, Evanston, Illinois

8.1   INTRODUCTION

The movie *Kinsey* opens with a scene of the great sex researcher being interviewed by one of his students. During the course of the interview, Kinsey not only reveals some surprising facts about his own sexual history, but also takes the opportunity to train the student in the fine art of interviewing. Much of the training concerns body language— eye gaze is important to indicate that you are listening, sitting far away creates a perception of distance, frowning will prevent the subject from relaxing. Kinsey's unfortunate student, however, has a hard time controlling his nonverbal behavior as the interview delves more and more deeply into the personal life of his mentor.

Survey researchers have long worried about the unconscious effects of interviewer appearance, including nonverbal behaviors such as these, on the responses of survey interviewees. For this reason (as well as to reduce costs), various communication technologies that allow partial or total self-administration of the interview have been adopted by survey researchers in attempts to objectify the survey process. Self-administered paper and pencil questionnaires, the telephone, computer-assisted self-interviews, and then the Web survey have been thought to hold the answer to the sorts of bias that might be introduced into surveys by the effects of the face-to-face contact of two humans in conversation. The telephone transmits only voice information about the interviewer, and other methods (in the text format most often used to present questionnaires) removes the interviewer altogether. A paradox exists,

Envisioning the Survey Interview of the Future, Edited by Frederick G. Conrad and Michael F. Schober
Copyright © 2008 John Wiley & Sons, Inc.

161

however, in the debate surrounding the use of these technologies. On the one hand, many researchers continue to hold to the belief that *rapport* is essential to the successful survey interview. On the other hand, it is also thought that the appearance of the interviewer as well as his/her unconscious nonverbal responses may affect respondents' answers, as the respondents seek to give the answers that they believe are expected. If, however, *rapport* is in fact primarily carried by nonverbal behavior—is induced by the very set of non-conscious behaviors that the telephone and Internet are thought to suppress—then there is a problem. For this reason the newest communication technology on the block, embodied virtual humans, are a possible answer to the survey interviewer's paradox. Embodied virtual humans might allow us to control the appearance of an interviewer as well as exactly which verbal and nonverbal behaviors are displayed, allowing us to then collect, in an atmosphere of utmost scientific accuracy, the information from citizens upon which the modern democracy depends.

In this chapter we examine the literature on survey interviewing, and the lessons we can learn about the effects of appearance and nonverbal behavior. We then turn to embodied virtual humans and the role that they might potentially play in better understanding these effects in survey interviewing, and perhaps even in mitigating them in practice.

Discussions of interviewer effects often obliquely implicate nonverbal communication between interviewer and respondent. The socioeconomic or racial divide between interviewers and respondents has produced worries about respondent deference, manifesting itself in socially desirable answers. Interviewers have been taught that they must establish "rapport" with respondents, but maintain a "professional distance" in order to get accurate information. Survey directors over the decades have speculated on what combination of demographic and attitudinal traits make for a "good interviewer." Interviewing lore is suffused with beliefs about how skin color, gender, dress, demeanor, attitude, and "body language" can influence responses for good or ill. But the contribution of these nonverbal factors to response error has not been identified in most cases, or even studied systematically. One often needs to "read between the lines" of methodological reports to infer how nonverbal communication may have played a role in the findings.

The following selective review looks at examples of research involving the physical characteristics of the interviewer and the idea of rapport in face-to-face interviews, before turning to an examination of research comparing face-to-face and other modes of respondent contact. The review concludes with some observations from studies that have closely examined within-interview interaction, before we turn to the topic of adapting embodied virtual humans to the task of the survey interview.

8.2   PHYSICAL APPEARANCE OF THE INTERVIEWER

The interviewing staffs employed by the polls come from a rather narrow range of the socio-economic scale. They are well above the median of the population in income, they are mostly people with some college education, they are better dressed, more academic in speech and more bourgeois in outlook than the lower income groups they interview. (Katz, 1942).

8.2.1 Social Class

A prominent theme in early studies of interviewer effects was the impact of the interviewer's social class on responses from their lower class counterparts. Katz's (1942) paper is an early example of such articles published between World War II and the 1960s. The basic hypothesis—implied in the quotation above—is that interviewers, who were universally middle to upper class, may obtain socially desirable answers from lower class respondents. Lower class respondents were thought to infer from the dress and mannerisms of interviewers what sort of answers they would like to hear and then "unconsciously" shade responses in that direction. Katz's study involved a quasi-experimental comparison of interviews administered by typical "white collar" Gallup interviewers and working class interviewers hired specially for the study. The "white collar" interviewers obtained responses that were less supportive of union activity than did their "blue collar" counterparts.

Lenski and Leggett's (1960) "Caste, Class and Deference in the Research Interview" furthers this theme in their examination of the social class and race differential between interviewers and respondents in the 1958 Detroit Area Study. Going beyond the notion that white middle class interviewers would elicit deferential responses on items related to unionism or race, these investigators documented that lower class and black respondents were more likely to give responses characteristic of acquiescence.

8.2.2 Age and Gender

Benney, Riesman, and Starr (1956) and Ehrlich and Riesman (1961) examined the effect due to age and gender differences between interviewers and respondents. In the former study, respondents gave different answers to sensitive sexual questions depending on the age and gender of the interviewer. In the latter case, adolescent girls were more likely to report disobedience to younger than older interviewers.

8.2.3 Race

Research on how the race of an interviewer affects responses, mainly focusing on black respondents, also has a lineage stretching back to the 1940s. Hyman (1954) reported the results of two experiments undertaken in 1942, one in Memphis and one in New York, in which white and black interviewers were randomly assigned to interview black respondents. Black respondents in the Memphis sample gave substantially different answers to white and black interviewers, even on innocuous items like automobile ownership. The differences observed in the New York experiment were notably smaller, leading to the belief that the oppressive segregationist environment in Memphis exacerbated the race-of-interviewer effect.

Robinson and Rohde (1945) conducted an experiment in which interviewers with varying degrees of "Jewishness" (appearance and names) administered interviews that included items measuring anti-Semitism. The more "Jewish" the interviewer (in appearance and name), the fewer anti-Semitic responses received.

Athey et al. (1960) compared responses obtained in interviews conducted by white, Asian, and black interviewers. Compared with whites, Asian interviewers got more positive answers to questions about Asian acceptance. Again, compared with whites, black interviewers received more positive views of housing integration. Williams (1964, 1968) reported the findings of a similar experiment in North Carolina in 1960. In this case, the white and black interviewers were randomly assigned to interview black respondents. The study found that respondents tailored racial attitudes in the direction of the interviewers' assumed preferences, particularly low SES respondents.

Another randomized experiment involving race of interviewer was conducted in Detroit in 1968 and the findings were reported by Schuman and Converse (1971). Hatchett and Schuman (1975) examined the impact of race of interviewer on *white* respondents in an experiment in Detroit in 1971. Schuman and Converse's oft-cited findings led to the conclusion in the methodological community that the primary site of race-of-interviewer effects is in responses to questions about racial matters, and that black interviewers obtain more valid responses on such items from respondents of the same race (though the authors caution that the picture is more complex and is apt to change over time as the nature of race relations changes). Hatchett and Schuman's study of white respondents showed a similar pattern of tailoring responses to what the respondent may perceive that interviewers want to hear. The study called into question the deference or fear explanation commonly offered for interviewer effects involving black respondents, suggesting that respondents—black or white—generally seek to get through a potentially difficult social exchange with the least conflict.

8.2.4   Political Affiliation

A unique study reported by Bischoping and Schuman (1992) was conducted during the campaign leading up to the Nicaraguan elections of 1990. This election, following years of civil war, pitted the ruling Sandinista party against a coalition of opposition parties under the UNO label. The campaign was bitter and took place in a volatile context in which political polls, frequently affiliated with one political camp or the other, varied markedly on the predicted outcome. They sought to explain how respondents came to tailor their responses toward the perceived preferences of interviewers who worked for even ostensibly neutral organizations. Schuman designed a randomized experiment in which the same cadre of interviewers, using the same questionnaire, sponsored by a local university, conducted interviews with only one variation in approach—the color of the pen they used to record responses. When interviewers used a red pen, a major color of the Sandinista party, reported political preferences tended to favor that group. When they used a blue pen, the color of the UNO coalition, the response pattern was reversed. This is the only study reviewed here in which a single nonverbal characteristic was the sole manipulation, and it demonstrates the extent to which putatively small changes in the interviewing situation can apparently lead to major differences in response.

8.2.5   Multiple Social Characteristics

Some investigations have studied interviewer–respondent disparities on multiple social characteristics. Dohrenwend, Colombotos, and Dohrenwend (1968) analyzed income and race differences between interviewers and respondents in a public health survey in New York. Weiss (1968) examined age, education, and socioeconomic status disparities between interviewers and respondents in a survey of welfare mothers. As noted previously, Williams (1964) examined race-of-interviewer effects for high and low SES black respondents.

What do we learn from this body of research into the effects of demographic characteristics on interviewing? With the exception of Katz's early quasi-experiment, studies examining the effect of social class differences between interviewers and respondents are based on post-hoc coding of interviewer and respondent characteristics. Katz's study, comparing white collar and blue collar interviewers, could not make clear inferences about interview effects because the interviewers were responsible both for recruiting respondents and questioning them. In none of the social class studies was there within-interview evidence of the process by which the social class differential produced the effect it was claimed to produce. Similarly, studies of age, gender, or race-of-interviewer effects infer a process of communication that is not documented through observation. Respondents may tell interviewers what they think they want to hear for a complex of reasons, including nonverbal cues, but the studies in this area were not designed to "unpack" how social class, age, gender, or race worked their effects in the interview, nor how the communication might vary in different interview contexts. Nevertheless, the studies presented earlier are highly suggestive of effects due to the appearance of survey interviewers.

8.3   RAPPORT

Along with interviewer demographic characteristics, "rapport" is a frequently cited factor that may affect the quality of data obtained in the interview. It is an elusive concept, as Weiss (1970) pointed out. A dictionary definition is "a relationship, especially one of mutual trust or emotional affinity." Establishing "rapport" was traditionally thought to be a prerequisite of gaining interview cooperation. Interviewing texts mention the importance of establishing trust so that respondents will be forthcoming in their answers, but, as Hyman (1954) famously noted, it can be overdone, leading to ingratiating rather than honest responses. For Cicourel (1964), rapport is incompatible with a task orientation to the interview. This concept has been the subject of much speculation but little research. If it is useful in the interview—and Weiss (1970) and Goudy and Potter (1975) suggest that it is not—nonverbal behaviors are likely to play a major role in establishing it. In her brief review, Weiss (1970) states—but does not cite the relevant studies—that nonverbal communication such as eye contact, smiles, and head nods contribute to rapport between interviewer and respondent.

Sheatsley's (1950, 1951) exhaustive description of the characteristics of the NORC field staff begins and ends with the concern that respondents who are questioned by dissimilar interviewers will provide erroneous answers. For Sheatsley, an interviewer's "rapport" with the respondent was the *same thing* as demographic similarity. See also Williams (1964), who states that "status characteristics directly affect communication, and similar status tends to reduce bias, especially by reducing inhibitions" (p. 340). Riesman and Glazer (1948) in a wide-ranging essay on "the meaning of public opinion," following the failure of the polls prior to the 1948 U.S. presidential election, also equate "rapport" in part with demographic similarity, noting that "the polls try to use an interviewer of [the respondent's] own ethnic group, though they are rarely able to use one of his own class" (p. 643).

Other investigators went beyond demographic similarity to measure "rapport" in different self-reported measures. Dohrenwend et al. (1968) classified interviewers according to whether they had preferences for interviewing particular kinds of respondents and whether they reported suffering embarrassment when asking about any of the topics in the interview. These scores, applied to the interviewers, were expected to capture how they may have behaved when talking with "unpreferred" people or when working through parts of the questionnaire they did not like to administer. A similar interviewer-based scoring system was employed by Williams (1968), who had interviewers complete measures designed to capture their general capacity for "rapport" and "objectivity." Weiss (1968), by contrast, had interviewers score each respondent on a scale measuring how cordial and disclosing she had been. This interview-specific measure, though not the product of an analysis of the interview interaction and based only on interviewer assessment, is a step closer toward the goal of capturing the affective flavor of an interview. Importantly (and ironically), Weiss found that, overall, the higher the rapport on average, the lower the validity of the responses (ascertained through matching responses with administrative records).

8.4   STUDIES OF INTERACTION IN THE FACE-TO-FACE INTERVIEW

The studies reviewed earlier identified factors involving nonverbal communication in face-to-face interviews that may lead to response error. The studies do not spell out a clear mechanism for the effects observed and do not offer interview observations of nonverbal communication. We do not know, for example, if white interviewers behaved differently nonverbally when asking questions of black respondents about racial matters than they did when addressing other topics. We don't know, similarly, if younger interviewers exhibited different nonverbal behaviors when they obtained more reports of disobedience from adolescent girls than did older interviewers. We know that interviewers must establish some sort of relationship with respondents if they are to complete an interview, but the studies of "rapport" leave us in the dark about the behaviors that produce the sort of "rapport" that leads to better reporting.

Although "race-of-interviewer" effect studies still appear in the literature now and then, there is little research on other aspects of interviewer physical appearance and "rapport" today. In part, the shift away from this line of research is the result of the very technological change we described earlier—away from face-to-face contacts and toward telephone, self-administered, and Web surveys. In part, the change also reflects lessening concern about some social fissures in American society (social class). Also, the field has come to believe, based on studies like Schuman and Converse (1971) and Hatchett and Schuman (1975), that we need only worry about response effects from interviewer appearance in special cases like racial attitudes. Finally, survey organizations recognized from the time of Sheatsley's report in the early 1950s that, due to the vagaries of the survey business and its labor pool, it was impossible on a general basis to physically match respondents with interviewers.

In concert with these observations, the study of interviewing shifted its focus to more microanalysis of interviewer–respondent interaction. Aided by the development of small, battery-powered audiotape recorders, records of question-asking and answering could be examined closely for characteristics related to better or worse reporting. The programmatic research by Cannell and colleagues, beginning in the 1960s and continuing for three decades, is the key exemplar of this approach. Like the earlier work on interviewer characteristics, it seldom addressed nonverbal behavior. But some of these studies offer clues into how nonverbal behavior might work in the interview and also offer a template for studies that take on the topic directly.

A formative investigation in this program of research was Cannell, Fowler, and Marquis (1968), in which reporting in health interviews was examined in relation to a wide variety of independent variables—interviewer and respondent demographic, knowledge and attitudinal variables, as well as all of the verbal behaviors that took place in the interviews. A key lesson of the study was that the behaviors predicted response quality better than the demographic variables that had characterized earlier interviewing studies. Interviewers and respondents tended to balance one another's verbal output. Some of the output produced useful answers—as when the interviewer asked a question correctly and the respondent gave an answer that met the question's objectives—and much verbal activity was devoted to other sorts of task-relevant and irrelevant objectives.

The focus on microanalysis of behavior offered the possibility of "unpacking" interviewer effects attributed to demographic differences with respondents. In a following study, Marquis and Cannell (1969) examined the exchanges between white interviewers with older and younger white and black respondents. They found, as in the earlier investigation (Cannell et al., 1968), that there was more verbal activity in total in the interviews with older respondents, and that there was more "task orientation" in the interviews with younger and black respondents—less overall verbal activity. The interviews with older and white respondents featured more nontask communication, including comments on the questions, "polite behavior," elaborations, suggestions to the interviewer, and refusals to answer. In other words, the white interviewers appear to have experienced something more like a "conversation" when interviewing white and older respondents. The addition of nonverbal observations would have allowed this hypothesis to be tested.

One more experiment in this research program suggests how the microlevel study of interviewing effects could benefit from observations of nonverbal behavior. Henson, Cannell, and Lawson (1973) compared two interviewing treatments for eliciting information about automobile accidents: a "personal" style emphasizing incipient friendship (like some earlier ideas of "rapport") and an "impersonal" approach focused on administering the questionnaire in a standard manner. The two approaches produced basically equivalent results in the accuracy of accident reporting. Van der Zouwen, Dijkstra, and Smit (1991) also report inconclusive results in a similar comparison.

8.5   LESSONS AND SHORTCOMINGS OF INTERVIEW INTERACTION STUDIES

Early studies of interviewing effects based on interviewer appearance begged the question of how physical characteristics of the interviewer worked the effects they were said to produce. More recent research, which has focused carefully on interviewer–respondent interaction, has improved on the "black box" of earlier studies, but its utility is limited by the absence of data on nonverbal behavior (with a few exceptions, e.g., Schober and Bloom (2004) on paralinguistic indicators of misunderstanding). We learn from studies of interviewer personal characteristics and "rapport" that respondents sometimes give answers that appear to be ones that would be well received by interviewers, rather than ones that seem more accurate—an effect that has been referred to as "social desirability." Rapport has been equated with similarity between the interviewer and respondent, or has been conceived of as a characteristic of the relationship between the two, in which case it may mediate or interact with physical similarity.

For the most part, research on rapport in the survey interview has not drawn from research in the field of interpersonal interaction. Thus, while there is a fairly substantial body of literature examining the verbal and nonverbal components of rapport, it has not influenced survey research. A particularly influential theory (Tickle-Degnen and Rosenthal, 1990) describes rapport as comprised of three components, which change in importance over time, and each of which is conveyed by both verbal and nonverbal means. *Positivity* is particularly important early in rapport-building and can be indicated by smiles. *Mutual attention* remains important throughout the development of rapport and is conveyed by eye gaze among other behaviors. *Coordination* becomes increasingly important over the course of a budding relationship and is marked by features such as quick turn-taking and frequent acknowledgment of the other's utterances. Other researchers have found important contributions in assessment of rapport played by posture (LaFrance, 1982) and certain kinds of nonverbal mimicry (Lakin et al., 2003). We do not know, however, the extent to which survey interviewers are consciously or unconsciously engaging in these behaviors and for this reason the process by which demographic similarity or rapport affects respondents' answers remains a "black box."

In sum, it is clear that the appearance of the interviewer, including interviewer demographics and moment-to-moment nonverbal reactions, have some effect on the interview, but neither the extent nor the mechanism is clear.

These questions became largely moot for many survey applications as the telephone replaced the face-to-face interview in commercial, academic, and increasingly, government surveys. The energy that might have gone into studying nonverbal behavior in the face-to-face interview was expended instead on aspects of the telephone contact—for example, how respondents would react to the lack of visual cues provided in person (Miller, 1984) or how the interviewer's tone of voice might affect nonresponse in telephone contacts (Oksenberg et al., 1986). Early consideration of the telephone interview produced the hypothesis that the audio-only mode might encourage more respondent disclosure because the interviewer was not physically present (Hochstim, 1967). But the issue of social desirability did not disappear with the growth of telephone interviewing. Instead, newer technology, operationalized in computerized self-administered questionnaires (computer-assisted self interviewing or CASI) or in audio computer-assisted self-interviewing (A-CASI) methods employed in telephone contacts, was introduced to remove the effects of the interviewer's presence and lessen social desirability bias. A-CASI has shown some notable results (Villarroel et al., 2006) and more mixed findings (Currivan et al., 2004; Moskowitz, 2004). In any case, the general trend in methodological research has been to examine ways in which the impact of the interviewer, whatever it may be, can be lessened or removed and self-administration can be increased, in the interest of obtaining honest answers from respondents.

8.6 EMBODIED CONVERSATIONAL AGENTS

Enter the embodied virtual human (henceforth called by the name by which it is most commonly known to researchers in human–computer interaction, *embodied conversational agent* or ECA). To be clear on our terms, an ECA is a graphical projection of a full-body human on a screen. The ECA may be life-size and projected onto a giant screen, or may be small enough to fit on a hand-held device such as a cellphone. The depiction of the human can fall anywhere between absolute realism and cartoon depiction. But for it to be called an embodied conversational agent it must be able to speak and also to display many of the nonverbal behaviors that humans do in face-to-face conversation, such as eye gaze, head nods, facial displays, body posture, and hand gestures. And it must be capable of simulating many of the same responses that humans give, such as happiness and sadness, attentiveness and boredom, desire to take the floor, and acknowledgment that the other's words have been understood. In fact, much research on ECAs is directed toward *autonomous* embodied conversational agents, where some or all of those responses are automatically generated on the basis of underlying models of human behavior (Cassell et al., 2000). That is, autonomous ECAs may nod when their speech recognition algorithm has actually understood what the real human interlocutor has said, and look happy when their model of emotion has detected the fulfillment of a hoped-for goal. During the development phase, however, the effects of ECAs on interaction are often studied using what is called the "Wizard of Oz mode," where a scientist hidden behind a curtain controls the behaviors of the ECA. In any case, and importantly for our purposes here, ECAs are implemented on the basis of research into human behavior, and they are most often evaluated with respect to a gold standard of real human behavior. They are also the result of decades of research in Artificial Intelligence (AI) and Natural Language Processing (NLP) that allows the construction of systems that understand language, reason on the basis of that understanding, and produce responses appropriate to the context. However, the ECA marked an important departure from previous work in AI and NLP, in that it recognized the importance of the social context in interaction—that reasoning and understanding are not sufficient for human conversation, which also requires a *display* of having understood, and the means to regulate the conversation through devices such as nods and lean-forwards. ECAs are attempts to make interaction with a computer more natural, intuitive, and like human interaction.

In this vein, the methodology underlying ECAs is quite specific. Researchers collect data on human–human conversation, analyze those data in such a way as to build a formal or predictive model of the human behavior observed, implement a computational system of a virtual human on the basis of the model, evaluate the system both by observing what it looks like themselves, and by observing how it interacts with real humans, evaluate the results, and go back to collect more data on human–human communication to fill the lacunae in the model.

As an example both of how ECAs are implemented and how they function, Fig. 8.1 shows an ECA named REA (for real estate agent) who was programmed on the basis of a set of detailed studies into the behavior of realtors and clients. Over a period of roughly five years, various graduate students, postdoctorates, and colleagues in Cassell's research group studied different aspects of house-buying talk, and then incorporated their findings into the ECA. Hao Yan looked at what features of a house description were likely to be expressed in hand gestures, and what features in speech (Yan, 2000). Yukiko Nakano discovered that posture shifts were correlated with shifts in conversational topic and shifts in whose turn it was to talk (Cassell et al. 2001). Tim Bickmore examined the ways in which small talk was employed to establish trust and rapport between realtor and client (Bickmore and Cassell, 1999). Earlier

FIGURE 8.1   REA, the virtual real estate agent.

work by Scott Prevost on intonation (Prevost, 1996) and by Obed Torres on patterns of eye gaze (Torres et al., 1997) also went into the implementation. As research into human conversation progressed, the group also came to better understand some of the overall properties of human conversation, and the system was iteratively redesigned to incorporate those insights.

The result was a virtual woman who would try to sell a home to whoever approached her. A small camera on top of the screen allowed her to detect the presence of real humans and initiate a conversation with them. Her knowledge of appropriate realtor conversation led her to ask questions about the person's housing needs and then nod, seem to reflect, and pull up data on appropriate properties, describing them using a combination of descriptive hand gestures, head movements, and spoken language.

A number of experiments evaluated REA's performance in terms of how well it compared to human–human interaction. Thus, Nakano's work looked at anecdotal evidence of how closely patterns of eye gaze during a conversation between a real person and REA resembled eye gaze between two humans (Nakano, Reinstein et al., 2003) and she found that implementing a model of nonverbal behavior in conversational acknowledgment (grounding) appeared to improve the extent to which the human–machine conversation resembled human–human conversation. Other experiments compared how well the ECA functioned with and without particular human-like "modules," such as facial displays of emotion, hand gestures, and small talk.

8.7   SOCIAL DIALOGUE IN EMBODIED CONVERSATIONAL AGENTS

The small talk study is particularly revealing of the kinds of complex effects that ECA researchers are likely to find. Cassell and Bickmore initially posited that small talk would be a particularly successful addition to the ECA for the very reasons described earlier: it can be used to provide such social cues as attentiveness, positive affect, and liking and attraction, and to mark shifts into and out of social activities (Argyle, 1988). And people who interact with ECAs seem to wish to engage in such social dialogue, as shown by a naturalistic study of an ECA embedded in an information kiosk, where roughly one-third of the 10,000 utterances from over 2 500 users were social in nature (Gustafson et al., 1999). In Cassell and Bickmore's work, two versions of the ECA were implemented, of which one was capable of using small talk in the way that researchers in conversational analysis have documented, to mitigate face threat and increase trust, while the other version simply had REA get down to business.

An initial study compared how likely users were to trust REA to sell them an apartment, when she used small talk versus task talk only. Since prior literature has suggested a mediating effect due to personality, the study further compared introverts and extroverts in their use of the two versions of REA. Results demonstrated that extroverts trusted the system more when it engaged in small talk, while introverts were not affected by the use of small talk (Cassell and Bickmore, 2002). Insofar as these results mirror research in social psychology, they seemed to indicate that people were reacting to REA as they would react to another human. They also lent support to our model of the role of small talk in task-oriented conversations such as selling real estate.

A subsequent study (Bickmore and Cassell, 2005) included the additional condition of medium, comparing users communicating with REA by standing in front of the screen on which she was displayed to users communicating with REA by telephone. In this second study, however, results concerning the role of small talk were strongly mediated by the role of the body in the interaction. That is, main effects showed that subjects in the *phone* condition felt that they knew REA better, liked her more, felt closer to her, felt more comfortable with the interaction, and thought REA was more friendly than those in the embodied condition. In addition, social dialogue was more fun and less tedious on the phone while only task-limited dialogue was judged to be more fun and less tedious when embodied. That is, subjects preferred to interact, and felt better understood, face-to-face when it was a question of simply "getting down to business," and preferred to interact, and felt better understood, by phone when the dialogue included social chit-chat.

Looking back at the implementation of REA, Cassell and Bickmore came to believe that these results were a condemnation of REA's nonverbal behaviors, which may have inadvertently projected an unfriendly, introverted personality that was especially inappropriate for social dialogue, and that was at odds with the model of small talk that had been implemented. REA's model of nonverbal behavior, at the time of this experiment, was limited to those behaviors linked to the discourse context, and had not been changed for the small-talk version. Thus, REA's smiles were limited to those related to the ends of turns, and she did not have a specific model of immediacy or other nonverbal cues for liking and warmth typical of social interaction (Argyle, 1988). The results obtained indicate that adding social dialogue to embodied conversational agents requires a model of social nonverbal behavior consistent with verbal conversational strategies. To this end, in more recent work Cassell and her students have begun to examine how rapport is signaled nonverbally with an eye toward revising REA's model of nonverbal behavior. Although some literature exists on this subject, it is incomplete. And so, in particular, they are interested in how friends differ from strangers, and people meeting for the first time differ from those same people meeting a third or fourth time; in situations where the participants can see one another (and thus nonverbal behaviors can play a role) and situations where there is no visual access. When these data are analyzed, REA's nonverbal behavior will be updated. But, until good process data can be collected on the user's behavior, REA will not be able to know whether her rapport-building behaviors are successful (see Person, D'Mello, and Olney, Chapter 10 in this volume).

But the results with REA also demonstrate the extent to which subjects quite unconsciously respond to REA as if she is a real person—for example, judging her as unfriendly when she does not display nonverbal behaviors linked to rapport. In this context, an analysis of the users' speech behavior in talking with REA is revealing. People tend to "hyperarticulate" their speech when they talk to computers (Oviatt et al., 1998), whereas their speech to other people contains more slurred speech, interruptions, and disfluencies. Looking at rates of disfluency in users communicating with REA showed that interactions with REA were more similar to human–human conversation than to human–computer interaction (Bickmore and Cassell, 2005). This subject's responses during debriefing made the same point:

> REA exemplifies some things that some people, for example my wife, would have sat down and chatted with her a lot more than I would have. Her conversational style seemed to me to be more applicable to women, frankly, than to me. I come in and I shop and I get the hell out. She seemed to want to start a basis for understanding each other, and I would glean that in terms of our business interaction as compared to chit chat. I will form a sense of her character as we go over our business as compared to our personal life. Whereas my wife would want to know about her life and her dog, whereas I really couldn't give a damn.

However, these results also serve as a warning of the extent to which *all* of the ECA's behaviors need to be considered in designing an experiment because "people cannot not interpret." That is, as Kinsey told his student, every behavior of an agent— an artificial agent just as much as a human being—is the basis for an inference about that person's personality, emotions, and stance toward the conversation. Reeves and Nass (1996) have demonstrated much the same thing. Replicating several decades of social psychological research, systematically substituting a computer for one of the participants, their findings show that computers can evoke behaviors quite similar to those evoked by another human, including behaviors of social desirability. Thus, for example, when subjects were asked to rate the performance of a piece of computer software, they were harsher in their assessment when the questionnaire was filled out on an adjoining computer, rather than the computer on which they had used the software—presumably because they didn't want to hurt the computer's feelings (Nass et al., 1999). And on the basis of these findings, they conclude that people treat computers as social actors rather than as tools.

With this caveat in place, how can ECAs be used in survey interviewing?

8.8   ECAs IN SURVEY INTERVIEWING

The ECA has already begun to be used as an survey tool, but most often to train survey interviewers (c.f., inter alia, Link et al., 2006). In this context, the ECA plays a patient or other interviewee, so that interviewers can practice their skills. ECAs have not yet come to serve as interviewers, or to replace CASI or A-CASI technologies, most probably for the reasons that we have just outlined. First, the human science that ECAs depend on is still in its relative infancy. Autonomous ECAs are built on models of human behavior, and we do not yet know enough about the nonverbal aspects of rapport, or the effects of certain kinds of questions on the interview, to build an ECA that can autonomously work its way through a questionnaire, responding to feedback appropriately, and making decisions about when to ask the same question again and when to go on to the next part of the interview.

Second, and more damningly, it may be difficult to *ever* use ECAs as interviewers because, as Tourangeau, Couper, and Steiger point out, if ECAs (and other computer systems) are treated as social actors, then they—and the interview methodologies that depend on them—may be as subject to social desirability effects as are real humans (Tourangeau et al., 2003). Tourangeau and colleagues go on to examine the effect of representations of humans on social desirability, by showing images of people (the authors of the study) above the text of the questions on an Internet-based survey. They find little social desirability in their results, but in fact what they may have found is that simple photos may not suffice as representations of the agency of the computer. Or, that when you show the authors of the study above the questions, the subjects will not attribute agency to the *computer* but to the author of the questions asked by the computer, which will not lead them to feel that the *computer* might be judging them (Sundar and Nass, 2000).

We started this chapter discussing the paradox of survey interviewing—that rapport is felt by some to be essential for a good interview, while others see it as underlying social desirability effects. The current twist on that paradox is that the more embodied conversational humans come to resemble real humans, the more they may evoke identical responses to human interviewers. If this is the case, they will more resemble human interviewers than CASI or A-CASI technologies, and we imagine the day when ECAs will need to learn how to wear a poker face . . . .

On the other hand, as we indicated previously, ECAs can serve as powerful tools in social science research. They can help us to understand fundamentals of human interaction that shape interview contacts. More concretely, they can allow us to manipulate aspects of interaction (e.g., nonverbal cues) that are difficult or impossible to examine systematically in human-to-human interaction. In this context, we would suggest that ECAs can help us tease apart the very effects of social desirability on survey interviewing, and provide the kind of results that will allow us both to better train human interviewers and to build more adequate interviewing technologies. Because nonverbal behaviors are largely unconscious, the vast majority of research on their deployment in survey interviewing has been post hoc and anecdotal. ECAs, on the other hand, in their Wizard of Oz mode, where an experimenter chooses each response, are infinitely controllable. For this reason, they may allow us to differentiate the effects of social class, race, gender, age, and conversational style on the interview context. In fact, analogously, some research has begun to look at the effects of ECA appearance when the ECA plays the role of a tutor. Thus, Baylor and colleagues (2006) found that young women's stereotypic views about women engineers changed after interacting with a female engineer ECA. However, motivation to actually study engineering was more likely to be changed when the young women interacted with a male engineer ECA. Likewise, Person had students receive tutoring from the four ECAs pictured in Fig. 8.2 (Person et al., 2002). She discovered that whereas college students preferred to learn from a tutor of their own race and gender, their learning gains were greatest when the ECA tutor was a white male.

As well as research on demographics, some research (see Person, D'Mello, and Olney, Chapter 10 in this volume) has also begun to target the effect of particular verbal and nonverbal styles of ECAs, once again on learning. In this experiment, some tutors were polite and some were rude. Surprisingly, rude tutors were judged as

FIGURE 8.2 Effects of race and gender on tutoring.

trustworthy, interesting, and better able to teach than were polite tutors, and learning gains were identical in the two conditions.

A similar paradigm could be used to study the effect on interview responses of rudeness, dialect, skin color, gender, and putatively rapport-evoking behaviors such as leaning forward and smiling. An identical set of questions asked by identical virtual humans, with just one single difference—gender or age or race or posture shifts—could give valuable information about the interviewing process.

In this chapter we have argued that researchers must walk between Scylla and Charybdis in using technology for survey interviews. On the one hand, technologies must be natural, intuitive, and easy to use—they must resemble human conversation to the extent possible so as to open their use to the broadest segment of the population. On the other hand, the more these technologies resemble human conversation, the more likely they are to evoke similar kinds of interviewee social desirability effects, and to be subject to similar constraints as human interviewing. For this reason, in parallel with the introduction of new technology, and attempts to understand the effects of these technologies on interviewing, there must be no slowdown in the attempt to understand human conversation, and how human interaction affects our use of technology.

REFERENCES

Argyle, M. (1988). *Bodily Communication*. New York: Methuen and Co.

Athey, E. R., Coleman, J. E., Reitmans, A. P., and Tang, J. (1960). Two experiments showing the effect of the interviewer's racial background in responses to questions concerning racial issues. *Journal of Applied Psychology*, 44, 381–385.

Baylor, A., Rosenberg-Kima, R., and Plant, A. (2006). Interface agents as social models: the impact of appearance on females' attitude toward engineering. *In Proceedings of the Conference on Human Factors in Computing Systems 2006*. Montreal, Canada.

Benney, M., Riesman, D., and Star, S. A. (1956). Age and sex in the interview. *American Journal of Sociology*, 62, 143–152.

Bickmore, T., and Cassell, J. (1999, Nov. 5–7). Small talk and conversational storytelling in embodied conversational characters. *In Proceedings of American Association for Artificial Intelligence Fall Symposium on Narrative Intelligence* (pp. 87–92). Cape Cod, MA: AAAI Press.

Bickmore, T., and Cassell, J. (2005). Social dialogue with embodied conversational agents. In J. v. Kuppevelt, L. Dybkjaer, and N. Bernsen (Eds.), *Natural, Intelligent and Effective Interaction with Multimodal Dialogue Systems*. New York: Kluwer Academic.

Bischoping, K., and Schuman, H. (1992). Pens and polls in Nicaragua: an analysis of the 1990 pre-election surveys. *American Journal of Political Science*, 36(2), 331–350.

Cannell, C. F., Fowler, F. J., and Marquis, K. H. (1968). The influence of interviewer and respondent psychological and behavioral variables on reporting in household interviews. *Vital and Health Statistics, Series 2*, 26, 1–65.

Cassell, J., and Bickmore, T. (2002). Negotiated collusion: modeling social language and its relationship effects in intelligent agents. *User Modeling and Adaptive Interfaces*, 12, 1–44.

Cassell, J., Nakano, Y., Bickmore, T., Sidner, C., and Rich, C. (2001, July 17–19). Non-verbal cues for discourse structure. In *Proceedings of Thirty-ninth Annual Meeting of the Association of Computational Linguistics* (pp. 106–115). Toulouse, France: Association for Computational Linguistics.

Cassell, J., Sullivan, J., Prevost, S., and Churchill, E. (2000). *Embodied Conversational Agents*. Cambridge, MA: MIT Press.

Cicourel, A. U. (1964). *Method and Measurement in Sociology*. New York: Free Press.

Currivan, D. B., Nyman, A. L., Turner, C. F., and Biener, L. (2004). Does telephone audio computer-assisted self interviewing improve the accuracy of prevalence estimates of youth smoking? *Public Opinion Quarterly*, 68, 542–564.

Dohrenwend, B. S., Colombotos, J., and Dohrenwend, B. P. (1968). Social distance and interviewer effects. *Public Opinion Quarterly*, 32, 410–422.

Ehrlich, J. S., and Riesman, D. (1961). Age and authority in the interview. *Public Opinion Quarterly*, 25, 39–56.

Goudy, W. J., and Potter, H. R. (1975). Interview rapport: demise of a concept. *Public Opinion Quarterly*, 39, 529–543.

Gustafson, J., Lindberg, N., and Lundeberg, M. (1999). The august spoken dialogue system. In *Proceedings of Eurospeech*. Budapest, Hungary.

Hatchett, S., and Schuman, H. (1975). White respondents and race-of-interviewer effects. *Public Opinion Quarterly*, 39, 523–528.

Henson, R., Cannell, C. F., and Lawson, S. (1973). *Effects of Interviewer Style and Question Form on Reporting of Automobile Accidents.* Ann Arbor, MI: Survey Research Center, University of Michigan.

Hochstim, J. R. (1967). A critical comparison of three strategies for collecting data from households. *Journal of the American Statistical Association*, 62, 976–989.

Hyman, H. (1954). *Interviewing in Social Research.* Chicago: University of Chicago Press.

Katz, D. (1942). Do interviewers bias poll results? *Public Opinion Quarterly*, 6, 248–268.

LaFrance, M. (1982). Posture mirroring and rapport. In M. Davis (Ed.), *Interaction Rhythms: Periodicity in Communicative Behavior* (pp. 279–298). New York: Human Sciences Press.

Lakin, J. L., Jefferis, V. E., Cheng, C. M., and Chartrand, T. (2003). The chameleon effect as social glue: evidence for the evolutionary significance of nonconscious mimicry. *Journal of Nonverbal Behavior*, 27(3), 145–162.

Lenski, G. E., and Leggett, J. C. (1960). Caste, class and deference in the research interview. *American Journal of Sociology*, 65, 463–467.

Link, M. W., Armsby, P. P., Hubal, R. C., and Guinn, C. I. (2006). Accessibility and acceptance of responsive virtual human technology as a survey interviewer training tool. *Computers in Human Behavior*, 22(3), 15.

Marquis, K. H., and Cannell, C. F. (1969). *A Study of Interviewer–Respondent Interaction in the Urban Employment Survey.* Ann Arbor, MI: Survey Research Center, University of Michigan.

Miller, P. V. (1984). Alternative question wording for attitude scale questions in telephone interviews. *Public Opinion Quarterly*, 48, 766–778.

Moreno, K. N., Person, N. K., Adcock, A. B., Van Eck, R. N., Jackson, G. T., and Marineau, J. C. (2002). Etiquette and efficacy in animated pedagogical agents: The role of stereotypes. *Proceedings of the 2002 AAAI Fall Symposium on Etiquette for Human Computer Work* (pp. 78–80). Menlo Park, CA: AAAI Press.

Moskowitz, J. (2004). Assessment of cigarette smoking and smoking susceptibility among youth. *Public Opinion Quarterly*, 68, 565–587.

Nakano, Y. I., Reinstein, G., Stocky, T., and Cassell, J. (2003, July 7-12). Towards a model of face-to-face grounding. In *Proceedings of Annual Meeting of the Association for Computational Linguistics* (pp. 553–561). Sapporo, Japan: Association for Computational Linguistics.

Nass, C., Moon, Y., and Carney, P. (1999). Are people polite to computers? Responses to computer-based interviewing systems. *Journal of Applied Social Psychology, 29*, 1093.

Oksenberg, L., Coleman, L., and Cannell, C. F. (1986). Interviewers' voices and refusal rates in telephone surveys. *Public Opinion Quarterly*, 50, 97–111.

Oviatt, S., MacEachern, M., and Levow, G.A. (1998, May). Predicting hyperarticulate speech during human–computer error resolution. *Speech Communication*, 24(2), 87–110.

Prevost, S. A. (1996). Modeling contrast in the generation and synthesis of spoken language. In *Proceedings of International Conference on Spoken Language Processing*. Philadelphia, PA: Association for Computational Linguistics.

Reeves, B., and Nass, C. (1996). *The Media Equation: How people Treat Computers, Televisions and New Media Like Real People and Places.* Cambridge, UK: Cambridge University Press.

Riesman, D., and Glazer, N. (1948). The meaning of opinion. *Public Opinion Quarterly*, 12, 633–648.

Robinson, D., and Rohde, S. (1945). A public opinion study of anti-Semitism in New York City. *American Sociological Review*, 10, 511–515.

Schober, M. F., and Bloom, J. E. (2004). Discourse cues that respondents have misunderstood survey questions. *Discourse Processes*, 38, 287–308.

Schuman, H., and Converse, J. M. (1971). The effects of black and white interviewers on white respondents in 1968. *Public Opinion Quarterly*, 35, 44–68.

Sheatsley, P. B. (1950). An analysis of interviewer characteristics and their relationship to performance. *International Journal of Opinion & Attitude Research*, 4 1950, 473–498.

Sheatsley, P. B. (1951). An analysis of interviewer characteristics and their relationship to performance; Part II. *International Journal of Opinion & Attitude Research*, 5 1951, 79–94.

Sundar, S. S., and Nass, C. (2000). Source orientation in human–computer interaction: programmer, networker or independent social actor? *Communication Research*, 27, 683–703.

Tickle-Degnen, L., and Rosenthal, R. (1990). The nature of rapport and its nonverbal correlates. *Psychological Inquiry*, 1(4), 285–293.

Torres, O. E., Cassell, J., and Prevost, S. (1997, July 14–16). Modeling gaze behavior as a function of discourse structure. *In Proceedings of First International Workshop on Human-Computer Conversation*. Bellagio, Italy: Intelligent Research, Ltd.

Tourangeau, R., Couper, M. P., and Steiger, D. M. (2003). Humanizing self-administered surveys: experiments on social presence in Web and IVR surveys. *Computers in Human Behavior*, 19, 24.

Van der Zouwen, J., Dijkstra, W., and Smit, J. H. (1991). Studying interviewer–respondent interaction: the relationship between interviewing style, interviewer behavior and response behavior. In P. Biemer et al. (Eds.), *Measurement Errors in Surveys*. Hoboken, NJ: John Wiley & Sons.

Villarroel, M. A., Turner, C. F., Eggleston, E., Al-Tayyib, A., Rogers, S. M., Roman, A. M., et al. (2006). Same gender sex in the United States: impact of T-ACASI on prevalence estimates. *Public Opinion Quarterly*, 70, 166–196.

Weiss, C. H. (1968). Validity of welfare mothers' interview responses. *Public Opinion Quarterly*, 32, 622–633.

Weiss, C. H. (1970). Interaction in the research interview: the effects of rapport on response. In *Proceedings of the American Statistical Association, Section on Social Statistics*, (pp. 18–19). Alexandria, VA: American Statistical Association.

Williams, J. A. (1964). Interviewer–respondent interaction: a study of bias in the information interview. *Sociometry*, 27, 338–352.

Williams, J. A. (1968). Interviewer role performance: a further note on bias in the information interview. *Public Opinion Quarterly*, 32(2), 287–294.

Yan, H. (2000). Paired speech and gesture generation in embodied conversational agents. Unpublished Masters of Science thesis, MIT, Cambridge, MA.

CHAPTER 9

Disclosure and Deception in Tomorrow's Survey Interview: The Role of Information Technology

Jeffrey T. Hancock
Cornell University, Ithaca, New York

9.1   INTRODUCTION

When do we answer a survey question with "the truth, the whole truth, and nothing but the truth"? Although we may like to believe that most responses to surveys are thoughtful, honest, and fully disclosive, a substantial body of research suggests otherwise. Under many conditions we fudge responses to make ourselves look better, or provide answers that are only reasonably accurate but the easiest to give. In some conditions we may even lie outright in order to protect our identity or privacy. But, there are other interview situations in which we are willing to bare our innermost secrets.

The question of when and how respondents provide truthful answers has long been a concern in survey research. When considering the role of new technologies in future survey interview methodologies, we need to revisit this concern. Our research in interpersonal communication suggests, for example, that people lie more frequently in some media than in others (Hancock, Thom-Santelli & Ritchie, 2004a). Given these findings from communication research, how might novel survey technologies affect whether a respondent will be more disclosive or more deceptive? What principles might guide survey technology design in an effort to create conditions that elicit more accurate answers? Can these principles help us understand the potential for response biases across different technologies? Finally, can techniques be devised that may facilitate the detection of deceptive responses?

Envisioning the Survey Interview of the Future, Edited by Frederick G. Conrad and Michael F. Schober
Copyright © 2008 John Wiley & Sons, Inc.

The present chapter examines these questions in light of recent research on deception and self-disclosure in digital contexts (for review, see Hancock, 2007). The chapter attempts to review and integrate research from the survey literature concerned with honesty, social desirability, and self-disclosure with the psychological and communication literatures concerned with how information technologies affect deception production and detection. A number of factors emerge that have important effects on disclosure and deception, including features of the communication environment, individual differences associated with technology and privacy, psychological effects associated with anonymity and self-awareness, and normative expectations.

The focus of the chapter is primarily on administration mode effects, which refers to differences in data quality between different communication environments that support the survey interview, including face-to-face, the telephone, and Web surveys. Typically, mode of administration effects have been considered undesirable, and attempts are made to eliminate or adjust for them. In the present chapter, however, we examine whether these mode effects might be usefully exploited to reduce deception and increase self-disclosure in an effort to improve data quality. A second focus is concerned with expanding the scope of consideration beyond simply the features of a technology or mode. For instance, while a given technology may engender a standard information environment, each user may perceive that environment differently. For instance, an important issue introduced by novel technologies is coverage error, namely, whether people with access to a particular mode may be different in their propensity to be dishonest from those who do not have access. Early adopters, who begin using a technology before the majority of the population, are necessarily more experienced with that technology than other respondents and may have different levels of privacy concerns. An important objective, therefore, is to identify what types of individual differences may interact with information technology to influence the propensity to deceive, and, ultimately, to develop systems that can help manage these individual differences.

9.2   DEFINING DECEPTION IN THE SURVEY CONTEXT

Because deception plays a central role in the following discussion, it is important to begin by defining this key concept and situating it within the survey interview context. Although whole books have attempted to conceptualize deception, one well-accepted definition from psychology offered by Vrij (2000, p. 6) will suffice: "a successful or unsuccessful deliberate attempt, without forewarning, to create in another a belief which the communicator considers to be untrue." The two key characteristics of deception identified in this definition and shared by most definitions are that (1) it is intentional and (2) it involves an attempt to create a false belief. So, providing mistaken information is not technically deception. Thus, the respondent who really believes that he spends $100/yr on movies when in fact he spends $200/yr is not lying, although he is providing false information. Similarly, a response told as a joke, as long as the respondent intended the interviewer to understand the joke, is not deception.

Obviously, not all deceptions are equal, and many distinctions can be made between different types of deception. There are several different taxonomies describing types of deception, but we can limit our concern to a few dimensions. First, lies range from the subtle to the serious and can take several forms. For instance, *falsifications* are outright lies and are contradictory to the truth; *distortions* present some modification of the truth that has somehow been altered to fit the liar's goals; *concealments* and lies of omission involve hiding the truth by deliberately withholding information. In the context of survey responses, each of these types of deception is possible. Perhaps the most common deception form in survey responses is distortion, such as socially desirable responding, in which participants' responses are biased in order to appear overly positive (Paulhus, 2002). As reviewed later, participants often distort their responses to sensitive topics, underreporting information that can be embarrassing or harmful to the respondent (e.g., drug use) or overreporting information that is socially valued (e.g., voting, income, status). Lies of omission can include not providing complete details asked for by an item, or even failing to respond to an item. Finally, falsification in survey responses includes any answers that contradict the truth rather than simply distort it. One example of falsification in survey responding is strong satisficing behavior, in which the respondent intentionally selects responses that will seem reasonable to the interviewer but without having done any assessment of whether that response actually represents the truth for the respondent.

A second dimension of deception is the benefactor of the lie. Lies can be self-oriented, in which the lie is designed to benefit the person telling the lie (e.g., telling a lie to make oneself appear better), or they can be other-oriented, in which the lie is designed to benefit the person lied to or some third entity (e.g., telling a lie to prevent the partner's feelings from being hurt). In a survey, deceptions are typically self-oriented and benefit the respondent. Socially desirable responding, for instance, is a distortion of the truth that is intended to make the respondent appear more positively.

Another important aspect of a deception is the nature of the relationship between the liar and partner. Psychological research suggests that we tend to lie less often to those with whom we are more intimate compared to strangers (DePaulo & Kashy, 1998). In the context of the survey interview, the interviewee is typically unacquainted with the interviewer. With the introduction of new technologies, however, the partner is sometimes not human. For instance, in interactive voice response (IVR) participants provide responses to a computerized voice over the telephone (see Bloom, Chapter 6 in this volume; Steiger 2006). With embodied virtual humans as interviewers, participants interact with a graphical avatar (see Cassell and Miller, Chapter 8 in this volume). As we will see, whether a respondent is interacting with a human or a computer system can have an impact on a respondent's decision to deceive or self-disclose, and new technology can blur this distinction.

Finally, deceptions are about different types of things. Some lies are about our opinions or attitudes. These kinds of lies are typically difficult to verify; only the opinion-holder really knows the truth (although our previous responses to related items may indicate whether we have been consistent or not). Other lies are about our behaviors, actions, or other details that are objectively verifiable, such as how many times you've visited the doctor (Hancock, Curry, Goorha, and Woodworth, in press). As Schuman and Presser (1981) observe, surveys typically involve questions about either opinions/attitudes or about behaviors and events; if this is the case, then deception may differ across these different types of questions.

With this understanding of deception and some of the important parameters that can affect its production in survey contexts, we now turn to a selective overview of how different modes of administration can influence deception and disclosure in responses.

9.3   ADMINISTRATION MODE EFFECTS AND DECEPTION IN SURVEY RESPONSES

Administration mode effects arise from differences in the methods in which data are collected. For example, one survey may collect information in face-to-face interviews while another may collect data over the telephone. Although the same questions may be asked, the responses can be considerably different. Indeed, a large body of research demonstrates that administration mode can have substantial effects on responses (for review, see Sudman and Bradburn, 1974; Tourangeau and Smith, 1996, 1998). These effects are particularly striking when collecting sensitive information or asking questions about threatening topics, such as illicit drug use, sexual activity, and abortion (Schaeffer, 2000).

Rather than focus on the characteristics of specific technology (see Couper, Chapter 3 in this volume, for an overview of differences between different survey technologies), a more useful approach is to examine some of the key factors that differ across administration modes. For example, Schwarz et al. (1991) identified nine differences across face-to-face and telephone interviews as well as self-administered questionnaires, including visual versus auditory presentation, sequential versus simultaneous presentation, time pressure, additional explanations from interviewer, perception of interviewer characteristics, perceived confidentiality, and external distractions (see p. 195). When considering how these various factors might affect deception in responding, Schwarz et al. (1991) observed that social desirability is reduced most by perceived confidentiality, and that perceived confidentiality is highest in self-administered surveys. They note that social desirability effects may be mediated by interviewer reactions. For instance, disapproving reactions on the part of the interviewer (say, in response to admissions of illicit drug behavior) are more likely to be noticed when the interviewer can be seen or heard, suggesting that visual and aural channels may increase socially desirable responding.

More recently, Tourangeau and Smith (1996, 1998) developed a model describing the relationship between features of different modes of survey administration, psychological variables, and data quality consequences. The most relevant data quality consequence for the present purposes is the accuracy of responding. In their model they describe two mode features and related psychological effects that increase accuracy. The first is self-administration, which refers to whether the survey is self-administered or interviewer-administered. As noted previously, this feature is perhaps the most widely studied and frequently observed administration mode factor to affect deception in surveys. Many studies have demonstrated that respondents are more willing to honestly disclose information about sensitive topics when the questions are self-administered (Currivan et al., 2004; Gribble et al., 2000; Tourangeau and Smith, 1996, 1998; Turner et al., 1996). For example, in one study comparing self-administered versus interviewer administered surveys, more illicit drug use and alcohol consumption was reported under self-administration (Aquilino, 1994). In another, more abortions were reported under self-administered conditions than interviewer-administered conditions. (Mosher and Duffer, 1994). Touragneau and Smith argue that these effects are mediated by the psychological variable of privacy. Self-administration increases a respondent's sense of privacy, which in turn increases the accuracy of responding.

The second feature in the model related to accuracy is the computerization of the administration mode. The impact of computerization on deception in surveys is not as consistent as that of self-administration but in general research suggests that computerization increases honest self-disclosure in survey responses. For example, a comparison of computer-assisted interviews with more traditional paper and pencil interviews revealed that respondents reported significantly more alcohol-related problems (Bradburn et al., 1991). Tourangeau and Smith argue that computerization leads to increased perceived legitimacy of the survey by bestowing professionalism or importance on the survey. For instance, laptops in the 1990s were still a novelty in many homes, and when used to administer a survey, participants may have perceived the survey as more important or legitimate. While this may no longer be the case with laptops in most of the United States, the kinds of novel technologies described in the present volume may enhance legitimacy. According to the Tourangeau and Smith model, enhanced legitimacy should in turn improve accuracy.

These two administration mode features, and their impact on accuracy as mediated by psychological variables, appear to be central to the expected evolution of the survey. Couper (Chapter 3 in this volume) describes two primary trends in survey administration: self-administration and computerization. Couper argues that the increased use of technology in the survey world typically functions either to supplement the role of the interviewer or to automate the data collection process. Given these trends, recent research on deception and technology can inform our understanding of how these technologies may impact the accuracy of survey responding.

9.4 DIGITAL FORMS OF DECEPTION

New forms of communication technology, such as the Web, often seem to be rife with deception. Examples abound in the popular press of cyber-cheating and cyber-scams. Each day e-mail users face a barrage of deceptive e-mails referred to as spam. More serious types of deception include identity fraud, in which one person appropriates another's identity by stealing his/her personal information. Among the millions of people when now engage in online dating, deception is a major concern. At the same time, however, there are many anecdotal examples of highly private disclosures online, with people providing intimate details about their love lives and financial circumstances in online diaries (known as weblogs, or blogs) accessible to anyone with access to the Internet. Psychological research also suggests that mediated interpersonal communication can be characterized by high levels of self-disclosure relative to face-to-face conversations (Joinson, 2001). What factors predict whether a technology will promote deception or enhance self-disclosure? And what implications might these factors have for accuracy in survey responses?

In an attempt to understand how communication technology affects deception, we carried out a series of diary studies (Hancock et al., 2004a, b), in which participants recorded all of their social interactions and lies for a seven day period. The diary procedure was a replication of a protocol developed by DePaulo et al. (1996), in which participants are first trained to identify their own deceptive behavior and then record them in a daily diary. In our procedure, participants recorded all of their lies and social interactions and identified in which medium the interaction and/or lie took place. Although there are limitations to this type of diary study, such as requiring participants to be honest about their lying behavior, the procedure has been used extensively in the psychology and communication literatures (see DePaulo et al., 1996). Our primary interest was to compare lying on the telephone, instant messaging, and e-mail relative to face-to-face interactions. In order to control for the fact that we use some media more than others, we compared the rate of lies (i.e., the number of lies divided by the number of social interactions) across the four media. The results revealed that lies were most common on the telephone (approximately 32%) and least common in e-mail (approximately 14%). The rate of lying face-to-face (24%) and instant messaging (21%) did not differ from one another. In addition, participants tended to tell lies about different things in different media. For instance, lies about one's actions were most common on the telephone, while lies about feelings were most common face-to-face. Explanation lies, in which false justifications and reasons are provided, tended to be made most often in e-mail.

It is perhaps interesting to note a few similarities with these findings and the findings reviewed earlier in the survey literature. First, e-mails, which are text-based interactions mediated through a computer, were the most honest of the media. This observation is similar to the observation that respondents tend to self-disclose more honestly when responding via text to a computer relative to responding to a human (Tourangeau and Smith, 1996, 1998). Similarly, the fact that our participants lied the most on the telephone is consistent with the observation in mode administration effect studies that the telephone often involves the most social desirable responding and satisficing (Holbrook et al., 2003).

Why did participants lie most on the telephone and least in e-mail? One theory that is often used to explain media effects is *social presence*. Social presence refers to the salience of the other in an interaction and the consequent interpersonal relationship. One argument, suggested originally by DePaulo et al. (1996), is that because lying makes us uncomfortable liars should seek to distance themselves from the target of the lie in order to reduce discomfort. According to this view, we should lie most often with media that reduce social presence. Given that e-mail creates less social presence than the other media (instant messaging is synchronous, telephone involves multiple interaction channels, and face-to-face presumably maximizes presence), this approach would predict the most lying in e-mail and the least face-to-face. This clearly was not the case as e-mail involved the fewest lies while the telephone involved the most.

There are a number of problems applying these sorts of higher level concepts to comparative media issues. The first is that concepts like social presence are often ambiguous and poorly defined. A second problem is the underlying assumption that different communication environments differ along one major dimension (e.g., level of social presence). It is clear, however, that there are more differences between the telephone and e-mail that may influence complex social behavior, such as deception, than socially presence. Indeed, in the research world social presence has fared poorly at predicting socially desirable responding, including one study in which increasing social presence items failed to affect response quality (Tourangeau et al., 2003).

So what factors might influence a decision to lie or be honest in a mediated communication? There are at least four factors to consider: (1) features of the communication environment, (2) psychological variables affected by the communication environment, (3) individual differences, and (4) normative expectations.

1. *Features of the Communication Environment.* The communication environment can be delineated along a number of dimensions that can affect language use (Herring, 2007). In the context of deceptive language use, we have argued that there are at least three features of the environment that influence the decision to lie or not, including recordability/permanence, synchronicity, and physical copresence (Hancock et al., 2004a). The first is the degree to which the medium is *recordable*. The more recordable or permanent a message is, the less likely a person should choose to lie in that context. E-mail is a highly recordable medium (copies are typically left on the liar's computer, intermediate servers, and the target's computer), and in our studies we consistently see fewer lies in e-mail than in other media. E-mails have also been the primary evidence in a number of high-profile criminal cases involving deception, such as Enron's accounting fraud.

   A second feature is whether the medium is synchronous, which refers to the degree to which messages are exchanged in real time. Lies in conversation tend to be spontaneous (DePaulo et al., 1996), and often arise from situations in which a truthful response may be problematic. For example, if Whitney asks Brent whether Brent likes Whitney's new shirt, which Brent hates, Brent may feel compelled to lie in order to avoid hurting the other's feelings. The time constraints of synchronicity can also make it more difficult to construct a response that is not an outright falsification but instead a more subtle distortion of the truth (e.g., Brent equivocates by saying that he likes Whitney's new haircut and doesn't mention the new shirt).

   Finally, physical copresence refers to whether two interactants are in the same physical space, such as face-to-face interactions. When physically copresent, a number of things are impossible to lie about, including who the liar is with, or what the liar is doing.

2. *Psychological Variables.* As noted earlier, a good deal of research now suggests that mediated communication often involves higher levels of self-disclosure

than face-to-face communication (for review, see Joinson and Paine, 2007). Perhaps the most frequent explanation for this phenomenon is the psychological effects associated with *perceived anonymity*. Perceived anonymity refers to a user's belief that she is unaccountable for her social actions within a specific medium. In online communication, perceived anonymity is facilitated by a lack of visual appearance and the delinking of one's real personal information from online identities. The basic argument is that the relationship between perceived anonymity and self-disclosure is one of disinhibition (Wallace, 1999). For example, Bargh, McKenna and Fitzsimmons (2002) have argued that anonymous Internet interactions resemble those identified in the "stranger on the train" phenomenon, in which people reveal their most intimate thoughts to strangers they meet on a train or plane (Thibaut and Kelly, 1959). The subjective experience of perceived anonymity, whether in an online interaction or in a conversation with a stranger on a train, promotes a focus on the self and reduces accountability concerns because the partner is unlikely to communicate with anyone in the user's social network.

A second, and related, psychological variable is the concept of self-awareness, which can be differentiated between private and public foci. Private self-awareness refers to the degree to which attention is focused on one's own feelings, beliefs, memories, and thoughts. Public self-awareness refers to the degree to which attention is focused on how others are viewing the self, including physical appearance, behaviors, and accent (Fenigstein et al., 1975). In a series of experiments, Joinson (2001) used private and public self-awareness to explain the elevated levels of self-disclosure in text-based online communication relative to face-to-face interactions. In particular, Joinson demonstrated that text-based communications tended to increase private self-awareness, due to increased focus on one's own thoughts, and decreased public self-awareness, due to the lack of a salient other in the interaction. The combination of enhanced private self-awareness and decreased public self-awareness were also the conditions that were most likely to increase self-disclosure, suggesting that one's self-awareness focus can affect whether a person is likely to disclose intimate information or withhold it.

3. *Individual Differences.* Not everyone comes to a technologically mediated interaction with the same set of skills, perceptions, experience, or sense of privacy. For instance, a lot of attention has been paid to the digital divide, which refers to the gap between those who are able to benefit from digital technologies and those who are not (for review, see Hargittai, 2004). While survey researchers have been concerned with differences in access rates across the population as a coverage concern, the digital divide also has implications for response biases. For instance, in our diary studies we observed that the more experience participants had with e-mail, the greater their propensity to lie in e-mail (Hancock et al., 2004a). These data suggest that as participants become experienced with a medium, they also become more facile at lying in that medium. This observation is potentially related to the legitimization issue raised by Tourangeau and Smith (1996, 1998), namely, that as participants

become familiar with a technology it becomes less imposing and mysterious, which in turn may affect response biases.

A second important individual difference related to deception and self-disclosure of information online is individual concerns with *privacy*. As Joinson and Paine (2007) note, self-disclosure is driven in part by how concerned an individual is with privacy. They outline Westin's (1967) categories of respondents with respect to privacy concerns: (1) *the privacy fundamentalists*, who value privacy especially highly and believe that their privacy has been substantially eroded and are resistant to further erosion incursions; (2) *the privacy pragmatists*, who also have strong feelings about privacy but tend to balance the value of providing personal information to society with the potential for its misuse; and (3) *the privacy unconcerned*, who have no real concerns about privacy or about how others use information about them. In general, the less concerned with privacy an individual is, the more likely the individual is to disclose information. Privacy, self-disclosure, and deception have been shown to be connected in online environments. For instance, Whitty (2002) observed that one of the motivations for lying in online environments, such as chat rooms, is to protect one's privacy.

A third individual difference is *trust*. Although trust is a difficult concept to define, it can be considered a "leap of faith" under conditions of vulnerability and uncertainty (Mollering, 2006). People differ in their willingness to take that leap, especially in the context of online disclosure. As Boyd (2003) notes, trust online may be more slow to develop and more fragile due to the reduced physical cues in online interactions. In fact, trust online seems to be a product of the previous individual differences described earlier, experience with a medium and concerns with privacy. Metzger (2004), for example, demonstrated that the likelihood that a person will disclose information in an online commerce context is driven by the person's experience with the Internet and low levels of privacy concerns, both of which increase levels of trust.

4. *Normative Expectations.* Regardless of the specific context, our behaviors are guided and constrained by relevant social norms, expectations, and conventions. This is no less the case in online, mediated social contexts. Indeed, a large body of research demonstrates that norms shape many online behaviors (for review, see Postmes et al., 2000). One of the important questions in this line of research is determining what kinds of cues can evoke normative behavior. One result that has consistently emerged is that in the relatively sparse information environment of online communication, subtle cues can have important impacts on the development of norms and adherence to them. For instance, subtle cues like the color of a user's avatar can impact the normative behavior of the user. In one study, when participants' avatars were dressed in black, participants in that group tended to hold more violent and destructive norms and attitudes than participants whose avatar was dressed in white (Merola et al., 2006).

Norms associated with deception emerge differently in various online environments. For example, in online games, such as Second Life, participants interact in a playful online version of life, complete with commerce, fashion, and parties. In these spaces, the normative expectation is that everyone is encouraged to play with their identity. That is, the norm in these sorts of online environments is one of joint pretense, in which participants are aware that each other is acting a role or playing a part. For example, one user in an online chat room full of male-presenting users was observed to say to the others "Would someone just please pretend to be a girl?" (Joseph Walther, personal communication). In other environments, though, deception is viewed as negatively as it is in face-to-face settings. For example, online support groups provide a safe environment for people with a wide range of issues to share their problems and experiences. When a deception in this context is revealed, it can lead to intense outrage, feelings of violation, and heavy sanctions on the perpetrator of the deception (Joinson and Dietz-Uhler, 2002).

9.5 IMPLICATIONS FOR DECEPTION AND DISCLOSURE IN SURVEY RESPONDING

The overview of the factors above suggests that psychological variables such as perceived anonymity and self-awareness, individual differences such as privacy concerns and experience, and normative expectations must be considered in addition to the specific features of the survey technology in an effort to understand how new technologies will affect deceptive responding. Consider a recent study examining deception in online dating profiles (Hancock, Toma, & Ellison, 2007). In this study, we compared online daters' profile information (i.e., what they disclosed in their online profile) with their actual characteristics as observed in the lab (e.g., their actual weight, height, and age). We found that each of the factors described above shaped the nature of the lies. In particular, we observed that participants lied frequently in their profiles (81% of respondents lied on their weight, height, or age) but by relatively small amounts (most deceptions were within 5% of their weight, height, or age).

The asynchronous and editable nature of the profiles allowed participants to exaggerate aspects of the self (e.g., men said they were taller than they were, women said they were thinner), but that the psychologically important anticipation of future interaction constrained the lies to subtle distortions (otherwise they would be perceived as dishonest when meeting their date face-to-face). Individual differences in privacy concerns emerged in the amount participants were willing to disclose in their profile. For example, approximately one-third of our sample did not include a photo in their profile, despite the drastic drop in hits for profiles without photos. In our interviews with the participants, privacy concerns were frequently cited for why information was not included in a profile, or even why a participant lied about a piece of information (e.g., general location). Finally, social norms emerged when we asked participants what kinds of lies were acceptable and which were not. Regardless of gender, almost all participants agreed that lying about relationship status and whether or not one had children was unacceptable, while most felt that lying about personal interests (e.g., reading, hiking, sunsets on the beach) were more acceptable. Taken together, this study reveals the importance of considering technology features, psychological variables, individual differences and normative expectations in understanding when a person is likely to deceive or disclose.

Given that each of the above-described factors needs to be considered, what are the implications for the survey context? There are at least three directions that survey designers might consider when developing technologies for the survey of the future.

1. *Recreate the "Stranger on the Train" Phenomenon.* One possibility is for survey design to attempt to recreate the "stranger on the train" phenomenon, in which people are highly disclosive and honest about their feelings, attitudes, and behaviors (Thibaut and Kelly, 1959). To do so, the survey interaction should promote private self-awareness while reducing public self-awareness (Joinson, 2001). Designs should include cues that cause participants to focus on their own thoughts, feelings, and beliefs while minimizing cues that shift attention to how others are viewing the self. As such, photos of humans or video conferencing should be avoided. One potential design technique to increase private self-awareness is the practice of *reflective human–computer interaction* described by Sengers and colleagues (2005), in which the interface is designed to stimulate the user's critical reflection on the current interaction with a technology. This technique may be adapted to stimulate user reflection on the self and the user's thoughts and feelings.

    One interesting question is whether embodied virtual agents, as described by Cassell and Miller (Chapter 8 in this volume), can facilitate the stranger on the train setting. There is no reason to believe they cannot; the agent is not likely to communicate with the respondent's social circle and therefore fits the status as stranger. A key question is whether the agent can instill a sense of perceived anonymity, which should enhance self-disclosure.

    Paradoxically, perhaps, the research above also suggests that making salient the permanence or recordability of responses may improve truthfulness. While recordability and perceived anonymity seem antithetical on the surface, they are not diametrically opposed. Consider a personal blog, in which the author relays highly intimate personal details. The author of the blog may be anonymous, but the blog itself is quite permanent and recordable. Is it possible to design surveys that replicate this type of setting? For instance, if a participant was made aware that his responses would be stored, anonymously but for the long term, and would potentially serve many surveys, this may increase a respondent's motivation to respond truthfully.

2. *Manage Individual Differences.* Future survey technologies should take advantage of individual differences related to deception by assessing experience with the survey technology and privacy concerns in order to determine that specific respondent's level of trust and willingness to disclose in that specific survey interview. One method for doing this would be to simply include privacy and experience assessment as part of the interview. Based on these responses, the survey's sensitive question asking may be curtailed or expanded for that participant depending on her responses, or the same set of standard questions

might be asked but the individual difference information may be used to weight the responses to sensitive questions. For example, a respondent identified as a privacy fundamentalist may not be asked highly sensitive questions, or the responses might be weighted accordingly.

A second approach may be to develop an identity or disclosure management system that allows a respondent to indicate how much personal information he is willing to disclose in that particular session. Participants willing to disclose would be asked more sensitive questions, while those indicating a preference for less disclosure would be asked fewer sensitive questions. Obviously, both of these approaches have implications for survey methodology, but one advantage is that the individual difference information would provide some insight into how to interpret responses to sensitive items, including socially desirable responding.

3. *Instill Normative Expectations of Honesty*. The relative scarcity of cues in technologically mediated survey interviews relative to face-to-face interviews suggests that cues to normative expectations and legitimization can be particularly powerful. Cues to honesty and legitimacy should be designed into the survey context, either through manipulation cues in the human–survey interface or norm development. Human–survey interface manipulations include displaying well-known and/or trustworthy brands (e.g., the Census Bureau), as well as primes of honesty, such as a picture of honest Abe Lincoln. At the same time, trust in mediated environments is fragile and easily disrupted (Boyd, 2003). As such, subtle cues in the interface, such as shoddy interface design, broken links, or other errors may disrupt a respondent's trust in that particular survey interaction.

A second, more comprehensive approach to instilling norms of honesty is to develop the normative expectations around a given survey brand. For example, in my own experience, the U.S. Census has achieved a norm of honest disclosure in the survey, through a mixture of emphasizing anonymity and highlighting the importance of the information (it is highly permanent data used in many contexts). Survey designers may be able to examine the success of the U.S. Census and develop a set of best practices for instilling normative expectations of disclosure rather than deception.

9.6   AUTHENTICATION COUNTERMEASURES

Regardless of the attempts described previously to elicit truthful responding, some percentage of responses are likely to be deceptive. Is it possible to use the novel technologies associated with surveys of the future to determine whether a respondent is lying on an item or not? One possible technique is the use of automated linguistic analysis programs to detect language patterns across deceptive and truthful responses. Several recent studies suggest that language use does change when someone is lying in a computer-mediated interaction (Hancock et al., in press; Zhou et al., 2004). For example, in one study of deception in synchronous computer-mediated communication, liars used more words overall, fewer first person singular references (e.g., "I"), more third person (e.g., "he," "she" "they"), and fewer exclusion terms (e.g., but, except), among other differences (Hancock et al., 2007). It may be possible to adapt these techniques to open-ended responses in survey contexts.

Linguistic analysis can also be used to identify the source of a message. Source authentication uses similar linguistic analysis techniques to determine whether two or more texts were written by the same source or not. In particular, stylistic features of a set of texts can be compared with one another to determine whether two pieces of text share sufficient stylistic features as to be likely to be authored by the same person. Other methods, involving security informatics, are also available for detecting other types of hacking attacks (for review, see Cybenko et al., 2002).

Finally, paradata, or the information that is associated with a response, such as the time it took the participant to respond to an item (see Person, D'Mello, and Olney, Chapter 10 in this volume), may also provide information about whether a participant is providing truthful or deceptive responses. I know of no research to date that has examined keyboard and mouse data in relation to deception, but it does not seem unreasonable to assume that there may be nonlinguistic patterns of responding that suggest deception.

9.7 LIMITATIONS TO THE SURVEY WORLD

A number of important potential limitations constrain how usefully the present discussion may apply to the survey context. For instance, given that much of the discussion is grounded in research in interpersonal communication, will the described mode features, individual differences, psychological variables, and normative expectations also apply to multiple-choice responding, or will they be limited to open-ended responses? It is certainly the case, at this point, that the countermeasures of linguistic analysis can only be applied to text-based, free-form responses. Similarly, the analysis described earlier may also be limited to questions about sensitive information, when participants have a motivation to deceive (e.g., in order to appear more socially desirable, or to avoid embarrassment).

Finally, although the future of technology is notoriously difficult to predict, an important trend is the movement toward ubiquitous and mobile computing (Weiser, 1991). The movement of computers off the desktop and into everyday parts of our world poses challenges and opportunities for deception and disclosure in surveys of the future. For instance, it may be difficult for designers to create the stranger on a train effect when the respondent could be anywhere when completing the survey. On the other hand, as computers become more context-sensitive and able to sense inputs from more than a keyboard or mouse, they may be more capable of sensing whether a respondent is preparing to deceive, disclose, or is just thinking hard. Nonetheless, research from both the survey world and the world of computer-mediated communication suggests that there are several factors that can help us understand when and why a respondent may choose to deceive.

REFERENCES

Aquilino, W. (1994). Interview mode effects in surveys of drug and alcohol use. *Public Opinion Quarterly, 58*, 210–240.

Bargh, J., McKenna, K. Y. A., and Fitzsimons, G. M. (2002). Can you see the real me? Activation and expression of the "true self" on the Internet. *Journal of Social Issues, 58*, 33–48.

Boyd, J. (2003). The rhetorical construction of trust online. *Communication Theory, 13*, 392–410.

Bradburn, N. M., Frankel, M. R., Hunt, E., Ingels, J., Schoeua-Glusberg, A., Wojcik, M., and Pergamit, M. R. (1991). A comparison of computer-assisted personal interviews with personal interviews in the National Longitudinal Survey of Labor Market Behavior – Youth Cohort. In *Proceedings of the US Bureau of the Census Annual Research Conference*. Washington DC: U.S. Bureau of the Census.

Currivan, D. B., Nyman, A. L., Turner, C. F., and Biener, L. (2004). Does telephone audio computer-assisted self-interviewing improve the accuracy of prevalence estimates of youth smoking? Evidence from the UMass Tobacco Study. *Public Opinion Quarterly, 68*, 542–564.

Cybenko, G., Giani, A., and Thompson, P. (2002). Cognitive hacking: a battle for the mind. *Computer, 35*, 50–56.

DePaulo, B. M., Kashy, D. A., Kirkendol, S. E., and Epstein, J. A. (1996). Lying in everyday life. *Journal of Personality and Social Psychology, 70*, 979–995.

DePaulo, B. M., & Kashy, D. A. (1998). Everyday lies in close and casual relationships. *Journal of Personality and Social Psychology, 74*, 63–79.

Fenigstein, A., Scheier, M. F., and Buss, A. H. (1975). Public and private self-consciousness: assessment and theory. *Journal of Consulting & Clinical Psychology, 43*, 522–527.

Gribble, J. N., Miller, H. G., Cooley, P. C., Catania, J. A., Pollack, L., and Turner, C. F. (2000). The impact of T-ACASI interviewing on reporting drug use among men who have sex with men. *Substance Use and Misuse, 80*, 869–890.

Hancock, J. T. (2007). Digital deception: when, where and how people lie online. In K. McKenna, T. Postmes, U. Reips, and A. N. Joinson (Eds.), *Oxford Handbook of Internet Psychology*, Oxford: Oxford University Press.

Hancock, J. T., Curry, L., Goorha, S., and Woodworth, M. T. (in press). On lying and being lied to: a linguistic analysis of deception. *Discourse Processes*.

Hancock, J. T., Thom-Santelli, J., and Ritchie, T. (2004a). Deception and design: the impact of communication technologies on lying behavior. *Proceedings, Conference on Computer Human Interaction*, Vol. 6, pp. 130–136. New York: ACM.

Hancock, J. T., Thom-Santelli, J., and Ritchie, T. (2004b). What lies beneath: the effect of the communication medium on the production of deception. Presented at the Annual Meeting of the Society for Text and Discourse, Chicago, IL.

Hancock, J. T., Toma, C., and Ellison, N. (2007). The truth about lying in online dating profiles. In *Proceedings of the ACM Conference on Human Factors in Computing Systems*.

Hargittai, E. (2004). Internet access and use in context. *New Media and Society, 6*, 137–143.

Herring, S. C. (2007). A faceted classification scheme for computer-mediated discourse. *Language@Internet*. Retrieved 16 February, 2007, from http://www.languageatinternet.de/articles/761

Holbrook, A. L., Green, M. C., and Krosnick, J. A. (2003). Telephone versus face-to-face interviewing of national probability samples with long questionnaires: comparisons of response satisficing and social desirability response bias. *Public Opinion Quarterly, 67,* 79–125.

Joinson, A. N. (2001) Self-disclosure in computer-mediated communication: the role of self-awareness and visual anonymity. *European Journal of Social Psychology, 31,* 177–192.

Joinson, A. N., and Dietz-Uhler, B. (2002). Explanations for the perpetration of and reactions to deception in a virtual community. *Social Science Computer Review, 20*(3), 275–289.

Joinson, A. N, and Paine, C. B. (2007) Self-disclosure, privacy and the Internet. In A. N. Joinson, K. McKenna, T. Postmes, and U.-D. Reips (Eds.), *Oxford Handbook of Internet Psychology.* Oxford: Oxford University Press.

Merola, N., Pena, J., and Hancock, J. T. (2006, June). Avatar color and social identity effects on attitudes and group dynamics in online video games. Paper presented at International Communication Association, Dresden, Germany.

Metzger, M. J. (2004) Privacy, trust and disclosure: exploring barriers to electronic commerce. Available online at http://jcmc.indiana.edu/vol9/issue4/metzger.html. Retrieved 3 August 2006.

Mosher, W. D., & Duffer, A. P., Jr. (May, 1994). Experiments in survey data collection: the national survey of family growth pretest. Paper presented at the meeting of the *Population Association of America,* Miami, FL.

Mollering, G. (2006). *Trust, Reason, Routine, Reflexivity.* Oxford, UK: Elsevier Ltd.

Paulhus, D. L. (2002). Socially desirable responding: the evolution of a construct. In H. Braun, D. N. Jackson, and D. E. Wiley (Eds.), *The Role of Constructs in Psychological and Educational Measurement* (pp. 67–88). Hillsdale, NJ: Lawrence Erlbaum.

Postmes, T., Spears, R., and Lea, M. (2000). The formation of group norms in computer-mediated communication. *Human Communication Research, 26,* 341–371.

Schaeffer, N. C. 2000. Asking questions about threatening topics: a selective overview. In A. A. Stone, J. S. Turkkan, C. A. Bachrach, J. B. Jobe, H. S. Kurtzman, and V. S. Cain (Eds.), The Science of Self-Report: Implications for Research and Practice. (pp. 105–122). Mahwah, NJ: Lawrence Erlbaum.

Schuman, H., and Presser, S. (1981). *Questions and Answers in Attitude Surveys: Experiments on Questions Form, Wording and Context.* New York: Academic Press.

Schwarz, N., Strack, F., Hippler, H. J., & Bishop, G. (1991). The impact of administration mode on response effects in survey measurement. *Applied Cognitive Psychology, 5,* pp. 193–212.

Sengers, P., Boehner, K., David, S., and Kaye, J. (2005). Reflective design. In *Proceedings of the 4th Decennial Conference on Critical Computing* (pp. 49–58).

Steiger, D. M. (2006). Interactive voice response (IVR) and sources of survey error. Paper presented at International Conference on Telephone Survey Methodology, Miami, FL.

Sudman, S., and Bradburn, N. M. (1974), *Response Effects in Surveys.* Chicago: Aldine.

Thibaut, J. W., and Kelly, H. H. (1959). *The Social Psychology of Groups.* Hoboken: NJ: John Wiley & Sons.

Tourangeau, R., and Smith, T. W. (1996). Asking sensitive questions: the impact of data collection mode, question format, and question context. *Public Opinion Quarterly, 60,* 275–304.

Tourangeau, R., and Smith, T. W. (1998). Collection sensitive information with different modes of data collection. In Mick P. Cooker, Reginald P. Baker, Jelke Bethlehem, Cynthia Z.

Clark, Jean Martin, William Nichols II, and James M. O'Reilly (Eds.), *Computer-Assisted Survey Information Collection*. Hoboken, NJ: John Wiley & Sons.

Tourangeau, R., Couper, M. P., and Steiger, D. M. (2003). Humanizing self administered surveys: experiments on social presence in Web and IVR surveys. *Computers in Human Behaviour, 19*, 1–24.

Turner, C. F., Miller, H. G., Smith, T. K., Cooley, P. C., and Rogers, S. M. (1996). Telephone audio computer-assisted self-interviewing (T-ACASI) and survey measurements of sensitive behaviors: preliminary results. In R. Banks, J. Fairgrieve, L. Gerrard, et al. (Eds.), *Survey and Statistical Computing 1996*. Chesham, Bucks, UK: Association for Survey Computing.

Vrij, A. (2000). *Detecting lies and deceit: The psychology of lying and the implications for professional practice*. Chichester, England: John Wiley & Sons.

Wallace, P. (1999). *The Psychology of the Internet*. Cambridge, UK: Cambridge University Press.

Weiser, M. (1991). The computer for the 21st century. *Scientific American, 265*, 94–104.

Westin, A., (1967). *Privacy and Freedom*. New York: Atheneum.

Whitty, M. T. (2002). Liar, Liar! An examination of how open, supportive and honest people are in chat rooms. *Computers in Human Behavior, 18*, 343–352.

Zhou, L., Burgoon, J. K., Nunamaker, J. F., and Twitchell, D. (2004). Automating linguistics-based cues for detecting deception in text-based asynchronous computer-mediated communication. *Group Decision and Negotiation, 13*, 81–106.

CHAPTER 10

Toward Socially Intelligent Interviewing Systems

Natalie K. Person
Rhodes College, Memphis, Tennessee

Sidney D'Mello and Andrew Olney
University of Memphis, Memphis, Tennessee

10.1 INTRODUCTION

Advances in technology are changing and personalizing the way humans interact with computers, and this is rapidly changing what survey researchers need to consider as they design the next generation of interviewing technologies. The nearly geometric growth rate of processing power, memory capacity, and data storage are allowing designers to endow software and hardware with capabilities and features that seemed impossible less than a decade ago. Among the innovations of greatest relevance for survey designers are intelligent systems: computers, robots, and software programs that are designed to exhibit some form of reasoning ability. The concept of an intelligent system does not necessarily imply that the system thinks independently; however, it does imply that the system has been programmed to respond in intelligent, adaptive ways to specific kinds of input.

In this chapter, we discuss how technology is already being used in interactive dialogue systems to optimize user input and to adapt to users in socially intelligent ways. Most of the methods and technologies that we will be discussing have been incorporated in systems other than automated survey systems; however, we believe it is only a matter of time before the social intelligence advances that have been made and implemented in other intelligent systems will be proposed for survey technologies—for better or worse. The extent to which what we discuss here ought to be extrapolated

Envisioning the Survey Interview of the Future, Edited by Frederick G. Conrad and Michael F. Schober
Copyright © 2008 John Wiley & Sons, Inc.

to survey systems depends crucially on the parallels between what these systems and potential interviewing systems do; we will return to this issue along the way and at the end, but there is sufficient overlap to make the comparisons valuable.

Thinking about automated interviewing systems with greater social intelligence is worthwhile given how increasingly important intelligent systems are becoming in our everyday experiences. Intelligent systems aid us in making travel arrangements, in finding phone numbers and addresses when we dial directory assistance, and in using our word processing applications on our personal computers. In some of our past work, we have been particularly interested in exploring how intelligent systems that manifest aspects of social intelligence affect users' behaviors and cognitive states. In particular, we have explored how systems that possess differing degrees of social agency affect language production when users are asked to disclose personal and emotionally charged information. We have also examined how users' affective states can be measured in intelligent tutoring systems, which are similar to automated survey interviews in certain important ways: both are interactive tasks in which the system ("tutor" or "interviewer") drives the dialogues, presents prompts and cues, and must motivate users (learners or respondents) to continue the activity and provide thoughtful responses. It is plausible that systems that can adapt or mirror users' affective states (see Cassell and Miller, Chapter 8 in this volume) will lead to more positive user perceptions, greater trust between the user and the system, and possibly greater learning gains. For example, an intelligent tutoring system that can detect when learners are frustrated, angry, or confused can adjust its teaching and communicative style accordingly. Similarly, a system that can sense when learners are pleased will know what and when particular actions and responses are desirable (Picard, 1997).

Unfortunately, detecting and responding to user affect are difficult endeavors for several reasons. First, humans themselves often have difficulty detecting the precise emotions of other humans (Graesser et al., 2006). This may be due to the tremendous variability between individuals in their outward expression of emotion. Second, although detecting user emotion is a difficult task, having a system select and deliver a socially appropriate response is a separate and equally difficult task. Simply linking canned responses to particular user affective states is probably not enough to enhance the quality of the interaction or have a significant effect on user performance. Third, it is unclear whether certain media are more effective in eliciting affective responses from users and responding to users than others. For example, users may respond very differently to animated agents that are visually and verbally responsive than to a system that only provides text responses.

In this chapter, three issues that are relevant to social intelligence and that have been studied in the context of interactive dialogue systems will be addressed. We have chosen to focus on issues that are receiving considerable attention in other research areas and that we feel are pertinent to the next generation of survey systems. The first issue involves *social agency*. We will discuss whether users communicate differently with systems that incorporate some form of social agency (e.g., animated agents) compared to those that do not. The second issue involves *detecting users' affective states* by monitoring bodily movements and using paradata, that is, data about the user's process in performing an interactive task. We will discuss some of the state-of-the art technologies and methodologies for detecting users' affective states during the course of a real-time dialogue. The third issue is concerned with designing systems that are *socially responsive to users*. Is it the case that dialogue systems should always adhere to social conventions and respond to users in polite and socially appropriate ways? It is our belief that interactive dialogue systems that possess forms of social intelligence will greatly enhance the interactive experience for users and will yield more informative data sets for researchers. As mentioned earlier, the methods and technologies that we will be discussing have yet to make their way into survey technologies; however, in each section of the chapter we will discuss how, with slight modifications, these methods and technologies could be used by the survey community.

10.2   SOCIAL AGENCY

Nass and colleagues have reported on numerous occasions that humans, mostly without realizing it, apply implicit rules of human-to-human social interaction to their interactions with computers. Computers that exhibit more human-like behaviors (e.g., those that include animated characters or voices) are subject to greater social expectations from their users than those with less human-like features [although Nass and colleagues argue that even line drawings can evoke social responses (Nass, 2004; Reeves and Nass, 1996)]. As a result, when computers violate rules of social etiquette, act rudely, or simply fail to work at all, users become frustrated and angry. Interestingly, users do not attribute their negative emotions to their own lack of knowledge or inadequate computer skills, but instead direct them toward the computer and its inability to sense the users' needs and expectations (Miller, 2004; Mishra and Hershey, 2004; Nass, 2004; Reeves and Nass, 1996). Such claims have numerous implications for the way dialogue systems are designed, especially those that attempt to create the illusion of social agency by including voices or animated agents with human-like personas.

Animated agent technologies have received considerable attention from researchers who are interested in improving learning environments and automated therapeutic facilities (Baylor et al., 2004; Marsella et al., 2000). However, it is still unclear whether such technologies improve user performance—for example, contribute to learning gains or offer substantial benefits (therapeutic or otherwise)—over environments without agents. It's worth noting that the ways in which animated agents are being used and studied in other domains could certainly inform their use in automated survey systems. For example, some intelligent tutoring systems use animated agents as tutors. These tutoring agents engage students in conversations in attempts to get students to provide information or construct knowledge. The tutor and student work persistently to negotiate meaning and establish common ground. The dialogue of tutoring interactions has some parallels to what transpires in interviewer-administered survey interviews. Respondents are often required to provide information to questions that can contain confusing or unfamiliar terminology (or even ordinary words that respondents are conceptualizing differently than the survey designers), and common ground has to be negotiated before the respondent can supply the information accurately (Conrad and Schober, 2000; Schober et al., 2004). To date, the results from most studies in which animated agent versus no agent comparisons have been made indicate that users tend to like environments with agents more than environments without agents. This effect is known as the *persona effect* (Andre et al., 1998; Lester et al., 1997). However, there is only preliminary evidence that animated agents actually contribute to the main goals (learning, therapeutic, etc.) of the systems they inhabit (Atkinson et al., 2004; Moreno et al., 2001).

The design, implementation, and maintenance of animated agents for any kind of environment (learning, therapeutic, survey, or otherwise) is expensive and time consuming. Before such resources are invested, it seems worthwhile to determine whether users do indeed communicate differently with systems that have some form of social agency versus those that do not. After all, it may be that the presence of a social agent (human or animated) is viewed as face-threatening. If this is the case, then in a survey about a sensitive topic an animated agent might collect less candid answers than a textual interface; this is why Cassell and Miller (Chapter 8 in this volume) conclude that agents are probably not promising as survey interviewers. Or the opposite may be true: users may be more inclined to learn and/or disclose more when some form of social agent is present. To the extent that a survey interviewer can reassure a respondent that many people give similar answers and all answers are confidential, so an animated agent might collect more candid responses than an interface without a human-like presence.

In a recent study following this line of inquiry, Person et al. (2005) investigated how different types of social agency affect language production in interviews with college students. Fifty-nine college students from two institutions participated in interviews in which they answered questions about their alcohol consumption behaviors, their attitudes and beliefs about alcohol use, their personal lives (e.g., What do they do in their free time? What kinds of things do they do when they are hanging out with friends?), and their personal and family histories of alcohol-related problems. The participants in the study were assigned to one of four social agency conditions: (1) animated agent (AA), (2) text-only (TO), (3) instant messaging (IM), and (4) human-to-human (HH). In the AA condition, an animated agent conducted the interview and the participants typed their responses. In the TO condition, the interview questions appeared one at a time on a computer monitor, and students typed their responses. In the IM condition, a human interviewer presented the questions to the participant via an instant messaging program (i.e., Instant Messenger), and participants in the IM condition were aware that they were communicating with another human. In the HH condition, a human conducted the interview in a face-to-face setting. The interviews in the HH condition were recorded and transcribed. The 12 interview questions were the same in all four conditions, and the interviewer in the social agency conditions (AA, IM, and HH) was always a male or a male persona (in the condition with the animated agent). The length of the interview sessions ranged from approximately 10 minutes to 45 minutes.

The participants' answers were analyzed with Wmatrix, a corpus analysis and comparison software tool (Rayson, 2003, 2005). Wmatrix provides a Web interface to the UCREL semantic analysis system (USAS) and CLAWS word-tagging corpus annotation tools, and to standard corpus linguistic methodologies such as frequency lists, word concordances, and grammatical and semantic category parsing. The results indicated differences between the conditions in terms of disclosure amount, semantic richness, and other linguistics features. For disclosure amount, although the means for the total numbers of words did not differ across conditions (TO = 660.6, AA = 622.8, HH = 606.2, and IM = 552.7), there were differences in disclosure amounts between the two computer conditions (TO and AA, Total words mean = 641.7) and the two human conditions (IM and HH, Total words mean = 578.5). That is, the participants disclosed more in the conditions in which the interviewer was a computer than they did in the conditions where the interviewer was a human. There were also differences between the conditions for particular semantic categories. Of interest were semantic categories related to drinking, family, relationships, religion, and emotion. Overall, it was found that participants used more semantically rich and emotion language in the AA and IM than in the TO and HH conditions.

One interpretation of these findings is that social agency is important for disclosing information, especially for information that is rich in content and emotion. However, such disclosure occurs less in human-to-human interviews, perhaps because the face-to-face interactions are too face-threatening. These findings indicate that animated agents may have a future in forthcoming survey systems given that, at least in the current case, animated agents do not seem to interfere with participants' willingness to disclose information (in contrast to Cassell and Miller's predictions, see Chapter 8 in this volume). In fact, animated agents may turn out to be preferable to both humans and textual interfaces for collecting personal and sensitive information in interviews.

10.3   USERS' AFFECTIVE STATES

Research in the relatively new area of affective computing is providing tremendous insight for the next generation of intelligent systems. Generally speaking, affective computing focuses on creating technologies that can monitor and appropriately respond to the affective states of the user (Picard, 1997). Evaluating the emotions of others is a skill that allows humans to function appropriately and effectively in social interactions. Mishra and Hershey (2004) reported that systems that interpret users' emotions and exhibit appropriate social responses are likely to result in increased usability and productivity on the part of the human user. In order to compensate for the lack of social skills of today's personal computers, affective computing has emerged as a new and exciting research area that attempts to bridge the gap between the emotionally expressive human and the socially challenged computer. Affective computing can be considered to be a subfield of human–computer interaction (HCI), where the affective states of a user (feelings, moods, emotions) are incorporated into the decision cycle of the interface in an attempt to develop more effective, user-friendly, and naturalistic applications (Bianchi-Berthouze and Lisetti, 2002; Prendinger and Ishizuka, 2005; Whang et al., 2003).

Although such interfaces are being developed for a wide variety of systems, including affect-sensitive robots (Rani et al., 2003), one particular area that has emerged as a leader in the development of affect-sensitive interfaces is the creation of intelligent tutoring systems and artificial peer learners (Conati, 2002; D'Mello et al., 2005; Kort et al., 2001; Litman and Forbes-Riley, 2004). Tutoring systems attempt to incorporate the affective states of a learner into their pedagogical strategies to (presumably) increase learning. This work parallels attempts to capture paradata in today's online surveys, in which information like timing of mouse clicks and keyboard entries tell something about the respondent's experience answering questions, either for subsequent additional analysis (Couper, 2000) or to change the system's behavior with the respondent (Conrad, Schober, and Coiner, 2007), although thus far the work on surveys has not focused on respondent affect.

In general, an affective interaction involves the immersion of a user into an affective loop or a *detect–select–synthesize cycle*. This involves the *identification* and then the real-time *detection* of the user's affective states relevant to the domain, the *selection* of appropriate actions by the system to optimize task efficiency, and the *synthesis* of emotional expressions by the system so that the user remains engaged and the interaction is not compromised. From a broad perspective, the players in the affective loop are a user and a computer interface. However, within the context of particular domains the roles of the players become more specific. For example, within the context of intelligent learning environments, the user becomes a learner and the interface becomes an artificial tutor (D'Mello et al., 2005) or a learning companion (Kort et al., 2001). Similarly, implementing the affective loop in a dialogue-based survey system would involve merging affect models for the respondent and interviewer. The respondent model would first be concerned with *identifying* the prominent affective states that a respondent experiences when participating in a survey interview. Of course, these affective states could vary depending on the nature of the information being collected in the survey or interview (e.g., ketchup brand preferences versus risky sexual behavior). Once this has been established, the manner in which respondents manifest these affective states can be investigated, thus providing the foundations for the development of an automatic affect *detection* system. The interviewer model would take into account how effective human interviewers *adapt* their behaviors to handle the emotions of the respondent (as when interviewers use a more empathetic tone or move on to the next question when a respondent seems overanxious or challenged). Finally, embodied conversational agents that simulate human interviewers would have to be programmed to *synthesize* affective elements through natural body movements such as the generation of facial expressions, the modulation of speech, and the inflection of posture.

It should be noted that the aforementioned description of an affective loop between an artificial interviewer and a human respondent is broad enough to encompass the entire gamut of the affective interaction. Of course, there may be circumstances during survey interviews in which overly expressive displays of emotions by artificial interviewers would be inappropriate or could bias the respondent's answer. Nevertheless, there is important information to be gained by simply monitoring the affective states of the respondent. For example, even if no reactive strategy is employed, simply recording the emotions of the respondent may be beneficial for offline analyses (Mossholder et al., 1995).

In the subsequent sections we highlight the major steps involved in designing an affect-sensitive interviewer. In order to appropriately ground the discussion, we consider a project at the University of Memphis that directly tackles this issue by attempting to transform an existing intelligent tutoring system, AutoTutor, into an affect-sensitive tutor. AutoTutor is a one-on-one tutoring system that helps learners construct explanations by interacting with them in natural language and by helping them solve problems in simulation environments (Graesser et al., 2001, 2005). The AutoTutor interface includes an animated conversational agent that speaks the content of AutoTutor's turns and displays simple facial expressions and rudimentary gestures. Students type their answers on the keyboard. AutoTutor provides *feedback* to the student (positive, neutral, or negative feedback), *pumps* the student for more information ("What else?"), *prompts* the student to fill in missing words, gives *hints*, fills in missing information with *assertions*, identifies and corrects *misconceptions* and erroneous ideas, *answers* students' questions, and *summarizes* topics. A full answer to a problem is eventually constructed by the student and AutoTutor and normally takes between 30 and 200 conversational turns.

The AutoTutor interface has five windows, as illustrated in Fig. 10.1. Window 1 (top of screen) has the main question that stays on the computer screen throughout the conversation that involves answering the question. Window 2 (bottom of screen) is affiliated with the learner's answer in any one turn and Window 3 (below agent)

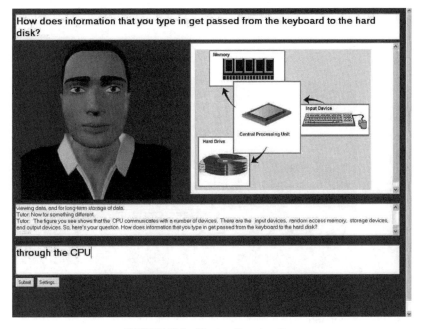

FIGURE 10.1 The AutoTutor interface.

echoes whatever the learner types in along with the responses of the tutor. Window 4 (left middle) hosts an animated conversational agent that speaks the content of AutoTutor's turns. The talking head has facial expressions and some rudimentary gestures. Window 5 (right middle) is either blank or has auxiliary diagrams.

AutoTutor has been tested in several experiments on approximately 1000 students in computer literacy and physics courses. Significant learning gains were obtained in all of these experiments (with an average effect size of .8), particularly at the level of deep explanations as opposed to shallow facts and ideas (Graesser et al., 2004). AutoTutor has also been evaluated on the conversational smoothness and the pedagogical quality of its dialogue moves in the turn-by-turn tutorial dialogue (Person et al., 2002). A *bystander Turing test* on the naturalness of AutoTutor's dialogue moves indicated that bystanders (i.e., research participants who read AutoTutor and human-to-human tutoring transcripts) were unable to discriminate AutoTutor dialogue moves from the dialogue moves of a real human tutor.

The general assumption behind the affect-sensitive AutoTutor project is that endowing the tutor with the ability to incorporate the learner's affective states into its pedagogical strategies (in addition to their cognitive states, which are already tracked by the tutor) will result in even greater learning gains for students. The project integrates state-of-the-art, nonintrusive, affect-sensing technology with the AutoTutor program in an endeavor to classify emotions on the basis of facial expressions, gross body movements, and conversational cues. Figure 10.2 illustrates the setup of a recent study involving learners ($N = 28$) interacting with AutoTutor in computer literacy while the sensors silently capture their bodily data (Graesser et al., 2006). Since many of the emotions tracked in the study may occur within an interviewer–respondent interaction, the methods and results achieved in the development of the affect-sensitive

FIGURE 10.2   Sensors used to track diagnostic data while learner interacts with AutoTutor. (*Note*: The left and right monitors are turned off during actual data collection.)

AutoTutor can potentially be used as a model for developing the first affect-sensitive interviewer.

10.3.1 Identifying the Affective States

Ekman and Friesen (1978) have proposed six *basic* emotions that are ubiquitous in everyday experience. These include fear, anger, happiness, sadness, disgust, and surprise. However, many have called into question the adequacy and validity of a theory of emotion that is based on these six basic emotions alone (Kort et al., 2001). For example, within the context of a survey it is unlikely that a respondent will experience particularly strong states of fear, happiness, or sadness. Similarly, these basic emotions rarely occur during learning, at least not with AutoTutor. In a series of exploratory studies that utilized online measures of affect such as observational and emote-aloud protocols, as well as offline judgments of emotions by multiple judges, Graesser and his colleagues have identified a set of affective states that typically *do* play a significant role in learning: boredom, flow (or engagement), confusion, and frustration (Craig et al., 2004; Csikszentmihalyi, 1990; D'Mello et al., 2006; Graesser et al., 2006; Graesser et al., in review).

It is reasonable to presume that the affective states of boredom, engagement, confusion, and frustration may also play important roles in the interviewer–respondent relationship. For example, boredom may occur if the survey involves an impersonal or uninteresting topic and may become more pronounced over an extended period of time. Confusion may be prevalent when respondents are not familiar with terminology or do not understand the question. Discomfort may transition into frustration if the questions are very sensitive in nature or are not deemed to be appropriate by the respondent. Ideally, a respondent could even be engaged in the survey and have a *flow* experience, where they are so engrossed that time and fatigue disappear (Csikszentmihalyi, 1990), although declining participation and completion rates in surveys argue against this being at all common.

10.3.2 Methods for Detecting User Emotion

An emotionally sensitive learning environment, whether it be human or computer, requires some degree of accuracy in classifying the learners' affect states. Over the last decade, efforts have been launched to automatically detect emotions by means of a variety of signal detection algorithms operating on a host of sophisticated sensors. Although the use of physiological signals has been relatively successful in emotion detection (Nakasone et al., 2005; Rani et al., 2003; Whang et al., 2003), these methods rely on obtrusive sensing technologies such as skin conductance and heart rate monitoring. Although acceptable in some domains, once users habituate to the presence of obtrusive sensors, they may not be optimal in learning or survey environments because this approach could cause distraction and task interference (not to mention they are very expensive and inaccessible to most end-users). Therefore, we recommend the use of nonintrusive bodily sensors such as cameras that track facial features and microphones that monitor speech contours (which are also less expensive and more accessible). In fact, the majority of current affect detection systems track facial features (Cohn and Kanade, 2007; Oliver et al., 1997) and acoustic-prosodic vocal features (Bosch, 2003; Fernandez and Picard, 2004; Shafran and Mohri, 2005). Additionally, some pioneering research has also focused on affect detection from posture patterns (Mota and Picard, 2003). The AutoTutor team has used nonintrusive sensors such as cameras and a posture sensor, along with discourse cues that may convey affect. In the following sections, we describe how affective states have been measured in AutoTutor via discourse cues, facial expressions of users, and a posture sensor.

10.3.2.1   Affect Detection from Conversational Cues

One of the first channels that we explored to detect affect was the discourse that occurred in AutoTutor's natural language, mixed-initiative dialogues. Although dialogue has traditionally been a relatively unexplored channel for affect/emotion detection, it is a reasonable information source to explore because dialogue information is recorded in virtually all human–computer conversations and is inexpensive to collect. In addition to detecting the emotions of a student or respondent, dialogue features can be used to infer additional task-specific information, some of which may be relevant to a tutoring or survey domain. In addition to affective computing, innovative uses of dialogue have surfaced from research on the identification of problematic points in human–computer interactions (Batliner et al., 2003; Carberry et al., 2002; Walker et al., 2002). For example, Carberry et al. (2002) proposed an algorithm to recognize doubt by examining linguistic and contextual features of dialogue in conjunction with world knowledge. Recognizing doubt is certainly important in tutoring, but recognizing doubt could allow a survey interviewing system to help the respondent understand a question or carry out the task, or to know when to reassure the respondent that he/she is doing a good job.

In AutoTutor the efficacy of tutorial dialogue as a predictor of the affective states of a learner was investigated by extracting a variety of dialogue features from AutoTutor's text-based log files and then connecting these with the emotions of the learner (D'Mello, Craig et al., in press). The features included temporal measures such as response time and time into the session, measures of response verbosity (number of words), and assessments of student's ability. Other measures that influenced the affective states of the learner were measures of tutor directness (e.g., how much information the tutor explicitly provides the learner), and tutor feedback (i.e., positive or negative), which is manifested in the verbal content, intonation, and a host of other nonverbal conversational cues of the embodied conversational agent that personifies the tutor.

Although the features of dialogue we analyzed were specific to AutoTutor, a similar set of features would presumably be expected in an interactive dialogue-based survey system. The lower level features specific to AutoTutor can be generalized to generic categories of dialogue features such as temporal assessments, response verbosity, respondent answer quality, answer attempts, and interviewer clarifications. Once these features are specified and the respondent affect model is developed, observed relationships between dialogue and affect can be applied to the survey domain. In AutoTutor, for example, boredom occurs later in the session and after multiple attempts to answer a question. Alternatively, confusion occurs earlier in the session, with slower and shorter responses, and with frozen expressions (e.g., "I don't know", "What did you say?"). Engagement seems to occur earlier on in the session and involves longer responses. Additional details on these and other patterns are reported in D'Mello, Craig et al. (2007).

Based on the relationships that emerged between dialogue features and emotions that occur during learning, AutoTutor researchers developed a system that could successfully differentiate the various affective states from a neutral baseline state. In particular, accuracy rates of 69%, 68%, 74%, 71%, and 78% in detecting boredom, confusion, delight, flow, and frustration, respectively, from a neutral affective state were obtained (D'Mello and Graesser, 2006; D'Mello et al., 2007). These results support the notion that dialogue is a reasonable source for measuring the affective states that a learner is experiencing.

10.3.2.2   Affect Detection from Posture Patterns

Monitoring posture to infer affect is interesting because it is rarely the case that posture is intentionally controlled by humans. Tracking posture patterns provides an added advantage when compared to facial expressions and gestures because these motions are ordinarily unconscious, unintentional, and thereby not susceptible to social editing. Ekman and Friesen (1969), in their studies of deception, used the term *nonverbal leakage* to refer to the unsuccessful attempts of liars to disguise deceit through less controlled channels such as the body movements compared to more controlled channels like facial expressions (for more on detecting deception, see Hancock, Chapter 9 in this volume).

There have been a few studies that have documented the importance of posture in expressing affect (Coulson, 2004; Schouwstra and Hoogstraten, 1995; Wallbott, 1998). Recently, D'Mello, Chipman, and Graesser (in review) have demonstrated that posture can be a viable channel to discriminate between the affective states of boredom (low engagement) and flow (high engagement). Being able to discriminate levels of engagement has utility in survey domains; presumably many breakoffs (premature terminations of the interview by respondents) are due to respondent boredom [for evidence see Conrad, Couper, Tourangeau, and Peytchev (2005)]. Establishing and maintaining the engagement of respondents is especially critical in situations with high degrees of user control, such as in automated telephone surveys and computer-based surveys where respondents can easily break off. For instance, with Web-based surveys, individuals are one-mouse-click-away (or one switch of Windows or departure from the room) from ending the session; desktop activities like clicking and typing won't be useful for detecting boredom because a bored respondent won't produce any. Therefore, detecting via other means when a respondent is transitioning into boredom and proactively engaging the respondent may help prevent respondent attrition.

D'Mello and colleagues (in review) used the Body Pressure Measurement System (BPMS), developed by Tekscan™ (1997), to track learners' posture when interacting with AutoTutor. The posture sensor is a thin pressure pad (or mat) that can be mounted on a variety of surfaces. In this study, it was placed on the seat and back of a chair that the learners sat on while interacting with AutoTutor. The pad is paper thin with a rectangular grid of sensing elements, each providing pressure in millimeters of mercury. High level posture features such as net pressure and arousal levels were tracked and their relationships to boredom and flow were determined by point biserial correlational analyses. The results indicated that boredom was manifested in two distinct forms. The first is consistent with the preconceived notion of boredom in which a learner stretches out, lays back, and simply disengages. However, a counterintuitive finding was that *boredom* was associated with a form of restlessness manifested by rapid changes in pressure on the seat of the chair. The affective state of *flow* was associated with a heightened pressure in the seat of the chair with minimal movement. This may imply that the learner is mentally engaged in absorbing the material and thereby devotes a smaller amount of cognitive processing toward trivial bodily motion.

Machine learning algorithms were also quite successful in segregating these emotions. In particular, a nearest neighbor classifier (Aha and Kibler, 1991), perhaps one of the simplest classification schemes, achieved an accuracy of 78% in discriminating between boredom and engagement. In addition, this and other classifiers were also moderately successful in segregating each of the other target emotions from a neutral state on the basis of the posture features alone. The accuracy scores were 70% for boredom, 65% for confusion, 70% for delight, 74% for flow, and 72% for frustration. Although the classification accuracy rates could be improved, this exploratory research has highlighted the efficacy of monitoring the posture of the user as a viable channel to infer complex mental states.

10.3.2.3 Affect Detection from Facial Features

Ekman and Friesen (1978) highlighted the expressive aspects of emotions with their Facial Action Coding System. This system specifies how "basic emotions" can be identified by coding specific facial behaviors and the muscle positions that produce them. Each movement in the face is referred to as an *action unit* (or AU). There are approximately 58 action units. Patterns of AUs were used to identify the emotions of happiness, sadness, surprise, disgust, anger, and fear. Developing systems to automatically detect the action units, however, is a challenging task because the coding system was tested primarily on static pictures rather than on expressions that change in real time.

AutoTutor researchers are currently exploring some of the technical challenges associated with the automated detection of the facial expressions. As an initial step, two trained judges coded a sample of the observed facial expressions using the action units. The coding yielded a set of 12 prominent action units that were correlated with the emotions, thus reducing the requirements of an automated facial expression measurement system (McDaniel et al., in press). This research has also elucidated important patterns in how learners display particular emotions. Confusion and delight are highly animated affective states and are easily detectable from facial expressions. From an evolutionary perspective, this might suggest that humans use their face as a social cue to indicate that they are confused, which helps them recruit information or resources to alleviate their confusion [see also Conrad, Schober and Dijkstra (2007) for discussion of facial confusion cues, including gaze aversion, in survey interviews]. Delight is another emotion that is also readily expressed on the face, perhaps because it is a positive emotion. However, it appears that learners do not readily display frustration, perhaps due to the negative connotations associated with this emotion. This finding is consistent with Ekman's theory of social display rules, which state that social pressures may result in the disguising of negative emotions such as frustration.

The accuracy scores for detecting affect on the basis of facial features alone are promising. Classification accuracies for detecting delight were the highest (90%), boredom the lowest (60%), and confusion (76%) and frustration (74%) were in between. These results support the conclusion that classifiers are more successful in detecting emotions that are manifested with highly animated facial activity, such as delight, than emotions that are more subtly expressed (e.g., boredom, which is easily confused with neutral).

It is important to acknowledge that the facial action units used in the emotion classification analyses described earlier were not automatically extracted as would be required for a real-time affect-sensitive system. Instead, they were rated by human coders, which adversely affects the applicability of these results. Fortunately, stemming from a NSF funded workshop in 1992 (Ekman et al., 1992), the last decade has been ripe with research geared toward developing automated facial expression recognition systems (Bartlett et al., 1999; Cohn and Kanade, 2007; Fasel and Luttin, 2000; Pantic and Rothkrantz, 2000). We are currently investigating the possibility of integrating these computational systems to aid in inferring the affective states of the learner.

10.4   SOCIALLY RESPONSIVE SYSTEMS

We have already addressed the importance of social agency and detecting users' affect states in the design of socially intelligent systems. In the previous section, we discussed in detail methods for detecting users' affective states and mentioned that part of the affective loop requires the system to provide a synthesized response to a user's current emotional and cognitive states. Simply put, systems should respond to users in socially and contextually appropriate ways. This begs the question of what actually constitutes a socially and contextually suitable response. Is it the case that a canned response to a user who is experiencing particular emotions (e.g., confusion and frustration) will be equally effective for all users experiencing the same emotions? Should computers avoid negativity and always respond in positive, upbeat ways? Gong (2002) reported that users tend to prefer systems that always express positive affect (e.g., are very polite or upbeat) versus ones that may display negative affect when it is contextually appropriate. This finding suggests that users are sensitive to face-threatening speech from computers. Person et al. (1994) reported that human tutors often provide ambiguous feedback to students to minimize face-threatening impositions. However, Person et al. (1994) also hypothesized that a tutor who caters solely to a student's emotions by strictly adhering to politeness and conversational norms may thwart the student's efforts to discard misconceptions and to interpret clearly the intention of the tutor's feedback. Similar trade-offs between maintaining rapport and collecting high quality responses can occur in survey interviews.

AutoTutor researchers conducted a study to determine whether altering Auto-Tutor's personality on politeness-related dimensions affects student learning gains (Person et al., 2003). An experimental version of AutoTutor, called RudeTutor, was designed to provide face-threatening feedback to users by flouting conversational and politeness norms and by only providing neutral or negative feedback to student responses. Below is an excerpt from a RudeTutor tutoring session.

RudeTutor: The sun exerts a gravitational force on the earth as the earth moves in its orbit around the sun. Does the earth pull equally on the sun? Explain why.

Student: The sun pulls more.

RudeTutor: I know that you know more than that. Say more.

Student: The sun pulls more on the sun than vice versa.

RudeTutor: Did you pay attention in class? How does Newton's third law of motion apply to this situation?

Student: The sun and earth are both in motion.

RudeTutor: No. Go back and answer the question completely. Can't you add anything?

Student: The earth pulls on the sun and the sun pulls on the earth?

RudeTutor: Do you know what you're talking about. I'm pretty sure you can't get this but, Newton's third law refers to the forces exerted by one body on another . . .?

Student: Body

RudeTutor: Another body. You really need some help. Try this. When two objects of different masses hit each other, which one exerts the greater force on the other, the heavy object, the light object, or are the forces equal?

Student: They are equal.

Participants in the study were randomly assigned to interact with either AutoTutor or RudeTutor. Learning gains were assessed for both groups and user perception data were collected. The results indicated that participants who interacted with RudeTutor rated the system very negatively with regard to politeness, effectiveness, and likeability compared to the participants who interacted with the regular AutoTutor. However, it was also the case that all students experienced equivalent positive learning gains, regardless of whether they interacted with RudeTutor or AutoTutor. That is, students who were subjected to face-threatening responses and who received excessive negative feedback learned just as much as those who interacted with the regular, polite AutoTutor. One particularly interesting result involved participants' overall enjoyment of the tutoring sessions: some participants enjoyed interacting with RudeTutor very much, while others seemed to be genuinely offended by its lack of politeness.

The results of the RudeTutor study generated more questions than they answered. For example, how are users' affective and cognitive states affected by the tutor's seemingly inappropriate responses? What is the nature of the individual differences that result in different enjoyment levels for the students? In human-to-human conversations, the affective states of the participants are determined not only by the affective nature of the conversational turns but also by the personalities of the participants. Human-to-computer interactions function much the same way because humans readily and subconsciously apply social expectations to computers irrespective of whether or not it is appropriate to do so (Reeves and Nass, 1996). Although many investigations have focused on how personality is related to a variety of learning-related factors (e.g., achievement, learning approaches, specific aptitudes), and personality researchers have theorized for years about the interaction between teacher and student personality (Mills, 1993, Zhang, 2003), little or no research has attempted to investigate the dynamics of tutor personality (or responsiveness), student personality, and affective states with a high level of experimental precision.

10.5 CONCLUSIONS

The work described here demonstrates that our vision of developing a socially intelligent, affect-sensitive tutor is not merely an exercise in science fiction, and that some components of social intelligence are already being implemented in tutoring systems. In our view, when our approaches are coupled with serious knowledge engineering efforts, several of the methods and computational algorithms that we have used or developed may be applicable for creating socially intelligent, affect-sensitive, dialogue-based, survey systems. Although tutoring interactions are not precisely the same as survey interactions, learner motivation to interact with a tutor is often quite different from respondent motivation to interact with an interviewer, and the flow of the tutorial dialogue follows different routes than survey dialogue. Nonetheless, the similarities are great enough that much can be extrapolated and made possible with existing (and increasingly less costly) technologies.

In particular, the use of dialogue and facial expressions could be used to effectively and reliably measure the emotions of a respondent. Although we have used an expensive, hand-crafted camera for facial feature tracking, recent technology makes this possible with a simple Web cam, thereby significantly reducing the associated equipment costs. In fact, certain laptop manufacturers now market their systems with integrated cameras that are optimized to perform on specialized hardware. Noise reduction microphones are also routinely shipped with contemporary laptops. Although we did not initially track acoustic-prosodic features of speech to infer the affective state of a learner, the literature is rich with such efforts [see Pantic and Rothkrantz (2003) for a comprehensive review].

In fact, the only system we have discussed that may not readily be useful for the survey domain would be the pressure-sensitive chair. However, with some innovation, some of its features can be approximated with a visual image captured by the camera. For example, the distance between the tip of the nose of a user to the monitor can be used to operationally define a leaning forward posture. The movement of the head can be used to approximate arousal. Also, some recent evidence suggests that posture features may be redundant with the facial expressions and dialogue features, thereby eliminating the need for an expensive pressure sensor (D'Mello and Graesser, 2007).

Likewise, agent technologies have matured considerably. Not only have agent technologies become easier to use, but they also have greatly increased in realism. Nowhere is this more evident than in the Microsoft Agent technology (Microsoft Corporation, 1998). Cutting edge ten years ago, Microsoft Agent's talking heads, and the infamous Clippy paperclip, have been eclipsed by the new wave of 3D full-body agent technologies from Haptek™ and the game Unreal Tournament 2004. These agent technologies are already in use by current tutoring systems (Graesser et al., 2005; Johnson et al., 2005) and have advanced tool suites that allow the agent's behavior and emotions to be customized. These agent technologies, already proven in the tutoring domain, can easily be extended to provide socially responsive agents in the survey domain. The real progress will come not in deploying existing affect detection and production models but instead in basic research that improves the accuracy of these models with respect to individual differences.

ACKNOWLEDGMENTS

We thank our research colleagues in the Emotive Computing Group and the Tutoring Research Group (TRG) at the University of Memphis (http://emotion.autotutor.org). Special thanks to Art Graesser. We gratefully acknowledge our partners at the Affective Computing Research Group at MIT. We thank Steelcase Inc. for providing us with the Tekscan Body Pressure Measurement System at no cost. This research was supported by the National Science Foundation (REC 0106965, ITR 0325428, and REC 0633918) and the DoD Multidisciplinary University Research Initiative administered by ONR under Grant N00014-00-1-0600. Any opinions, findings, and conclusions or recommendations expressed in this chapter are those of the authors and do not necessarily reflect the views of NSF, DoD, or ONR.

REFERENCES

Aha, D., and Kibler, D. (1991). Instance-based learning algorithms. *Machine Learning, 6,* 37–66.

André, E., Rist, T., and Müller, J. (1998). WebPersona: a life-like presentation agent for the World-Wide Web. *Knowledge-Based Systems, 11,* 25–36.

Atkinson, R. K., Mayer, R. E., and Merrill, M. M. (2004). Fostering social agency in multimedia learning: examining the impact of an animated agent's voice. *Contemporary Educational Psychology.* New York: Elsevier, Inc.

Batliner, A., Fischer, K., Huber, R., Spilker, J., and Noth, E. (2003). How to find trouble in communication. *Speech Communication, 40,* 117–143.

Bartlett, M. S., Hager, J. C., Ekman, P., and Sejnowski, T. J. (1999). Measuring facial expressions by computer image analysis. *Psychophysiology, 36,* 253–263.

Baylor, A. L., Shen, E., and Warren, D. (2004). Supporting learners with math anxiety: the impact of pedagogical agent emotional and motivational support. In *ITS 2004 Workshop Proceedings on Social and Emotional Intelligence in Learning Environments.* Maceio, Brazil: Springer-Verlag.

Bianchi-Berthouze, N., and Lisetti, C. L. (2002). Modeling multimodal expression of users affective subjective experience. *User Modeling and User-Adapted Interaction, 12*(1), 49–84.

Bosch, L. T. (2003). Emotions, speech, and the ASR framework. *Speech Communication 40*(1–2), 213–215.

Carberry, S., Lambert, L., and Schroeder, L. (2002). Toward recognizing and conveying an attitude of doubt via natural language. *Applied Artificial Intelligence 16*(7), 495–517.

Cohn, J. F., and Kanade, T. (2006). Use of automated facial image analysis for measurement of emotion expression. In J. A. Coan and J. B. Allen (Eds.), *The Handbook of Emotion Elicitation and Assessment. Oxford University Press Series in Affective Science.* New York: Oxford University Press.

Conati C. (2002). Probabilistic assessment of users emotions in educational games. *Journal of Applied Artificial Intelligence, 16*, 555–575.

Conrad, F. G., and Schober, M. F. (2000). Clarifying question meaning in a household telephone survey. *Public Opinion Quarterly, 64,* 1–28.

Conrad, F. G., Couper, M. P., Tourangeau, R., and Peytchev, A. (2005). Effectiveness of progress indicators in web survey: First impressions matter. *Proceedings of SIGCHI 2005: Human Factors in Computing Systems.* Portland, OR.

Conrad, F. G., Schober, M. F., and Coiner, T. (2007). Bringing features of human dialogue to web surveys. *Applied Cognitive Psychology, 21,* 165–188.

Conrad, F. G., Schober, M. F., and Dijkstra, W. (2007). Cues of communication difficulty in telephone interviews. To appear in J. M. Lepkowski, C. Tucker, M. Brick, E. de Leeuw, L. Japec, P. Lavrakas, M. Link, and R. Sangster (Eds.), *Advances in telephone survey methodology.* New York: Wiley.

Couper, M. P. (2000). Usability evaluation of computer assisted survey instruments. *Social Science Computer Review, 18*(4), 384–396.

Coulson, M. (2004). Attributing emotion to static body postures: recognition accuracy, confusions, and viewpoint dependence. *Journal of Nonverbal Behavior, 28,* 117–139.

Craig, S. D., Graesser, A. C., Sullins, J., and Gholson, B. (2004). Affect and learning: an exploratory look into the role of affect in learning. *Journal of Educational Media, 29,* 241–250.

Csikszentmihalyi, M. (1990). *Flow: The Psychology of Optimal Experience.* New York: Harper-Row.

D'Mello, S. K., Craig, S. D., Gholson, B., Franklin, S., Picard, R., and Graesser, A. C. (2005). Integrating affect sensors in an intelligent tutoring system. In *Affective Interactions: The Computer in the Affective Loop Workshop at 2005 International Conference on Intelligent User Interfaces* (pp. 7–13). New York: ACM Press.

D'Mello, S., and Graesser, A. C. (2006). Affect detection from human–computer dialogue with an intelligent tutoring system. In J. Gratch et al. (Eds.), *IVA 2006, LNAI 4133* (pp. 54–67). Berlin: Springer-Verlag.

D'Mello, S., and Graesser, A. C. (2007). Mind and body: dialogue and posture for affect detection in learning environments. In *Proceedings of the 13th International Conference on Artificial Intelligence in Education (AIED 2007).*

D'Mello, S. K., Craig, S. D., Sullins, J., and Graesser, A. C. (2006). Predicting affective states through an emote-aloud procedure from AutoTutor's mixed-initiative dialogue. *International Journal of Artificial Intelligence in Education, 16,* 3–28.

D'Mello, S. K., Chipman, P., and Graesser, A. C. (in press). Posture as a predictor of learner's affective engagement. *Proceedings of the 29th Annual Meeting of the Cognitive Science Society.* Nashville, TN.

D'Mello, S. K., Craig, S. D., Witherspoon, A. M., McDaniel, B. T., and Graesser, A. C. (in press). Automatic detection of learner's affect from conversational cues. *User Modeling and User Adapted Interaction.*

Ekman, P., and Friesen, W. V. (1969). Nonverbal leakage and clues to deception. *Psychiatry, 32,* 88–105.

Ekman, P., and Friesen, W. V. (1978). *The Facial Action Coding System: A Technique for the Measurement of Facial Movement.* Palo Alto, CA: Consulting Psychologists Press.

Ekman, P., Huang, T. S., Sejnowski, T. J., and Hager, J. C. (1992, July 30 to August 1). *Final Report to NSF of the Planning Workshop on Facial Expression Understanding,* Washington DC: NSF. http://face-and-emotion.com/dataface/nsfrept/overview.html.

Fasel, B., and Luttin, J. Recognition of asymmetric facial action unit activities and intensities. In *Proceedings of the International Conference on Pattern Recognition (ICPR 2000),* Barcelona, Spain.

Fernandez, R., and Picard R. W. (2004). Modeling driver's speech under stress. *Speech Communication, 40,* 145–159.

Gong, Li. (2002). Towards a theory of social intelligence for interface agents. Paper presented at Virtual Conversational Characters: Applications, Methods, and Research Challenges, Melbourne, Australia.

Graesser, A. C., Chipman, P., Haynes, B. C., and Olney, A. (2005). AutoTutor: an intelligent tutoring system with mixed-initiative dialogue. *IEEE Transactions in Education 48,* 612–618.

Graesser, A. C., Chipman, P., King, B., McDaniel, B., and D'Mello, S. (in review). *Emotions and Learning with AutoTutor.*

Graesser, A. C., Lu, S., Jackson, G. T., Mitchell, H., Ventura, M., Olney, A., and Louwerse, M. M. (2004). AutoTutor: a tutor with dialogue in natural language. *Behavioral Research Methods, Instruments, and Computers, 36,* 180–193.

Graesser, A. C., McDaniel, B., Chipman, P., Witherspoon, A., D'Mello, S. K., and Gholson, B. (2006). Detection of emotions during learning with AutoTutor. In R. Son (Ed.), *Proceedings of the 28th Annual Meetings of the Cognitive Science Society* (pp. 285–290). Mahwah, NJ: Lawrence Erlbaum.

Graesser, A. C., Person, N. K., Harter, D., and the Tutoring Research Group. (2001). Teaching tactics and dialogue in AutoTutor. *International Journal of Artificial Intelligence in Education, 12,* 257–279.

Johnson, L., Mayer, R., Andre, E., and Rehm, M. (2005). Cross-cultural evaluation of politeness in tactics for pedagogical agents. In *Proceedings of the 12th International Conference on Artificial Intelligence in Education.*

Kort, B., Reilly, R., and Picard, R. (2001). An affective model of interplay between emotions and learning: reengineering educational pedagogy—building a learning companion. In T. Okamoto, R. Hartley, Kinshuk, and J. P. Klus (Eds.), *Proceedings IEEE International Conference on Advanced Learning Technology: Issues, Achievements and Challenges* (pp. 43–48). Madison, WI: IEEE Computer Society.

Lester, J., Converse, S., Kahler, S., Barlow, T., Stone, B., and Bhogal, R. (1997). The persona effect: affective impact of animated pedagogical agents. In *Proceedings of CHI '97* (pp. 359–366). New York: ACM Press.

Litman, D. J., and Forbes-Riley, K. (2004). Predicting student emotions in computer–human tutoring dialogues. In *Proceedings of the 42nd Annual Meeting of the Association for Computational Linguistics* (pp. 352–359). East Stroudsburg, PA: Association for Computational Linguistics.

Marsella S., Johnson, W. L., and LaBore, K. (2000). Interactive pedagogical drama. In *Proceedings of the 4th International Conference on Autonomous Agents*.

McDaniel, B. T., D'Mello, S. K., King, B. G., Chipman, P., Tapp, K., and Graesser, A. C. (in press). Facial features for affective state detection in learning environments. *Proceedings of the 29th Annual Meeting of the Cognitive Science Soceity*. Nashville, TN.

Microsoft Corporation. (1998). *Microsoft Agent Software Development Kit with CD-ROM*. Redmond, WA: Microsoft Press.

Miller, C. (2004). Human computer etiquette: managing expectations with intentional agents. *Communications of the ACM, 47,* 31–34.

Mills, C. (1993). Personality, learning style, and cognitive style profiles of mathematically talented students. *European Journal for High Ability, 4,* 70–85.

Mishra P., and Hershey, K. (2004). Etiquette and the design of educational technology. *Communications of the ACM, 47,* 45–49.

Moreno, R., Mayer, R. E., Spires, H. A., and Lester, J. C. (2001). The case for social agency in computer-based teaching: Do students learn more deeply when they interact with animated pedagogical agents? *Cognition and Instruction, 19*, 177–213.

Mossholder, K. W., Settoon, R. P., Harris, S. G., and Armenakis, A. A. (1995). Measuring emotion in open-ended survey responses: an application of textual data analysis. *Journal of Management, 21*(2), 335–355.

Mota, S., and Picard, R. W. (2003). Automated posture analysis for detecting learner's interest level. In *Workshop on Computer Vision and Pattern Recognition for Human-Computer Interaction*. CVPR HCI.

Nakasone, A., Prendinger, H., and Ishizuka, M. (2005). Emotion recognition from electromyography and skin conductance. In *Proceedings of the Fifth International Workshop on Biosignal Interpretation* (pp. 219–222). Tokyo, Japan: IEEE.

Nass, C. (2004). Etiquette and equality: exhibitions and expectations of computer politeness. *Communications of the ACM, 47,* 35–37.

Oliver, N., Pentland, A., and Berand, F. (1997). LAFTER: a real-time lips and face tracker with facial expression recognition. In *Proceedings of the IEEE Conference on Computer Vision and Pattern Recognition* (pp. 123–129). San Juan, Puerto Rico: IEEE.

Pantic, M., and Rothkrantz, M. (2000). Expert system for automatic analysis of facial expression. *Image and Vision Computing, 18,* 881–905.

Pantic, M., and Rothkrantz, L. J. M. (2003). Towards an affect-sensitive multimodal human–computer interaction. *Proceedings of the IEEE, Special Issue on Multimodal Human-Computer Interaction, 91*(9), 1370–1390.

Person, N. K., Burke, D. R., and Graesser, A. C. (2003). RudeTutor: a face-threatening agent. In *Proceedings of the Society for Text and Discourse Thirteenth Annual Meeting*, Madrid, Spain.

Person, N. K., Graesser, A. C., and The Tutoring Research Group (2002). Human or computer: AutoTutor in a bystander Turing test. In S. A. Cerri, G. Gouarderes, and F. Paraguacu (Eds.), *Intelligent Tutoring Systems 2002 Proceedings* (pp. 821–830). Berlin: Springer-Verlag.

Person, N. K., Kreuz, R. J., Zwaan, R., and Graesser, A. C. (1994). Pragmatics and pedagogy: conversational rules may inhibit effective tutoring. *Cognition and Instruction, 2,* 161–188.

Person, N. K., Petschonek, S., Gardner, P. C., Bray, M. D., and Lancaster, W. (2005). Linguistic features of interviews about alcohol use in different conversational media. Presented at the 15th Annual Meeting of the Society for Text and Discourse. Amsterdam, The Netherlands.

Picard, R. W. (1997). *Affective Computing.* Cambridge, MA: MIT Press.

Prendinger, H., and Ishizuka, M. (2005). The empathic companion: a character-based interface that addresses users' affective states. *International Journal of Applied Artificial Intelligence, 19*(3,4), 267–285.

Rani, P., Sarkar, N., and Smith, C. A. (2003). An affect-sensitive human–robot cooperation: theory and experiments. In *Proceedings of the IEEE Conference on Robotics and Automation* (pp. 2382–2387). Taipei, Taiwan: IEEE.

Rayson, P. (2003). Wmatrix: a statistical method and software tool for linguistic analysis through corpus comparison. PhD thesis, Lancaster University.

Rayson, P. (2005) Wmatrix: a Web-based corpus processing environment. Retrieved from Lancaster University Computing Department Web site: http://www.comp.lancs.ac.uk/ucrel/wmatrix/.

Reeves, B., and Nass, C. (1996). *The Media Equation: How People Treat Computers, Television, and New Media Like Real People and Places.* New York: Cambridge University Press.

Schober, M. F., Conrad, F. G., and Fricker, S. S. (2004). Misunderstanding standardized language. *Applied Cognitive Psychology, 18,* 169–188.

Schouwstra, S., and Hoogstraten, J. (1995). Head position and spinal position as determinants of perceived emotional state. *Perceptual and Motor Skills, 81,* 673–674.

Shafran, I., and Mohri, M. (2005). A comparison of classifiers for detecting emotion from speech. In *Proceedings of the International Conference on Acoustics, Speech, and Signal Processing* (pp. 341–344). Philadelphia, PA: IEEE.

Tekscan (1997). *Tekscan Body Pressure Measurement System User's Manual.* South Boston, MA: Tekscan Inc.

Walker, M. A., Langkilde-Geary, I., Hastie, H. W., Wright, J., and Gorin, A. (2002). Automatically training a problematic dialogue predictor for a spoken dialogue system. *Journal of Artificial Intelligence Research, 16,* 293–319.

Wallbott, N. (1998). Bodily expression of emotion. *European Journal of Social Psychology, 28,* 879–896.

Whang, M. C., Lim, J. S., and W. Boucsein, W. (2003). Preparing computers for affective communication: a psychophysiological concept and preliminary results. *Human Factors, 45,* 623–634.

Zhang, L. (2003). Does the big five predict learning approaches? *Personality and Individual Differences, 34,* 1431–1446.

CHAPTER 11

Culture, Computer-Mediated Communication, and Survey Interviewing

Susan R. Fussell
Carnegie Mellon University, Pittsburgh, Pennsylvania

Qiping Zhang
Long Island University, Brookville, New York

Frederick G. Conrad
University of Michigan, Ann Arbor, Michigan

Michael F. Schober
New School for Social Research, New York, New York

Leslie D. Setlock
Carnegie Mellon University, Pittsburgh, Pennsylvania

11.1   INTRODUCTION

As survey designers test and implement new interviewing technologies, the grow-ing body of evidence on cultural differences in computer-mediated communication (CMC) is becoming increasingly relevant. People from different cultures can differ in patterns and styles of communication and interpretation, and this can affect how they interact with new technologies for communicating with human partners and with computer systems. This is likely to be the case for current and future interviewing systems. Although there is as yet little direct evidence from studies of survey inter-views, studies of CMC help lay the groundwork for understanding and predicting effects of culture in technologically mediated survey interviews.

Consider the sample dialogue in Table 11.1. These conversations come from pairs of American and Chinese students negotiating a jointly agreed upon order of priority for items in the Arctic Survival Task (Setlock et al., 2004; Stewart et al., in press) either face-to-face or over instant messaging (IM). In the selected excerpts, the pairs are trying to agree on the most important item in the set. When American pairs do this task they discuss each item in a cursory manner (6–7 speaking turns) regardless of communication medium, and they are quick to acquiesce to their partners' suggested rankings (see the last turn of each utterance). When Chinese pairs do this task face-to-face, they discuss each item in depth, asking each other questions and working through the survival scenario (e.g., "… the most important thing we need to fight is the coldness. Right?"). Discussion of a single item can take many speaking turns, 42 in this excerpt. The most striking aspect of these dialogues is the way in which the Chinese pairs' conversation shifts when they talk over IM: These conversations are similar to those of the American pairs in terms of brevity and acquiescence, and quite unlike the lengthy discussions of the Chinese pairs in a face-to-face setting.

CMC studies such as the one that produced these dialogues suggest design considerations—though not yet prescriptions—for designers of new interviewing systems. Designers of future interviewing systems will be able to choose system features that are not available to designers of current surveys and these choices may differently affect the behavior of respondents from different cultures. For example, a designer might choose to display an interviewer's or interviewing agent's facial cues in the user interface and these cues may affect respondents from context-dependent cultures differently than respondents from context-independent cultures (see later discussion). Similarly, the dialect of the interviewer's or interviewing agent's voice seems likely to affect respondents who are speakers of that dialect differently than re-spondents from other linguistic communities. Such differences can potentially affect people's willingness to participate in the interview, to provide thoughtful answers, to provide honest answers to sensitive questions, and the likelihood that they will complete the interview. But these effects will only be evident if the medium commu-nicates interviewer dialect. Such differences would not be evident in an IM interview, for example. In a globalizing world with increasing migration, survey interviews are increasingly intercultural (with interviewers and respondents coming from different cultural backgrounds), which heightens the need for understanding these issues.

A growing body of literature about the impact of culture on survey data now ex-ists. A substantial component of this work addresses the practical issues of conduct-ing cross-cultural survey research such as translating questionnaires (e.g., Harkness, 2003). Closer to the concerns of the current chapter, other work addresses cultural sources of measurement error, that is, the discrepancy between what a respondent reports and the true value of the answer. For example, Johnson and van de Vijver (2003) report that survey respondents from *collectivist* societies, that is, societies in which people prioritize the benefit of the larger group over their own benefit, are more likely to give socially desirable answers than their counterparts from *individualistic* societies, that is, societies in which these priorities are reversed, presumably because there is greater pressure to conform to social norms in the former than the latter type of society. Even closer to the topic of the current chapter, there is some evidence that

TABLE 11.1   Speech Between American and Chinese Dyads in Arctic Survival Task Carried Out Face-to-Face (Row 1) and Through Instant Messaging (Row 2)

Medium	American Dyad	Chinese Dyad
Face-to-Face	A: Ok um what do you have for number one? B: Um I thought that personally, I thought the most important thing to have was, I'm checking to make sure, the matches? A: Yeah. B: So- A: Yeah those are pretty important. A: I put that for number two, but it was interchangeable with number one.	B: What do you feel most important? A: So I- I choose the first. Number one. The gallon can of the maple syrup. B: Why? Eh that's food right? That's xxx A: That's food, but that's can keep your body warm. B: Oh but I feel that the wood match would be most important because you need the fires, you need- A: That's right, but you can not take that, you can not take like a burning wood with you when you are walking. A: It's hard. A: When you sit down you can take a raft, you can use that to burn some wood and then you- you you become warm. A: But when you are walking B: But when you are walking you- you are walking so you are warm I feel like. A: mhm. A: So- so the let's first to make sure that the most important thing we need to fight is the coldness. Right? [continues another 30 turns]
IM	A: I put the water tablets first B: right B: next is tough A: yeah B: I put the ax because I was thinking B: cut wood B: then matches to start a fire A: that's fine, my next three are pretty interchangeable and they include those two.	A: What did you have as the most important? B: Hand ax, then matches A: I had matches then hand ax B: and then? [discussion of lower ranked items] A: would you agree with ax, matches syrup as the first three? (but not in order) B: yes A: ok now for the order A: I think you had axe, matches, syrup A: and I had matches, axe, syrup A: either way works for me

cultural differences in socially desirable reporting are moderated by the mode of survey administration. Acquilino (1994) found that the mode effect on reports of using drugs and alcohol (i.e., more use reported in self-administered questionnaires than in face-to-face interviews) was larger for African-American and Hispanic respondents than for whites. Johnson and van de Vijver suggest that this may be related to greater privacy concerns among members of vulnerable minority groups when they are asked to report socially undesirable behaviors than among members of the predominant cultural group.

These findings suggest that survey responses *might* be differently affected by mode across cultural groups, but the evidence is just suggestive. The definitive controlled studies have not been done. Moreover, there are no studies to our knowledge that investigate whether culture interacts with mode for cutting edge survey modes like video, text chatting, speech dialogue systems, and Web questionnaires with embedded animated agents. Nonetheless, we can derive predictions from studies of CMC and culture about how culture might affect survey interviews across different media. For example, much of the interaction that has been observed in survey interviews involves "paradigmatic" sequences (see Schaeffer and Maynard, Chapter 2 in this volume): that is, the interviewer asks the question, the respondent gives a problem-free answer, and the interviewer acknowledges this answer, sometimes by simply asking the next question. But the example exchanges in Table 11.1 suggest that belief among survey methodologists that paradigmatic sequences are the norm (Ongena, 2005) may be more culture specific than we have realized. Perhaps in Chinese or more generally East Asian interviews, it is typical for the face-to-face interaction between interviewers and respondents to involve more turns and more checking that the parties understand each other (grounding) than in American (or Western) interviews. If so, this seems likely to promote more accurate understanding and, as result, accurate responding in the East Asian than Western interviews as grounding has been shown to affect response accuracy (e.g., Conrad and Schober, 2000; Schober and Conrad, 2002; Schober et al., 2004). However, the example also suggests that when interviews are conducted through a medium like IM, these cultural differences would go away and that both East Asian and Western interviews would be brief and more likely to follow the paradigmatic pattern. The reduction in grounding this would imply for Asian respondents could signal reduced comprehension accuracy but at levels similar to the Western counterparts. While this is just our best guess about what might happen in the interview domain, this is the kind of connection we will attempt to establish in this chapter: we will consider the implications of the results from CMC and culture studies for survey interviews across different media with respondents and interviewers from different cultures.

We encourage the reader to keep in mind survey interviewing through computers is similar to and different from the kinds of communication in CMC studies. When a respondent completes a Web-based questionnaire, he/she communicates with the survey researchers through a computer, but this "conversation" does occur in real time and may never occur on an individual basis as it does in the collaborative tasks that characterize CMC research. When an interviewer is part of the data collection, the conversation is more individualized—the interviewer asks questions and records the respondent's answers; however, when the interviewer enters the responses into a computer, it is the interviewer, not the computer, that is the intermediary (Clark and Schober, 1992). Despite these differences, we believe the interview is similar enough to most CMC tasks so that what is known about CMC and culture can at least stimulate thinking about the role of culture in the use of future interviewing technologies and the quality of the data they are used to collect.

One more caveat before we begin our discussion of CMC research. Culture is obviously a complex and nuanced construct (e.g., Miller, 2002). At least in the early stages of research on culture and communication, culture is operationalized with broad brush strokes (e.g., collectivist versus individualistic societies) that may feel overly simple to many readers. This is in part a result of conducting relatively small-scale laboratory studies in which it simply isn't possible to enlist enough participants to span the range of cultural diversity that may be necessary to do justice to some distinctions. However, even rather broad distinctions seem to have some measurable effects on the way participants communicate through different media so, at least as a starting point, studies of culture at this level seem appropriate. Indeed, the early work on culture and survey responding mentioned earlier has proceeded with similarly broad distinctions. So despite the relatively high level at which culture is characterized in current CMC research, the effects observed in that literature may well transfer to communication through computational media in survey data collection tasks.

11.2 INTRODUCTION TO CMC AND CULTURE

A number of well-developed theories, based on evidence from Western participants, can be used to generate predictions about which media might work best for a given set of people performing a given set of tasks (e.g., Clark and Brennan, 1991; Daft and Lengel, 1984; Postmes, et al., 2002; Short et al., 1976; Walther, 1992, 1995), and a number of investigators have begun to examine cultural effects on CMC (e.g., Anderson and Hiltz, 2001; Kayan et al., 2006; Reinig and Mejias, 2003, 2004; Setlock et al., 2004, 2007; Zhang et al., 2006) in a variety of technologies and cultures, using a variety of research methods. The results to date suggest that people's use of CMC tools is influenced by their cultural background.

Adding issues of culture into the CMC mix complicates matters in interesting and important ways. Cultures vary along a number of dimensions that may affect group processes and outcomes, such as *individualism versus collectivism* (e.g., Hofstede, 1983; Triandis, 1995), *low versus high context of communication* [how much contextual information is required for communication (Hall, 1976)], and *task versus relationship orientation* [whether people focus on getting work done or on establishing rapport with their partners (e.g., Triandis, 1995)]. These and other cultural dimensions may interact with features of media, such as the availability of visual cues, to create different effects on interaction and data quality in interviewer- and self-administered interviews.

We first present a conceptual framework to investigate how culture and CMC shape communication processes and task outcomes in general. Then we review research on each component of this framework, highlighting findings that we believe have

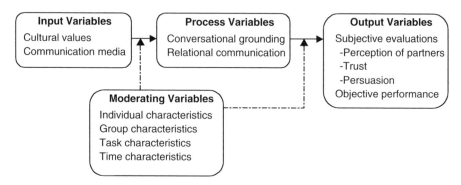

FIGURE 11.1   Basic Input–Process–Output framework.

broad implications for adopting new technologies in surveys. Note, however, that the findings thus far are rarely from studies carried out in interviewing settings but come from other arenas; our primary contribution is in raising important questions that deserve future research in the survey context.

11.3   BACKGROUND

The theoretical framework we use to examine relationships between culture and CMC is an Input–Process–Output (I-P-O) model (Hackman, 1987), shown in Fig. 11.1.

Here, culture and media are inputs that people bring to collaboration. These inputs, both alone and in interaction, influence communication processes and, in turn, subjective and objective outcomes. There are also a number of moderating variables that may influence relationships between inputs and processes and between processes and outcomes. Although the I-P-O framework is a simplification, it can help us conceptualize how culture and CMC interact by explicating relationships between input, process, outputs, and moderating variables. In the survey interview we conceive of the I-P-O sequence at the level of individual question–answer exchanges, where the input includes the question as well as culture and particular media; the process is the dialogue between respondent and interviewer (or interviewing system); and the outcome is the response, which in turn becomes one of the inputs to the process of answering the next question.

11.4   DIMENSIONS OF CULTURAL VARIABILITY

In cross-cultural research (both on intracultural groups and on intercultural interaction), there has been substantial debate about the definition of culture, as well as about the number, size, and significance of dimensions along which cultures vary (e.g., Hofstede, 1983; Oyserman et al., 2002; Schwartz, 1992; Triandis, 1995). For the purposes of this chapter, we define culture as *a set of norms, roles, and values emphasized by a culture and adopted, to greater or lesser degrees, by members of that culture through such processes as imitation and teaching.* We focus on three cultural dimensions—individualism/collectivism, high versus low context of communication, and task versus relationship focus—that affect processes central to collaborative work (of which interviewing is a special case). These dimensions are not intended as an exhaustive description of how cultures differ but rather as a way of focusing our investigation on those dimensions most likely to influence what happens in new interviewing technologies.

11.4.1 Individualism Collectivism

Virtually all dimensional culture theories distinguish between individualistic cultures, in which people tend to identify themselves as individuals and focus on their own personal gain, and collectivistic cultures, in which people identify themselves as a member of a collective and focus on the betterment of that collective (e.g., Hofstede, 2001; Triandis, 1995). Nisbett (2003) describes a wide range of cognitive processes affected by membership in individualistic versus collectivistic cultures, including reasoning styles and memory processes. Hofstede's (2001) analyses of survey responses from a global sample of IBM employees show how individualism/collectivism is associated with preferences for business practices, child-raising, and many other aspects of culture. Markus and Kitayama (1991) show that individualism/collectivism is associated with people's concept of themselves as either independent of or interdependent with other individuals.

11.4.2 High Versus Low Context of Communication

Hall (1976) proposed that cultures vary in how much contextual information is required for communication. Low context, typically Western communication is verbally explicit, to the point, with relatively little attempt to mask one's feelings. In contrast, high context, typically Eastern communication is indirect, often ambiguous, and sensitive to the context in which it occurs (e.g., the relationship between speaker and addressee, nuances of facial expressions or tone of voice). Much of the research on communication styles has used a self-report methodology, in which people respond to questions such as "I catch on to what others mean even when they do not say it directly" or "My speech tends to be very picturesque" (Gudykunst and Ting-Toomey, 1988; Gudykunst et al., 1996). These studies typically show that while people in all cultures use both styles, low context communication is preferred in individualistic societies and high context communication is preferred in collectivistic societies (Gudykunst and Ting-Toomey, 1988; Gudykunst et al., 1996). As we will discuss further later, cultural differences in use of indirectness in speech have been further supported through analyses of actual conversations in face-to-face and mediated settings.

11.4.3 Task Versus Relationship Focus

A third dimension of cultural variation pertinent to new forms of interviewing is task versus relationship orientation (Triandis, 1995). Task-oriented cultures such as the United States, Canada, and Australia focus on getting work done, whereas relationship-oriented cultures such as Japan, Korea, and China focus on establishing rapport with one's partners. The task versus relationship focus is only quasi-independent of the other dimensions: cultures identified by Triandis as task-oriented overlap substantially with those categorized as individualistic (Hofstede, 2001) and those described as low context communicators (Hall, 1976; Gudykunst and Ting-Toomey, 1988). Similarly, cultures identified by Triandis as relationship-oriented overlap with those identified by other researchers as collectivistic and high context communicators.

Dimensions like individualism collectivism are often applied at the national level (Hofstede, 1983, 2001; Triandis, 1989), but it has become increasingly obvious that to understand links between culture and communication, it is necessary to examine how national values are related to individuals' personal values (Schwartz, 1992) and their construals of themselves as interdependent versus independent (Markus and Kitayama, 1991; Singelis and Brown 1995). Both individual values and individuals' self-concepts are influenced by national culture but not entirely determined by it. Gudykunst et al. (1996) and Oetzel (1998), among others, have shown that national values, individual values, and self-concepts each have an impact on self-reported communication behavior. Generally, these studies have looked at the impact of these variables in the abstract, independent of any given communicative domain. Thus, the question of how the results would pertain to the interviewing context remains open for investigation.

11.5 AFFORDANCES OF MEDIA

To build a theoretical model of how culture interacts with features of communication media, it is essential to characterize media at the right level of analysis. A number of theories distinguish media along a single dimension such as media richness (Daft and Lengel, 1984). For our purposes, single-dimension theories do not differentiate clearly enough among media. Instead, we draw on Clark and Brennan's (1991) influential theory of *media affordances,* which provides a finer-grained analysis of the resources media provide for communication (see Table 11.2). For example, telephone calls and video conferencing provide audibility, and thus afford the use of speech, whereas IM does not. In this framework, communication over different media will entail different costs for producing messages, receiving and understanding messages, changing speakers, and repairing misunderstandings.

Newer modes for survey interviewing will certainly require extensions of the above framework. For example, in Web-based surveys and audio-CASI, no conversational partner is explicitly present during the interaction but the dynamics of the medium may enable the survey researchers to impose a sense of presence. This kind of effect has been demonstrated by Bradner and Mark (2001), who found social presence effects were as strong in a computer-based math task when people used an application-sharing tool as when they were observed via two-way video (even though they could not be

TABLE 11.2   Some Affordances of Communication Media and Their Typical Presence (Y), Partial Presence (P), or Absence (N) in Face-to-Face (FTF) Communication, Video Conferencing, Telephone, and Instant Messaging (IM)

Affordance	Definition	FTF	Video	Phone	IM
Audibility	Participants hear other people and sounds in the environment.	Y	Y	Y	N
Visibility	Participants see other people and objects in the environment.	Y	Y	N	N
Co-presence	Participants are mutually aware that they share a physical environment.	Y	P	N	N
Cotemporality	Participants are present at the same time.	Y	Y	Y	Y

Source: Adapted from Clark and Brennan (1991).

sure anyone was really watching). In these cases there might be a grey "Y" in the co-temporality row of the table.

11.6   APPLYING THE FRAMEWORK TO INTERVIEWING

Starting from Clark and Brennan's model, dimensions of cultural variability may alter the perceived importance of affordances such as audibility, visibility, and co-presence. For example, Gudykunst and Kim (1997) suggest that nonverbal cues may be more important for communication in high context cultures because the meaning of messages resides in the situational context, not in the words themselves. Thus, we might anticipate that visibility will be more important for successful communication among members of high context cultures than for members of low context cultures. If this is the case, then one should expect that new interviewing technologies that afford visibility of partners—the respondent being able to see the interviewer, the interviewer being able to see the respondent, during question asking, during answers— should matter differently for members of different cultural groups. In fact, as we will see later, there are a few pieces of evidence from current interviewing modes that are consistent with this idea. Let us first examine how culture and media have been shown to influence conversational processes and team outcomes in noninterviewing arenas.

11.7   COMMUNICATION PROCESSES

Culture may, alone or in interaction with features of media, influence group processes, particularly processes of communication. Here, we focus on two aspects of mediated communication that we view as essential for successful understanding in interviews— conversational grounding and relational communication—and review prior work on the ways in which culture and media affect these two communication processes.

11.7.1   Conversational Grounding: The Basis for Question Comprehension

Questions in survey interviews consist of words that respondents need to interpret, and the cognitive and interactive resources that respondents use are those that they use to understand what their partners say more generally (Schober, 1999). *Conversational grounding* refers to the interactive process by which communicators exchange evidence in order to reach mutual understanding (Clark and Brennan, 1991; Clark and Schober, 1992; Clark and Wilkes-Gibbs, 1986). Speakers and listeners work together by asking questions, providing clarifications, and other procedures to ensure that messages are understood as intended. Grounding is easier, and conversation more efficient, when collaborators share *common ground*—mutual knowledge, beliefs, and so on (Clark and Marshall, 1981). This common ground can arise from co-membership in social groups (e.g., Fussell and Krauss, 1992; Isaacs and Clark, 1987), through the process of exchanging messages (*linguistic copresence*), or by sharing a physical setting (*physical copresence*).

In Clark and Brennan's (1991) framework, affordances of media influence the strategies people use to ground their utterances. For example, face-to-face settings afford visibility and physical co-presence, so speakers can use gestures to refer efficiently to task objects (e.g., Bekker et al., 1995; Clark and Krych, 2004). In media that lack visibility and physical co-presence, speakers must use lengthier verbal descriptions of the same objects (e.g., Doherty-Sneddon et al., 1997; Kraut et al., 2003). A substantial body of research supports the conjecture that features of media influence grounding. For example, conversation is more efficient when technology provides a shared view of the workspace (e.g., Gergle et al., 2004; Kraut et al., 2003) and when tools allow people to gesture in that workspace (Fussell et al., 2004; Kirk and Stanton-Fraser, 2006).

How these findings apply to interviews is an important question, as in interviews it is rare that the questions refer to what is in the immediate physical environment or shared workspace. Nonetheless, the evidence from survey interviews thus far is that the ability to ground interpretation of question concepts in telephone interviews can indeed lead to more accurate question interpretation (Conrad & Schober, 2000; Schober and Conrad, 1997; Schober et al., 2004); when interpretation can be clarified the interaction is less efficient (takes longer) but can lead to better answers and thus better data quality. The evidence is also that the effects of being able to ground can extend to both text-based self-administered interviewing (Conrad et al., 2007) and to speech-based self-administered interviews (Ehlen, Schober, and Conrad, in press). Visibility seems at least indirectly related to grounding in survey interviews. In a comparison of face-to-face and telephone interviews (Conrad et al., 2007) respondents provided spoken cues of comprehension difficulty (*uhs* and *ums*) more on the phone than face-to-face, presumably to compensate for the absence of visual cues of uncertainty (e.g., facial evidence that the respondent is confused like a furrowed brow or looking away from the interviewer while answering). So how culture and media interact to affect the ability to ground understanding is particularly relevant when considering adopting new interviewing technologies.

And there is indeed some evidence that cultures can vary in their strategies for grounding meaning in conversation (Li, 1999a,b). Hall (1976) proposed that audibility and visibility may be more important for grounding in high context cultures than in low context cultures, because awareness of how others are reacting to one's messages is an important aspect of high context communication. This notion is supported indirectly by Veinott and Colleagues (1999), who examined how well pairs could perform a map-based task in which one person gave directions and the other had to draw the identical route on his/her own map. Veinott and colleagues found that nonnative English speakers, many of whom were Asian, benefited from video over audio conferencing, whereas native English speakers did not. They infer that the richer cues to mutual understanding provided by visibility (e.g., quizzical looks, raised eyebrows) were especially valuable for nonnative speakers. However, this study confounded native language with intercultural communication, so we don't know which factor accounts for the results.

In one of our own studies (Setlock et al., 2004), we compared American, Chinese, and mixed American-Chinese dyads performing scenario-based negotiation tasks face-to-face or via IM. The goal of these tasks is to rank salvaged items from a crash in order of importance. Pairs first rank the items individually, and then negotiate until they come to agreement on a joint ranking. We hypothesized that the lack of visual cues in IM would make it poorly suited for communication among members of high context cultures but not affect communication among members of low context cultures. Consistent with this hypothesis, we found no difference between media in terms of how much grounding American pairs required to complete the task, but a large impact of medium for Chinese pairs who spoke much more face-to-face (see the example interactions in Table 11.1). This culture by medium interaction is displayed in Fig. 11.2.

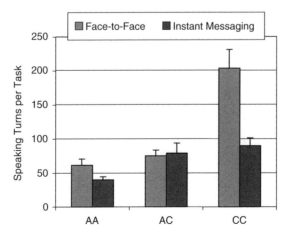

FIGURE 11.2 Mean speaking turns per task by culture group and medium (AA = American only, AC = mixed American-Chinese, CC = Chinese only).

As we showed in Table 11.2, face-to-face interaction has many affordances not present in IM, including audibility and visibility. To assess which of these two affordances was more important, Setlock et al. (2007) compared the same cultural groups interacting via audio or video conferencing. No main effect of culture, nor a culture by medium interaction, was found. Thus, the increased speech in face-to-face interactions for Chinese dyads seems to result from the presence of auditory information but is not further benefitted by adding visual cues via video conferencing. These results conflict with those of Veinott et al. (1999) and suggest the need for a more detailed examination of factors that differed across the two studies (e.g., tasks, specific cultural backgrounds of participants). Possibly the most relevant difference between the two studies for our current purposes is whether a physical artifact was part of the task (i.e., a map). In this respect the verbally based negotiation task used by Setlock and her colleagues is more similar to a typical interviewing situation where the questions refer to events and attitudes that are not present or visible. Thus, it would seem that telephone interviews should increase conversational grounding for Chinese respondents relative to IM interviews (or interviews in any mode with Westerners) but that face-to-face interviews are unlikely to lead to more grounding.

11.7.2   Relational Communication and Rapport

In addition to the cognitive aspects of survey interviews, socioemotional aspects also matter, affecting not only respondents' willingness to participate, but their motivation to provide thoughtful and accurate answers, to answer sensitive questions, and to complete interviews. These aspects of interviewing are related to the larger literature on relational aspects of communication, which are concerned not with what information is conveyed but with *how* that information is conveyed and what this indicates about the relationship between speaker and addressee(s). Much research on relational communication has focused on nonverbal cues such as eye gaze (e.g., Argyle and Cook, 1976), facial expressions (e.g., Ekman, 1982), and posture (e.g., Mehrabian, 1967), which can be used to indicate intimacy, trust, and attraction. In addition, messages themselves can be formulated in different ways to establish, maintain, and/or build closeness with a partner. In Table 11.3, we list some of the ways that nonverbal, paralinguistic, and verbal cues can add socio-affective meaning to people's messages.

A key aspect of relational communication is *face maintenance,* or ensuring that one does not cause another person to lose respect (Goffman, 1967). *Linguistic politeness*

TABLE 11.3   Basic Categories of Relational Communication

Category	Examples
Nonverbal behavior	Eye gaze, posture, facial expression
Paralinguistic behavior	Intonation patterns, speech rate, loudness
Verbal behavior	Form of address (e.g., *John, Mr. Jones*), pronouns (e.g., *I, we*), hedges (e.g., *sort of*), intensifiers (e.g., *very very*), indirect requests (e.g., *would you mind . . .*), swear words

refers to a range of strategies by which people demonstrate concern for their own and others' faces (Brown and Levinson, 1987). For example, indirect requests such as, "Could you close the door?" are more polite than directives such as, "Close the door!" (Holtgraves, 1997). Similarly, hedging an opinion (e.g., "I think you might be wrong") is more polite than directly stating that opinion (e.g., "You are wrong"). Misunderstandings can arise when one partner places less emphasis on relational concerns than another partner.

Features of media have been shown to affect relational communication (e.g., Herring, 1994; Kiesler et al., 1988; see Whittaker, 2003 for a review). In some cases, greater negative emotion and "flaming" has been found in text communication (e.g., Kiesler et al., 1988), a finding that was attributed by Kiesler and colleagues to the lack of social context cues in text communication but which others have attributed to the difficulty of producing politeness markers in typed discourse (e.g., Brennan and Ohaeri, 1999). Other studies have found more relational communication in face-to-face settings than in text communication (e.g., Hiltz and Turoff, 1978). Early studies also suggest that relational aspects of communication are reduced when conversations take place over the phone versus face-to-face (Rutter, 1987; Stephenson et al., 1976). Less research has compared relational communication in audio versus video conferencing, although many media theories (e.g., Daft and Lengel, 1984; Short et al., 1976) suggest that video will better support relational communication.

The cultural theories outlined earlier suggest that high context, relationship-oriented cultures place more emphasis on relational communication than do low context, task-oriented cultures (Ting-Toomey et al., 1991). This hypothesis has been supported in studies of conversational indirectness, which show that high context cultures such as China and Korea use more indirectness than low context ones such as the United States (Ambady et al., 1996; Holtgraves, 1997). Chinese speakers are also more likely to use "we" pronouns and social language than American speakers, both face-to-face and via IM (Setlock et al., 2004). Further support comes from cross-cultural research on negotiation, which has shown that *relational* strategies (e.g., compromising) are favored by high context negotiators whereas *informational* strategies (e.g., dominating the conversation) are favored by low context cultures (e.g., Adair and Brett, 2005; Adair et al., 2001; Ting-Toomey et al., 1991). Such differences have been attributed to cultural variation in concern for one's own face versus the other person's face (Ting-Toomey, 1988).

Although it has not to our knowledge been directly tested, we anticipate interactions between the cultural background of communicators and the affordances of a medium on the amount and valence of relational communication. High context communicators may especially rely on facial expressions and tone of voice when producing and interpreting relational cues, whereas low context communicators may find verbal substitutes such as hedges and indirect requests to be adequate substitutes. This leads to a set of testable predictions for new interviewing technologies: high context respondents should be likely to build rapport with interviewers using interviewing media that support visibility and audibility, while low context respondents' motivation and satisfaction should be less affected by visual and auditory affordances. For example, we might expect reduced benefit or even a cost from A-CASI for participants from high cultures versus low context cultures. Respondents from low context (i.e., Western) cultures seem to feel more private answering questions posed via A-CASI than by face-to-face interviewers, based on the increase in their reports of sensitive behaviors in the former versus the latter mode (e.g., Tourangeau and Smith, 1996; Turner et al., 1995). However, for high context cultures, the fact that a human voice is displayed under A-CASI may overwhelm the sense of privacy and lead to more socially desirable (i.e., less candid) responding. Similarly, an emotive animated agent may increase rapport with low context cultures, but it may lead to an overinterpretation of affective cues by high context cultures: for example, if the interviewing agent's smiles are poorly timed or inappropriate this might distract high context respondents or lead them to feel they are not performing adequately (see Person, D'Mello, and Olney, Chapter 10 in this volume, for a discussion of affective agents and rapport). Furthermore, one can imagine that without some way to ground interviewer affect in a textual medium such as IM (e.g., the use of "emoticons" like ☺), high context respondents may ascribe affect to the interviewer that is not warranted, much like leaving the interpretation of concepts undefined in standardized interviews leads to more variation in how the terms are interpreted, including unintended meanings (see Schober and Conrad, 2002; Suessbrick et al., 2005).

11.8   OUTCOME MEASURES

In our I-P-O model in Fig. 11.1, inputs (culture and features of technology) impact communication processes, which in turn impact a range of outcome measures. Here, we touch briefly on several outcome measures that are especially important in intercultural teamwork and that are likely to affect the quality of survey data: subjective evaluations of one's partners, persuasion, trust, and objective performance measures. This set of outcome measures, common in CMC research, maps only partially to the outcomes of interest in survey research. For example, in CMC studies, subjective outcomes like partner perception and trust are typically measured at the end of the conversation, under the assumption that such outcomes will have bearing on future interactions between the same individuals. In survey research, future interactions between the surveyor and respondent may be quite unlikely. Instead, we might conceptualize person perception and trust as outcome measures at the end of each question–answer pairing, outcomes that build up over the course of the survey interview. In addition, CMC studies have rarely if ever considered whether the answers people provide are valid or reliable, so additional research will be needed to understand how the inputs and processes in our I-P-O model affect such outcomes.

11.8.1   Perception of Partners

Features of media can affect collaborators' impressions of each other. For example, early studies found greater liking for partners when using video versus audio alone (Short et al., 1976; Williams, 1977). Hancock and Dunham (2001) suggest that the lack of social cues in text CMC creates ambiguity that affects one's impressions of a partner. When less is known about a remote collaborator's immediate experiences, problems such as delays and awkward expressions are more likely to be attributed to internal, dispositional factors (e.g., rudeness) rather than external causes (e.g., network delays) (Cramton, 2001). Such effects may be especially strong in intercultural interaction, in which people share less initial background knowledge. Consistent with this, Setlock et al. (2004, 2007) found that members of intercultural dyads rated each other more negatively than homogeneous American or Chinese dyads. However, the tendency to attribute behavior to dispositional factors is in part culturally specific: East Asians are more likely to consider situational explanations for behaviors than Westerners (Choi et al., 1999; Morris and Peng, 1994). We anticipate that media that reduce or eliminate visual and auditory cues would have a greater impact on partner perceptions in high context than low context cultures. Thus, as social cues are reduced across interviewing modes (e.g., face-to-face => telephone => IM => textual Web-based questionnaires), respondents from high context cultures may increasingly (mis)attribute lack of interest or disapproval to the interview (or interviewing system). The danger is this kind of perception could lead high context respondents to terminate the interview, whereas low context respondents may experience none of this.

11.8.2 Persuasion

The relatively high response rates in face-to-face interviews (versus telephone interviewers versus Web survey participation) are often attributed to the interviewer's persuasive abilities, which are more effectively applied when the interviewer is physically present and the respondents cannot make the interviewer disappear by hanging up the phone or ignoring an e-mail invitation. How might this differ cross-culturally, especially with new and emerging interview technologies? The relevant studies have not been conducted but the CMC literature is again instructive.

Persuasion in CMC refers to the extent to which one team member can convince others that his/her viewpoint is correct. Early studies indicated that persuasion varied as a function of medium (e.g., Chaiken and Eagly, 1976; Guadagno and Cialdini, 2002; Heim et al., 2002; Morley and Stephenson, 1977), but many of these studies used artificial role-playing paradigms in which grounding and relational communication couldn't be measured. Other studies have compared consensus using text-based group decision support systems (GDSS) and found less consensus after GDSS than face-to-face negotiations (Reinig and Mejias, 2003; Watson et al., 1994). There is little consensus as to whether cultural differences in negotiation styles influence persuasion, either alone or in interaction with features of technology. Some studies (e.g., Reinig and Mejias, 2003; Watson et al., 1994) have found main effects of culture and medium on persuasion but no interaction between the two. Others (e.g., Setlock et al., 2004) have found main effects of culture but no effects of medium. Adair et al. (2001) suggest that persuasion may be reduced when there are mismatches in negotiators' styles, although they did not investigate interactions with medium. Anderson and Hiltz (2001) found that in face-to-face communication culturally heterogeneous groups showed the most consensus and in asynchronous communication culturally homogeneous groups showed the least consensus after group discussion. Taken together, these results lead to the following hypothesis: interviewers might be more effective in persuading potential respondents to participate if the respondent can see the interviewer, as in a video introduction (see Fuchs, Chapter 4 in this volume, for a discussion of video introductions in mobile Web surveys) and that this might be particularly effective for high context groups.

11.8.3   Trust

Trust is an individual's confidence in the goodwill of others and the expectation that others will reciprocate if one cooperates (e.g., Ring and Van de Velden, 1994). McAllister (1995) differentiates two broad foundations for trust in organizational settings: cognitive and affective. Cognitive trust is built on people's intelligence, competence, and reliability, whereas affective trust is built on people's emotional bond and relationship. Several studies suggest that establishing trust is more difficult in remote collaborations than in face-to-face, and more difficult with leaner text media than with richer media like audio or video conferencing (Bos et al., 2002), although having initial face-to-face interaction before working at a distance seems to help (Jensen et al., 2000; Rocco, 1998). We would expect affective trust to be weighted more heavily in high context, relationship-oriented cultures and cognitive trust to be weighted more heavily in low context, task-oriented cultures. For high context cultures, establishing affective trust in leaner media may be especially difficult. Consistent with this, Zhang, Olson, and Olson (2004) found that Chinese pairs showed higher affective trust when negotiating by video conferencing than by audio conferencing, whereas American pairs showed no differences on either affective or cognitive trust in both media.

In the domain of survey interviewing, trustworthiness of an interviewer can strongly affect respondents' willingness to report sensitive or less socially desirable behaviors, and their tendency to overreport socially desirable behaviors. There is a small body of evidence suggesting that culture and (existing) media interact on this score (e.g., Acquilino, 1994; Johnson & van de Vijver, 2003). One can extend this to emerging technologies such as animated interviewing agents. Imagine an agent that communicates warmth (e.g., it smiles and is polite) but lacks conversational ability (e.g., cannot clarify the questions it asks) and imagine an agent with the opposite characteristics (i.e., lacks warmth but possesses the ability to clarify question meaning). The CMC evidence on trust and culture just discussed would lead us to expect that high context respondents would trust the first interviewing agent more than the second, answering sensitive questions asked by the former with greater candor than the same questions asked by the latter. In contrast, respondents from low context cultures might be more likely to disclose sensitive facts to the second agent.

11.8.4   Objective Performance

The I-P-O framework ultimately concerns task performance. In surveys, there are various straightforward indicators of objective performance, which align with the CMC research to varying degrees. Response accuracy and reliability are the most obvious indicators. The extent to which answers are influenced by question wording, question ordering, or response options are another possible indicator, with the assumption that less influenced answers are more likely to be accurate and reliable. There are various pieces of evidence on how media affordances affect people's performance on joint tasks in noninterviewing domains though, again, relatively little in surveys. We turn to the CMC results first.

Unsurprisingly, the number of utterances spoken during a task is often significantly correlated with task completion time (Gergle et al., 2004; Kraut et al., 2003). Thus, performance times are generally shorter when a medium allows for more efficient communication. When performance is measured in other ways, however, the effects of media are less clear. For example, Doherty-Sneddon et al. (1997) found no differences in the accuracy of map routes described in video versus audio conferencing; Jackson et al. (2000) found no effect of video frame rate on the quality of poster designs; and Straus and McGrath (1994) found no performance differences between text-based and face-to-face interaction on idea generation, intellective, or judgment tasks. Theories of CMC suggest that tasks involving negotiation and persuasion will be more influenced by communication medium than tasks requiring less interpersonal finesse (e.g., Daft et al., 1987; Short et al., 1976). Even so, many studies using these sorts of tasks have found no differences in performance quality (e.g., Hiltz et al., 1986; see Whittaker and O'Conaill, 1997, for a review). Few studies have looked at how culture influences performance quality. Li (1999a,b) found no differences in accuracy of information transmission in Canadian and Chinese dyads, but significantly poorer transmission in mixed Canadian-Chinese dyads. Others (e.g., Adair et al., 2001) have found that intercultural teams perform more poorly on negotiation tasks.

There really is little evidence on how culture and media interact to affect performance in interviews but there are several demonstrations that culture can affect answers. For example, Ji, Schwarz, and Nisbett (2000) found that a well-known response scale effect—observed with Western respondents—only partly replicates with Chinese respondents. The original finding (e.g. Menon et al., 1995) was that there is a tendency for respondents to treat the middle scale value as reflecting the average frequency in the population and position themselves relative to what they see as the "average." However, when Ji, Schwarz, and Nisbett (2000) replicated the study with both American and Chinese respondents, the Americans reproduced the earlier finding but the Chinese did not. The Chinese only showed the effect for unobservable behaviors such as having nightmares but not for observable behaviors such as coming to class late. The authors suggest that in an interdependent culture such as Chinese culture, respondents are far better attuned to the behavior of others than in a dependent culture such as American and German culture. As a result, Chinese respondents know a great deal about the frequency of behaviors they can observe and thus are less tempted to consult the response scale for distributional information than are their American counterparts.

These findings come from self-administered questionnaires. It is possible that introducing an interviewer whose appearance provides clues relative to the frequency of the relevant behavior (e.g., the interviewer asks about frequency of exercise and looks relatively fit) might have more of an impact for American than Chinese respondents because the Americans have less idea about what is average. This underscores the idea made throughout this volume that in designing interviewing systems that depict the interviewer, such as the embodied agents discussed by Cassell and Miller (Chapter 8) and by Person et al. (Chapter 10), the designers' choice of interviewer features may not only affect answers but may do so differently across cultures.

11.9   NEW INTERVIEWING TECHNOLOGIES THAT CONSIDER CULTURE

Considering culture in developing new interviewing systems raises a large number of never-asked questions, and new interviewing technologies will require answers about basic theoretical questions on culture and media in order to make the necessary design choices. Say it turns out to be true that visual cues are particularly helpful for increasing data quality, trust, and interview satisfaction for one cultural group, while not mattering much or even harming data quality for another group. This would suggest that allowing respondents to have visual access to the interviewer (via the successors to today's video conferencing, videoSkype, or iChat technologies) should be a central consideration in designing the interviewing system, and perhaps that visual access should be a feature that can be turned on and off for different respondents—either at their request, or based on the survey system's diagnosis of the respondent's cultural background.

The same questions turn out to be relevant for self-administered interviewing systems. With current interviewer-administered surveys, various aspects of the interviewer are unalterable, and the only place to choose an interviewer's acceptability (on the unalterable dimensions) are in hiring, and (on any alterable dimensions) in training and monitoring. But imagine, for example, designing a new interviewing medium that includes spoken interaction with an interviewing system. A number of culturally relevant choices need to be made, including:

- What accent and pronunciation should the recorded or text-to-speech "interviewer" use? Should the same accent be used for all respondents? Should the accent and vocal tone adjust to the cultural background of the respondent, in order to increase trust and comfort with the interviewer?
- What kinds of interruption in the dialogue should be allowable? There is substantial evidence on cultural variability in interruption styles (Schiffrin; Tannen), with certain high context cultures that not only allow but expect substantial overlap in speech as signs of interest and attention, and other low context cultures that find overlapping speech offensive and intrusive. Should an interviewing system ever interrupt a respondent? Should a respondent be allowed to interrupt the interviewer?
- If the effects of facial cues on data quality are shown to be high for respondents from a particular culture but low for others, should the system add a face for the respondents who will be helped by one?

For example, a new interviewing tool might automatically modify messages to be more appropriate for the recipient's cultural background, similar to real-time translation software (Yamashita and Ishida, 2006). Alternatively, interviewing tools might seek to educate interviewers about respondents' cultural backgrounds, or respondents about interviewers' backgrounds, for example, by informing the sender of a message as to why it might be inappropriate given the recipient's culture.

11.10 CONCLUSION

In this chapter we have presented an Input-Process-Output framework for understanding how culture influences CMC and considered how this might inform the design of future interviewing systems. We described three dimensions of cultural variability—individualism/collectivism, high versus low context of communication, and task versus relationship orientation—and discussed how these dimensions may interact with features of communication media to influence the outcomes of tasks using those media. A review of the literature to date suggests that cultural factors do indeed shape how people use technology to communicate and this may well be true of communication through survey interviewing systems. More specifically, people from cultures that emphasize nonverbal and contextual aspects of communication are more affected by the visual and auditory affordances of communication media than are people from cultures that emphasize the verbal aspects of communication. Designers of interviewing systems are confronted with choices and none are culturally neutral.

REFERENCES

Adair, W. L., and Brett, J. M. (2005). The negotiation dance: time, culture and behavioral sequences in negotiation. *Organizational Science, 16,* 33–51.

Adair, W. L., Okumura, T., and Brett, J. M. (2001). Negotiation behavior when cultures collide: the United States and Japan. *Journal of Applied Psychology, 86,* 371–385.

Ambady, N., Koo, J., Less, F., and Rosenthal, R. (1996). More than words: linguistic and nonlinguistic politeness in two cultures. *Journal of Personality and Social Psychology, 70,* 996–1011.

Anderson, W. N., and Hiltz, R. S. (2001). Culturally heterogeneous vs. culturally homogeneous groups in distributed group support systems: effects on group process and consensus. In *Proceedings of the 34th Hawaii International Conference on System Sciences.* Los Alamitos, CA: Computer Society Press.

Argyle, M., and Cook, M. (1976). *Gaze and Mutual Gaze.* London: Cambridge University Press.

Aquilino, W. S. (1994). Interview mode effects in surveys of drug and alcohol use: a field experiment. *Public Opinion Quarterly, 58,* 210–240.

Bekker, M. M., Olson, J. S., and Olson, G. M. (1995). Analysis of gestures in face-to-face design teams provides guidance for how to use groupware in design. In *Proceedings of DIS 95 Conference on Design in Computing Systems.* New York: ACM Press.

Bos, N., Olson, J., Gergle, D., Olson, G., and Wright, Z. (2002). Effects of four computer-mediated communications channels on trust development. In *Proceedings of the CHI 2002 Conference on Human Computer Interaction* (pp. 135–140). New York: ACM Press.

Bradner, E., and Mark, G. (2001). Social presence in video and application sharing. In *Proceedings of Conference on Supporting Group Work (GROUP '01)* (pp. 154–161), New York: ACM Press.

Brennan, S. E., and Ohaeri, J. O. (1999). Why do electronic conversations seem less polite? The costs and benefits of hedging. In *Proceedings, International Joint Conference on Work Activities, Coordination, and Collaboration* (WACC '99) (pp. 227–235). San Francisco, CA: ACM Press.

Brewer, E., Demmer, M., Ho, M., Honicky, R. J., Pal, J., Plauche, M., and Surana, S. (2006). The challenges of technology research for developing regions. *Pervasive Computing*, April-June, 15–23.

Brown, P., and Levinson, S. (1987). Universals in language usage: politeness phenomena. In E. Goody (Ed.), *Questions and Politeness* (pp. 56–289). Cambridge: Cambridge University Press.

Carmel, E. (1999). *Global Software Teams: Collaborating Across Borders and Time Zones.* Upper Saddle River, NJ: Prentice Hall.

Chaiken, S., and Eagly, A. H. (1976). Communication modality as a determinant of message persuasiveness and message comprehensibility. *Journal of Personality and Social Psychology, 34,* 606–614.

Choi, I., Nisbett, R. E., and Norenzayan, A. (1999). Causal attribution across cultures: variation and universality. *Psychological Bulletin, 125,* 47–63.

Clark, H. H., and Brennan, S. E. (1991). Grounding in communication. In L. B. Resnick, R. M. Levine, and S. D. Teasley (Eds.), *Perspectives on Socially Shared Cognition* (pp. 127–149). Washington DC: APA Press.

Clark, H. H., and Krych, M. A. (2004). Speaking while monitoring addressees for understanding. *Journal of Memory & Language, 50,* 62–81.

Clark, H. H., and Marshall, C. E. (1981). Definite reference and mutual knowledge. In A. K. Joshi, B. L. Webber, and I. A. Sag (Eds.), *Elements of Discourse Understanding* (pp. 10–63). Cambridge: Cambridge University Press.

Clark, H. H., and Schober, M. F. (1992). Asking questions and influencing answers. In J. M. Tanur (Ed.), *Questions About Questions: Inquiries into the Cognitive Bases of Surveys* (pp. 15–48). New York: Russell Sage Foundation.

Clark, H. H., and Wilkes-Gibbs, D. (1986). Referring as a collaborative process. *Cognition, 22,* 1–39.

Conrad, F. G., and Schober, M. F. (2002). Clarifying question meaning in a household telephone survey. *Public Opinion Quarterly, 64,* 1–28.

Conrad, F. G., Schober, M. F., and Coiner, T. (2007). Bringing features of human dialogue to web surveys. *Applied Cognitive Psychology, 21,* 165–187.

Cramton, C. D. (2001). The mutual knowledge problem and its consequences for dispersed collaboration. *Organizational Science, 12,* 346–371.

Daft, R. L., and Lengel, R. H. (1984). Information richness: a new approach to managerial behavior and organizational design. In L. L. Cummings and B. M. Staw (Eds.), *Research in Organizational Behavior Volume 6* (pp. 191–233). Homewood, IL: JAI Press.

Daft, R. L., Lengel, R. H., and Trevino, L. K. (1987). Message equivocality, media selection and manager performance: implications for information systems. *MIS Quarterly, 11,* 355–366.

Doherty-Sneddon, G., Anderson, A. H., O'Malley, C., Langton, S., Garrod, S., and Bruce, V. (1997). Face-to-face and video mediated communication: a comparison of dialogue structure and task performance. *Journal of Experimental Psychology: Applied, 3,* 105–125.

Ehlen, P., Schober, M. F., and Conrad, F. G. (in press). Modeling speech disfluency to predict conceptual misalignment in survey speech interfaces. Discourse Processes.

Ekman, P. (1982). *Emotions in the Human Face.* Cambridge: Cambridge University Press.

Fussell, S. R., and Krauss, R. M. (1992). Coordination of knowledge in communication: effects of speakers' assumptions about what others know. *Journal of Personality and Social Psychology, 62,* 378–391.

Fussell, S. R., Setlock, L. D., Yang, J., Ou, J., Mauer, E. M., and Kramer, A. (2004). Gestures over video streams to support remote collaboration on physical tasks. *Human–Computer Interaction, 19,* 273–309.

Gergle, D., Kraut, R. E., and Fussell, S. R. (2004). Language efficiency and visual technology: minimizing collaborative effort with visual information. *Journal of Language and Social Psychology. 23,* 491–517.

Goffman, E. (1967). *Interaction Ritual: Essays in Face-to-Face Behavior.* New York: Pantheon Books.

Guadagno, R. E., and Cialdini, R. B. (2002). Online persuasion: an examination of gender differences in computer-mediated interpersonal influence. *Group Dynamics: Theory, Research and Practice, 6,* 38–51.

Gudykunst, W. B., and Kim, Y. Y. (1997). *Communicating with Strangers: An Approach to Intercultural Communication,* 3rd ed. New York: McGraw-Hill.

Gudykunst, W. B., and Ting-Toomey, S. (1988). *Culture and Interpersonal Communication.* Newbury Park, CA: Sage.

Gudykunst, W. B., Matsumoto, Y., Ting-Toomey, S., Nishida, T., Kim, K., and Heyman, S. (1996). The influence of cultural individualism–collectivism, self-construals, and individual values on communication styles across cultures. *Human Communication Research, 22,* 510–543.

Hackman, J. R. (1987). The design of work teams. In J. W. Lorsch (Ed.), *Handbook of Organizational Behavior* (pp. 315–342). Englewood Cliffs, NJ: Prentice Hall.

Hall, E. (1976). *Beyond Culture.* New York: Doubleday Anchor Books.

Hancock, J. T., and Dunham, P. J. (2001). Impression formation in computer-mediated communication revisited: an analysis of the breadth and intensity of impressions. *Communication Research, 28,* 325–347.

Harkness, J. A., van de Vijver, F. J. R., and Mohler, P. P. (2003). Cross-cultural survey methods, New York: John Wiley & Sons.

Heim, J., Asting, T., and Schliemann, T. (2002). Medium effects on persuasion. In *Proceedings of NordiCHI 2002* (pp. 259–261) New York: ACM Press.

Herring, S. C. (1994). Politeness in computer culture: why women thank and men flame. In M. Bucholtz, A. Liang, L. Sutton, and C. Hines (Eds.), *Cultural Performances: Proceedings of the Third Berkeley Women and Language Conference* (pp. 278–294). Berkeley, CA: Berkeley Women and Language Group.

Hiltz, S. R., and Turoff, M. (1978). *The Network Nation: Human Communication via Computer.* Reading, MA: Addison-Wesley.

Hiltz, S. R., Johnson, K., and Turoff, M. (1986). Experiments in group decision making: communication process and outcome in face-to-face versus computerized conferences. *Human Communications Research, 13,* 225–252.

Hofstede, G. (1983). Dimensions of national cultures in fifty countries and three regions. In J. Deregowski, S. Dzuirawiec, and R. Annis (Eds.), *Explications in Cross-Cultural Psychology.* Lisse, Switzerland: Swets and Zeitlinger.

Hofstede, G. (2001). *Culture's Consequences: Comparing Values, Behaviors, Institutions, and Organizations Across Nations.* Thousand Oaks, CA: Sage.

Holtgraves, T. (1997) Styles of language use: individual and cultural variability in conversational indirectness. *Journal of Personality and Social Psychology, 73,* 624–637.

Isaacs, E., and Clark, H. H. (1987). References in conversation between experts and novices. *Journal of Experimental Psychology: General, 116,* 26–37.

Jackson, M., Anderson, A. H., McEwan, R., and Mullin, J. (2000). Impact of video frame rate on communicative behaviour in two and four party groups. In *Proceedings of CSCW 2000* (pp. 11–20). New York: ACM Press.

Jensen, C., Farnham, S. D., Drucker, S. M., and Kollock, P. (2000). The effect of communication modality on cooperation in online environments. In *Proceedings of the Conference on Human Factors and Computing Systems (CHI'00),* (pp. 470–477). New York: ACM Press.

Ji, L., Schwarz, N., and Nisbett, R. E. (2000). Culture, autobiographical memory, and behavioral frequency reports: measurement issues in cross-cultural studies. *Personality and Social Psychology Bulletin, 26,* 586–594.

Kayan, S., Fussell, S. R., and Setlock, L. D. (2006). Cultural differences in the use of instant messaging in Asia and North America. In *Proceedings of CSCW 2006.* New York: ACM Press.

Kiesler, S., Siegel, J., and McGuire, T. W. (1988). Social psychological aspects of computer-mediated communication. *American Psychologist, 39,* 1123–1134.

Kirk, D. S., and Stanton-Fraser, D. (2006). Comparing remote gesture technologies for supporting collaborative physical tasks. In *Proceedings of CHI 2006.* New York: ACM Press.

Kraut, R. E., Fussell, S. R., and Siegel, J. (2003). Visual information as a conversational resource in collaborative physical tasks. *Human–Computer Interaction, 18,* 13–49.

Li, H. Z. (1999a). Grounding and information communication in intercultural and intracultural dyadic discourse. *Discourse Processes, 28,* 195–215.

Li, H. Z. (1999b). Communicating information in conversations: a cross-cultural comparison. *International Journal of Intercultural Relations, 23,* 387–409.

Markus, H. R., and Kitayama, S. (1991). Culture and the self: implications for cognition, emotion, and motivation. *Psychological Review, 20,* 568–579.

McAllister, D. J. (1995). Affect- and cognition-based trust as foundations for interpersonal cooperation in organizations. *Academy of Management Journal, 38,* 24–59.

Mehrabian, A. (1967). Orientation behaviors and nonverbal attitude communication. *Journal of Communication, 16,* 324–332.

Menon, G., Raghubir, P., and Schwarz, N. (1995). Behavioral frequency judgments: an accessibility-diagnosticity framework. *Journal of Consumer Research, 22,* 212–228.

Miller, J. (2002). Bringing culture to basic psychological theory: beyond individualism and collectivism. Comment on Oyserman et al. (2002). *Psychological Bulletin, 128*(1), 97–109.

Morley, I. E., and Stephenson, G. M. (1977). *The Social Psychology of Bargaining.* London: Allen & Unwin.

Morris, M. W., and Peng, K. (1994). Culture and cause: American and Chinese attributions for social and physical events. *Journal of Personality and Social Psychology, 67,* 949–971.

Nisbett, R. E. (2003). *The Geography of Thought: How Asians and Westerners Think Differently … and Why.* New York: The Free Press.

Oetzel, J. G. (1998). Explaining individual communication processes in homogeneous and heterogeneous groups through individualism–collectivism and self-construal. *Human Communication Research, 25,* 202–223.

Olson, G. M., and Olson, J. S. (2000). Distance matters. *Human–Computer Interaction, 15,* 139–179.

Olson, G. M., Atkins, D. E., Clauer, R., Finhold, T. A., Jahanian, F., Killen, T. L., Prakash, A., and Weymouth, T. (1998). The Upper Atmospheric Research Collaboratory (UARC). *Interactions of the ACM, 5,* 48–55.

Ongena, Y. (2005). Interviewer and respondent interaction in survey interviews. PhD dissertation, Free University of Amsterdam.

Oyserman, D., Coon, H. M., and Kemmelmeier, M. (2002). Rethinking individualism and collectivism: evaluation of theoretical assumptions and meta-analyses. *Psychological Bulletin, 128,* 3–72.

Postmes, T., Spears, R., and Lea, M. (2002). Intergroup differentiation in computer-mediated communication: effects of depersonalization. *Group Dynamics: Theory, Research and Practice, 6,* 3–16.

Reinig, B. A., and Mejias, R. J. (2003). An investigation of the influence of national culture and group support systems on group processes and outcomes. In *Proceedings of HICSS 2003.* Los Alamitos, CA: Computer Society Press.

Reinig, B. A., and Mejias, R. J. (2004). The effects of national culture and anonymity on flaming and criticalness in GSS-supported discussions. *Small Group Research, 35,* 698–723.

Ring, P. S., and Van der Velden, A. (1994). Developmental processes of cooperative interorganizational relationships. *Academy of Management Review, 19,* 90–118.

Rocco, E. (1998). Trust breaks down in electronic contexts but can be repaired by some initial face-to-face contact. In *Proceedings of Conference on Human Factors and Computing Systems* (pp. 496–502). New York: ACM Press.

Rutter, M. (1987). *Communicating by Telephone.* Oxford: Pergamon Press.

Schober, M. F. (1999). Making sense of questions: an interactional approach. In M. G. Sirken, D. J. Hermann, S. Schechter, N. Schwarz, J. M. Tanur, and R. Tourangeau (eds.), *Cognition and survey research* (pp. 77–93). New York: John Wiley & Sons.

Schober, M. F., and Conrad, F. G. (1997). Does conversational interviewing reduce survey measurement error? *Public Opinion Quarterly, 61,* 576–602.

Schober, M., and Conrad, F. G. (2002). A collaborative view of standardized survey interviews. In D. W. Maynard, H. Houtkoop-Steenstra, N. C. Schaeffer, and J. van der Zouwen (Eds.), *Standardization and Tacit Knowledge: Interaction and Practice in the Survey Interview* (pp. 67–94). Hoboken, NJ: John Wiley & Sons.

Schober, M. F., Conrad, F. G., and Fricker, S. S. (2004). Misunderstanding standardized language in research interviews. *Applied Cognitive Psychology, 18,* 169–188.

Schwartz, S. (1992). Universals in the content and structure of values: theoretical advances and empirical tests in 20 countries. *Advances in Experimental Social Psychology, 25,* 1–65.

Setlock, L. D., Quinones, P. A., and Fussell, S. R. (2007). Does culture interact with media richness? The effects of audio vs. video conferencing on Chinese and American dyads. In *Proceedings of HICSS 2007.* Los Alamitos, CA: Computer Society Press.

Setlock, L. D., Fussell, S. R., and Neuwirth, C. (2004). Taking it out of context: collaborating within and across cultures in face-to-face settings and via instant messaging. In *Proceedings of the CSCW 2004 Conference on Computer-Supported Cooperative Work* (pp. 604–613). New York: ACM Press.

Setlock, L. D., Fussell, S. R., and Shih, Y. Y. (2006, July). Effects of culture, language and communication medium on conversational grounding. In *Annual Meeting of the Society for Text and Discourse,* Minneapolis, MN.

Short, J., Williams, E., and Christie, B. (1976). *The Social Psychology of Telecommunication.* London: John Wiley & Sons.

Singelis, T. M., and Brown, W. J. (1995). Culture, self, and collectivist communication: linking cultures to individual behavior. *Human Communication Research, 21,* 354–389.

Stephenson, G., Ayling, K., and Rutter, D. (1976). The role of visual communication in social exchange. *British Journal of Social and Clinical Psychology, 15,* 113–120.

Stewart, C. O., Setlock, L. D., and Fussell, S. R. (in press). Conversational argumentation in decision-making: differences across cultures and communication media. *Discourse Processes.*

Straus, S., and McGrath, J. (1994). Does the medium matter? The interaction of task type and technology on group performance and member reactions. *Journal of Applied Psychology, 79,* 87–97.

Suessbrick, A., Schober, M. F., and Conrad, F. G. (2005). When do respondent misconceptions lead to survey response error? In *Proceedings of the American Statistical Association, Section on Survey Research Methods.* Alexandria, VA: American Statistical Association.

Ting-Toomey, S. (1988). Intercultural conflict styles: a face-negotiation theory. In Y. Y. Kim and W. Gudykunst (Eds.), *Theories in Intercultural Communication* (pp. 213–235). Newbury Park, CA: Sage.

Ting-Toomey, S. Gao, G., Trubisky, P., Yang, Z., Kim, H. S., Lin, S.-L., and Nishida, T. (1991). Culture, face maintenance, and styles of handling interpersonal conflict: a study in five cultures. *International Journal of Conflict Resolution, 2,* 275–296.

Triandis, H. C. (1989). The self and behavior in different cultural contexts. *Psychological Review, 96,* 506–520.

Triandis, H. C. (1995). *Individualism and Collectivism.* Boulder, CO: Westview.

Tourangeau, R., and Smith, T. W. (1996). Asking sensitive questions: the impact of data collection mode, question format and question content. *Public Opinion Quarterly, 60,* 275–304. DOI: 10.1086/297751.

Turner, C. F., Leighton, K., and Frenya, L. (1995). Impact of ACASI on reporting of male–male sexual contacts. In R. Warnecke (Ed.), *Preliminary Results from the 1995 National Survey of Adolescent Males*, Breckenridge, Colorado. Washington DC: DHHS, pp. 171–176.

Veinott, E., Olson, J., Olson, G., and Fu, X. (1999) Video helps remote work: speakers who need to negotiate common ground benefit from seeing each other. In *Proceedings of the CHI 1999 Conference on Human–Computer Interaction* (pp. 302–309). New York: ACM Press.

Walther, J. B. (1992). Interpersonal effects in computer-mediated interaction: a relational perspective. *Communication Research, 19,* 52–90.

Walther, J. B. (1995). Relational aspects of computer-mediated communication: experimental observations over time. *Organization Science, 6,* 186–203.

Watson, R. T., Ho, T. H., and Raman, K. S. (1994). Culture: a fourth dimension of group support systems. *Communications of the ACM, 37*(10), 44–55.

Whittaker, S. (2003). Theories and methods in mediated communication. In A. Graesser, M. Gernsbacher, and S. Goldman (Eds.), *The Handbook of Discourse Processes* (pp. 243–286). Mahwah, NJ: Lawrence Erlbaum Associates.

Whittaker, S., and O'Conaill, B. (1997). The role of vision in face-to-face and mediated communication. In K. Finn, A. Sellen, and S. Wilbur (Eds.), *Video-Mediated Communication* (pp. 23–49). Mahwah, NJ: Lawrence Erlbaum Associates.

Williams, E. (1977). Experimental comparisons of face-to-face and mediated communication: a review. *Psychological Bulletin, 84,* 963–976.

Yamashita, N., and Ishida, T. (2006). Effects of machine translation on collaborative work. In *Proceedings of CSCW 2006.* New York: ACM Press.

Zhang, Q. P., Olson, G. M., and Olson, J. S. (2004). Does video matter more for long distance collaborators? In *Proceedings of the XXVIII International Congress of Psychology.* East Sussex, UK: Psychology Press.

Zhang, Q. P., Sun, X., Chintakovid, T., Ge, Y., Shi, Q., and Zhang, K. (2006). How Culture and media influence personal trust in different tasks. Paper presented at the HCIC Winter Consortium, Feb. 2006.

CHAPTER 12

Protecting Subject Data Privacy in Internet-Based HIV/STI Prevention Survey Research

Joseph A. Konstan, B. R. Simon Rosser, Keith J. Horvath, Laura Gurak, and Weston Edwards
University of Minnesota, Minneapolis, Minnesota

12.1 INTRODUCTION

Researchers conducting Internet studies face a myriad of ethical and logistic challenges as they attempt to protect subject privacy (see also Marx's discussion on research ethics, Chapter 13 in this volume). Because of the increased opportunity for anonymity, remote monitoring, and distant attacks, Internet researchers face even greater challenges than when conducting offline research. The more sensitive the topic area, the greater the ethical obligation to provide data protection. Data security considerations will become more obvious as the Internet is increasingly used as a research tool to assess a variety of health behaviors (Lenert and Skoczen, 2002; Ross et al., 2000). This chapter is a case study of an online HIV risk assessment survey research project and presents a decision-making model for policy and procedure factors affecting subject privacy in such studies. We do not presume to present a single correct answer, but rather identify the issues, discuss the trade-offs, and provide a framework to help researchers identify the data security considerations most appropriate for their research. While our case study is specific to sex questionnaires and HIV prevention, we anticipate these issues and principles may have wide application.

Envisioning the Survey Interview of the Future, Edited by Frederick G. Conrad and Michael F. Schober
Copyright © 2008 John Wiley & Sons, Inc.

12.2   BACKGROUND

Computer technology has revolutionized in many ways the research process (see Couper, Chapter 3 in this volume), with the effect of altering how participants may respond to survey items. Participants' perception of data confidentiality in Internet-based studies may influence response and completion rates (Sills and Song, 2002; Stanton, 1998), with outcomes varying with race, income, and sex (O'Neil, 2001). The sense of community that individuals build on the Internet, as well as the deliberate use of the Internet for escape, suggests that individuals encountering a survey on the Internet may "have their guard down" and thereby depend even more strongly on those conducting research to protect their interests (Reeves, 2001). In the most comprehensive treatment we have seen of issues relating to privacy protection in Internet surveys, Cho and LaRose (1999, p. 421) argue that "online surveyors commit multiple violations of physical, informational, and psychological privacy that can be more intense than those found in conventional survey methods" and recommend a set of techniques for researchers to overcome user reluctance to participate, including offering online incentives, separating consent from the survey instrument, avoiding "trolling" for e-mail addresses in online communities, and posting results of the research online.

When this project started, we were conducting one of the earliest Internet-based sex surveys and accordingly had to learn a great deal from practice in other domains (e.g., e-commerce). While Internet-based studies of sexual behavior are more prevalent (Bull et al., 2001; Horvath et al., 2006), we have been disappointed to find that many Internet-based survey studies fail to report steps taken to protect the privacy of subjects, while others appear to lack the expertise in the team to address the complex issues involved. The first goal of this chapter is to provide practical information for researchers to plan privacy policies. The second goal is to report the logistical details of configuring systems to implement that policy. To achieve these goals, we address the protection of subjects from external monitoring, the survey's Web and database servers from tampering, identified survey information, anonymized survey information, and subjects during phone or e-mail contacts with research staff. We conclude with policy and process recommendations for researchers undertaking other Internet-based studies and reflections on applying these recommendations. It should be noted that, in our experience, few simple solutions to ensuring subject data privacy exist and, therefore, we caution against using the suggestions we provide as a "cookbook" approach to data security.

12.3   CASE DESCRIPTION

The *Men's INTernet Study (MINTS)* is a national, Internet-based HIV prevention study of online and offline sexual behavior of 1026 Latino Men who use the Internet to seek Sex with Men (MISM). Sexual surveys of populations at risk for HIV may be among the most sensitive areas of research and, hence, provide excellent examples for issues of protecting data privacy. This online study advertised for Latino

MISM through banner ads at the gay.com Web site, and offered participants $20 to complete a lengthy survey (455 specific items were recorded) of their Internet use, demographics, sexual experiences, and alcohol and drug use. The survey instrument was developed initially in English, then forward and back translated into Spanish, with translation equivalency and reliability estimated through pilot test–retesting with bilingual respondents.

The structure of the research team included a principal investigator in HIV prevention and sexual health research and a multidisciplinary group of coinvestigators (computer science, e-communication, ethics, sexology and methodology/statistics) who together supervised a project coordinator, a webmaster, and a research assistant. These roles became important as we defined "need-to-know" and decisions to limit the number of individuals with access to the "delimited" or full database. It is also important to note that this study was carried out in Minnesota. Under our state's open records laws, nearly all research records, phone logs, e-mail messages, and computer files are open to public request. We addressed this challenge in part by obtaining a Certificate of Confidentiality (a U.S. government document issued by the National Institutes of Health to protect confidential data about research subjects from subpoena for law enforcement and other purposes). However, because the power of such a certificate has not been tested through Minnesota court cases, we felt it prudent to take other precautions in case such a certificate did not fully protect subject identity and data.

12.4   DATA SECURITY PROTOCOL PLANNING

The project team had extensive training in sensitive handling of sexual data issues from previous offline and clinical studies. Based on this experience, the following security goals were identified in order of priority:

- *Protecting Subjects Against External Knowledge of the Fact of or Content of Their Participation.* Our first priority was to ensure that participation in the study could not become known to some third party (e.g., employers, family members, the general public, or law enforcement).
- *Protecting the Integrity of the Data Collected Against External Attacks.* Given that the purpose of data collection in our case was to learn how to develop effective HIV-prevention interventions, compromised data could have undermined that mission and rendered subject participation pointless. Unfortunately, there were, and continue to be, technically savvy individuals whose aim may be to disrupt studies such as this one.
- *Protecting Subjects Against Unnecessary Discoverable Knowledge of the Fact or Content of Their Participation.* Given the size of the project team, and the potential for open records requests, we felt it was important to limit as many investigators as possible from having access to data that might identify subjects.

While translating these goals into specific privacy protection protocols, we recognized several important trade-offs, including:

- *Anonymity Versus Compensation.* Given that this was a long survey, we believed it was necessary to compensate participants for their effort. However, it was a challenge to ensure participant anonymity during the compensation process. Even when the level of compensation fell below that required for tax reporting, noncash payments usually necessitated evidence trails.
- *Anonymity Versus Data Integrity.* Positive identification of subjects makes it much easier to verify inclusion criteria or detect duplicate enrollments. At the same time, identification may discourage some participants from participating or reduce the likelihood of honest responses.
- *Ease of Operation Versus Security.* All security systems compromise some level of ease of use. We were concerned for both subject ease-of-use (which could affect recruitment and retention) and administrative/research ease-of-use (which could undermine project efficiency).
- *Privacy Versus Cost Efficacy.* We strove to address the challenge of recognizing how much security is sufficient. At some point, increasing security measures yields diminishing participant benefit and increasing project costs.

Finally, we established seven pragmatic principles to guide security protocol:

- *Identity Can Be Inferred from Computer Log Data.* While certain data (e.g., an IP address that identifies a computer or service provider) does not directly identify a subject, it can be combined with login records or customary usage patterns to identify individuals. Accordingly, we chose to protect computer log data in the same manner as we protect birthdates, e-mail addresses, payment contact information, and other partially identifying data.
- *Security Measures Should Be Proportional to the Sensitivity of the Data Being Protected.* Anonymous responses do not require the same level of protection that is required for subject-identifying data.
- *Computer Backups Occur Regularly, and Often Without the Knowledge and/or Control of the Data Holder.* Automatic backup processes for individual machines are executed often without the owner's awareness. Accordingly, sensitive data should be loaded onto computers in an encrypted form.
- *We Should Facilitate Subject Awareness of Security Steps They Could Take.* Thinking through security concerns in an Internet-based study requires considering security measures at all stages of the study. We saw it as part of our role to help concerned subjects to identify alternatives that provided them with extra security and educated them about avoiding the most severe subject-end security risks.
- *Technological Means Alone Are Insufficient for Protecting Private Subject Data.* As long as the data files are kept, they are susceptible to subpoena and open records laws. Accordingly, we sought and obtained a Certificate of Confidentiality to add legal defenses to our technological ones.

- *Identifying Information Must Be Restricted to Need-to-Know Access.* It was our responsibility to limit as many investigators as possible from having access to or knowledge of privacy-compromising data.
- *A System Is Only As Secure As Its Weakest Link.* We took steps to identify points in which security could be most easily compromised throughout the study process.

12.4.1   Security Protocol Logistics

12.4.1.1   Protecting Subjects from External Monitoring
For our study, we identified three points where subjects were at risk for external monitoring:

- Monitoring of the subject's communication link with our survey.
- Monitoring of the subject's computer or e-mail.
- Detecting the subject's participation through the payment mechanism (e.g., information identifying the study on a check envelope or e-payment e-mail).

To minimize these potential risks, we conducted the survey over secure connections (using SSL via a secure HTTP connection). This encrypts the communication between the subject's computer and our server such that even the subject's ISP or corporate network administrators cannot detect the content being transmitted. A URL name that was not easily identifiable was chosen to avoid external monitoring. In addition, we provided participants with a set of educational messages during the sign-up process advising them to consider one or more of these measures:

- Use an anonymizer service (such as www.the-cloak.com) to further disguise their visit. In this way, their ISP or site administrator only sees that they've visited the anonymizer, not the specific site they visit through it.
- Create and use a free Web-based e-mail account for communication with the study. Services such as Microsoft's Hotmail (www.hotmail.com) and Yahoo! Mail (mail.yahoo.com) provide such e-mail accounts, allowing the subject to avoid any communication to his regular accounts.
- Complete the survey in a private setting and exit the browser after completing the survey (to prevent someone else from going back to previous pages).

Finally, we provided a variety of safeguards accompanying four compensation options. Subjects could choose to be paid by check, be paid using PayPal (a service that delivers money to an e-mail account), have a donation made to charity on their behalf, or forego compensation entirely. For each option, we took the following security measures:

- The payment-by-check process was kept completely separate from actual survey data results. The check and envelope contents were documented and monitored

TABLE 12.1   Comparison of Payment Choice by Mean Income of Participants and Completion Rate ($N = 1026$ men who use the Internet to seek sex with men)

Payment Choice	Annual Gross Income		p^a	Completion Rate[b]
	Mean	SD		
Mailed check	$32,367	29,675	—	73%
E-money (Paypal)	$41,232	59,999	.008	67%
Charitable donation	$43,595	28,376	.023	48%
No payment	$52,294	28,376	$p < .05$	39%

[a] Least significant difference (LSD) post hoc comparisons.
[b] $X^2(1) = 68.74, p < .001$.

to ensure that no information about the specific purpose was evident either from the outside or to someone viewing the check (e.g., a bank teller).

- PayPal payments initially required no identifying information other than subject e-mail address. We provided this option as a means for subjects to receive payment entirely electronically. We later discovered that PayPal returned information to us about the name of the person "cashing" the payment. While this form of payment compromised anonymity, it had the benefit of alerting us to the fact that some participants were creating multiple e-mail addresses to receive and complete the survey multiple times.

- The charity and no payment options required no identifying information, although we requested an e-mail address to contact subjects in the event of problems.

In the MINTS, choice of compensation significantly impacted completion rate, with lower income MISM choosing a mailed check while higher income participants tended to choose donation to charity or no payment (see Table 12.1). Those who chose payment were also significantly more likely to complete the survey. Hence, for studies wanting lower SES participants to participate and studies wanting high completion rates, monetary compensation, preferably by check, appears optimal. Interestingly, providing a donation to charity option appeared significantly more attractive to participants and achieved a higher completion rate than the no payment option. Taken together, these findings illustrate the complexity and importance of considering payment options within broader discussions of data privacy and survey completion rates.

12.4.1.2 *Protecting the Survey's Web and Database Servers from Tampering*
An Internet study is conducted through one or more computers connected to the Internet. Computers receive raw data from subjects and store that data into one or more databases. An attack on these computers could disable the study, corrupt the stored data, and/or lead to disclosure of sensitive data. The study's Web and database servers are vulnerable through five channels:

- *Denial of service* attacks that aim to shut down the study, temporarily or permanently. Such attacks could be as direct as cutting power or communications to the building with the server, or as indirect as having hundreds or thousands of users flood our system with simultaneous (and likely fake) surveys.

- *Physical access* attacks that attempt to compromise the servers through direct, physical access. Physical access could permit an attacker to remove disk drives with databases, intercept data, or potentially change the functioning of the system.

- *Network intrusion* attacks in which an attacker attempts to compromise the servers through network access. The goal of such attacks may be to gather or alter data, or change the system.

- *Virus or Trojan horse* attacks, where a broadly spread piece of malicious software may damage or compromise data or software. Computer viruses spread frequently, mostly through electronic mail. Trojan horses are malicious programs embedded in otherwise innocuous-seeming software (including in some cases images). In both cases, the likely attack would be incidental—broadly distributed rather than targeted at this specific system—and therefore the greater risk is destruction of all or part of the data rather than targeted alterations of the data.

- *Spoofing* attacks in which the attacker redirects Internet traffic to a different site or creates "replicate" ads and links that point people to the other site. The risk in this case is that people may believe they are providing data to a legitimate study when the data are in fact intercepted and misused. A recently documented case involved e-mails alleged to be from PayPal that led people to a real-looking site that collected credit card information "to update records."

Some of these potential attacks were best managed through the environment in which we operated our study. The University of Minnesota network provided a certain level of protection. For example, unlike a new top-level domain, a ".umn.edu" subdomain cannot be redirected except through the University. Also, since the University's computers are regularly under attack by hackers, extensive security monitoring protocols are in place, as well as some firewall protection that prevents certain dangerous outside access. We also decided to advertise solely within a site catering to our target population, as to avoid attracting attention and possible attacks.

Although it is nearly impossible to prevent denial of service attacks, we instituted a regular pattern of monitoring that would ensure that we could detect unusual access patterns (e.g., an unusually high or low number of surveys being responded to during a time interval, or an unusual percentage of survey recipients from a particular geographic area) and respond when problems occurred. We did not experience such attacks during the survey.

Of greater threat were physical access attacks, which we addressed through specific measures. Our first decision was whether to locate the machine in a "computer room" space that would not be in our control or in space under our sole control. We chose the latter, mostly due to the lack of advantages in having the machine physically out of our control, and because it enabled us to turn off remote access if need be. Since our environment was relatively low risk, we chose not to store the machine in a secure, off-site location. We would recommend that a high profile site, such as one that has already attracted protesters or sabotage, consider such an alternative.

Additional steps that we took included:

- Locking the server in the office of the system administrator, where access requires keys to two locked doors.
- Password protection on all server accounts, to prevent easy access to system data and software by someone with temporary physical access to the machine.
- The server was configured to run a screen saver with a password required to unlock the system when left idle for a period of time.

We were far more concerned about network intrusions and viruses/Trojan horse attacks, to which the following precautions were taken:

- Before the study began, the server was configured so that remote access was only available through "ssh" (a secure, encrypted remote access protocol). Access through unencrypted "rsh," "ftp," and "telnet" protocols was disabled.
- During the operation of the study, even "ssh" was turned off. The only remote access to the system was through the Web server. Our administrative Web interface had very limited functionality; it could display summary statistics and it allowed the user to shut down the survey. No subject-specific data could be viewed or changed.
- The server was configured with a commercial operating system and Web server with all of the latest security patches installed.
- The server was configured with antivirus software that automatically updated virus definitions weekly.

We considered two further security measures against network intrusions and virus/Trojan horse attacks. We could have set up a separate hardware firewall to protect the server, but, because of the experience of our webmaster in securing computers, we determined that the cost was not justified by the added security it might provide. We also considered whether to keep Web logs of access to the system to help track down the source of any unusual activity or potential attack. Ultimately, we decided that the presence of these logs, which would have provided potentially identifying information on subjects in a nonencrypted and backed-up form, created too great a risk to subject privacy. However, these steps might be taken depending on the nature of the study.

Finally, we did not take special measures to prevent spoofing attacks, since our domain name was fairly secure through the University, and we would have noticed any sudden elimination of traffic. Furthermore, the effort needed for someone to clone our site and lure people to the site seems excessive for the small amount of potential data they could have gathered. Indeed, it would have been as easy for such hackers to invent a survey from the institution of their choice. As far as we know, we were not spoofed.

12.4.1.3   *Protecting Identified Survey Information*

The most sensitive data were the records of potentially identifying information about survey respondents. We have information on birth date, zip code, e-mail address, IP address, and in many cases more specific payment information including name and address. This information is stored in a database separate from the actual survey responses. We refer to this as the *subject database*.

The subject database is used during payment processing and the data validation process (to ensure that subjects were consistent in certain key demographic questions). Accordingly, protocols were needed to address the possibility of compromise on the server, between the server and other machines, and when stored on the machines of our project coordinator and research assistant.

We focused protection of identified survey information in the following three areas:

- Encryption of the data and protection of the encryption key.
- Preservation of the data in the event of database damage or loss.
- Destruction of the data after its use is completed.

In deciding on the level of data encryption necessary for this study, we balanced two concerns: the strength of the encryption and the ease of transferring or using the encrypted data. We used the Microsoft Access database and its password-protected database feature, which stores the database in an encrypted format that prevents access to the data fields by simply reading the raw file. We believed that layering stronger encryption would have made transferring or using the encrypted data unmanageable. If we perceived a greater risk, we could have used other commercial databases with even stronger encryption. Our specific database encryption and password control steps were as follows:

- The database was stored as a password-protected database in Microsoft Access.
- Only three project members have the password: the system administrator, the project coordinator, and the research assistant. The latter two must access the subject database to process payments and for regular reviews to detect potential abuses of the system.
- The password was not written down and only communicated in person.
- In the event that any of the password-holding project staff left the project, the principal investigator was responsible for determining whether the circumstances of departure warranted changing the password.

To further protect the subject database during data collection, we established the following three parallel backup procedures:

- Twice a day, we transmitted the subject database from the project server computer to the project coordinator's computer, which is located in a different building on a separate campus.

- We scheduled a process to e-mail the database every two hours to another secure computer at a third location. The last two versions were to be retained for backup purposes by a coinvestigator, but nobody at that site had the encryption password. Two-hour backups eventually overloaded the server and, therefore, we performed such incremental updates less frequently as the study progressed.
- We also performed daily backups onto tape. The two most recent backups were retained and stored on site. Older backup tapes were reused for subsequent backups.

Recognizing that it is nearly impossible to destroy all copies of electronic data, we implemented the following procedures to help ensure that such data were not usable after the completion of the study:

- The protected subject database was excluded from backups under our control.
- No unencrypted copy of the subject database was ever made and all access to the database was through Microsoft Access.
- After the study was completed and the subject data processed, all online copies were deleted and all known backup copies were destroyed.
- The password used to protect the protected subject database was not to be reused.

12.4.1.4 *Protecting Anonymized Survey Information*
The *survey database* contained records identified only by a subject number, which in this case was primarily information on sexual behavior and Internet use. Because such data are likely to be available to researchers for meta-analyses and other purposes, we focus our protection on data integrity and ensuring appropriate anonymity. We adopted the following procedures for the survey response database:

- The survey database contains no identifying information. We allowed age and zip code for demographic analysis, but excluded birthdates because they could be used in combination with zip code to identify a subject. (This study was completed prior to Health Insurance Portability and Accountability Act regulations, which now only permit the first three digits of the zip code to be collected.)
- The copy of the survey database on the database server was deemed the primary copy. (This is not in a strict sense a security measure, but rather an important data control measure, as it helps identify where database changes must occur for them to be considered official and propagated to other copies.)
- As with the subject database, the survey database was transmitted twice daily to the project coordinator and stored in a password-protected file system. Copies were also scheduled to be sent every two hours to a third site, although system capacity limited the frequency of such transfers. Tape backups were made daily and two days of backups were stored off-site.
- The survey database was excluded from backups under our control.

- At the close of the survey, the primary copy of the survey database was transferred to the project coordinator. Two copies of this uncleaned database were preserved on CD-ROM and stored in separate, secure, locked environments. An additional copy was preserved on tape. These three copies are retained for seven years and are accessible to the principal investigator or any investigator wishing to check data integrity. All other copies of the uncleaned database were destroyed after data cleaning.

- Three permanent copies of the cleaned database were made on CD-ROM. One was stored on-site, another off-site, and a third at another university campus.

- The principal investigator is required to maintain copies of the cleaned and uncleaned data for seven years or longer if required by NIH.

12.4.1.5   *Protecting Subjects During Phone or E-mail Contacts with the Researchers*

One of our more significant challenges was the maintenance of subject confidentiality when a question or problem arose that led a subject to contact us. Research practices required us to maintain logs of these contacts, yet we needed to ensure that these logs had as little identifying material as possible. In addition to the study Web site providing advice on how to minimize risk of identification (e.g., do not use your full name and do not call where you will be overheard), we developed security measures for handling and recording e-mail contacts, handling and recording telephone contacts, communication among the researchers in response to subject contacts, and securing the records of the contacts.

Because the e-mail server that usually serves program staff is backed-up, regularly, we used a separate mail server on a site operated by our webmaster that did not back up e-mail. This slightly increased the risk of losing messages in the event of a problem (although none occurred), but it greatly reduced the risk that the messages would appear unexpectedly on a backup tape. For our records, we kept a paper record of all messages. Project staff retrieving e-mail messages made a single print out of each message and reply. The electronic copies of each message were deleted as soon as possible. Project staff were trained on how to delete messages, since many e-mail programs do not actually delete messages directly but instead save them in a folder of deleted items.

For telephone calls, we were concerned about identifying information that could be provided automatically, such as the participant's telephone number. We trained our staff handling calls to avoid requesting identifying information (e.g., first name only) and to ask permission if a call-back was needed. A project toll-free number was available for subject convenience, however, many toll-free number services provide calling number identification with the call or as part of billing. As such, we:

- Requested a toll-free phone service that did not provide calling number identification either to us or to the University's telecom department. While the telephone company necessarily has this data, they are legally required to protect it.

- Did not use an ordinary phone number. While this would be a requirement for studies involving international participation (since toll-free numbers generally serve only North America), we decided it did not add any protection over a toll-free number.
- Advised concerned subjects to call us from a public pay phone if they did not want any identifying telephone records.
- Logged telephone calls using paper records.

Most calls and e-mails could be handled directly by either the project coordinator or research assistant. When it was necessary to involve the principal investigator who could only be reached at a distance (e.g., he was attending a conference), we established a policy that e-mails of subject queries were kept to the minimum possible and contained as little identifying information as possible. Messages that included an e-mail address, IP address, phone number, or last name were considered particularly sensitive. They were not relayed by e-mail if another alternative existed (e.g., telephone, e-mail requesting a call-back, or waiting until the principal investigator returned). Paper records of e-mail and telephone calls were kept in a locked file cabinet until the end of their analysis, after which they were disposed of by shredding or burning.

12.5   RECOMMENDATIONS

1. Involve a multidisciplinary team in developing your privacy policy to achieve a broad perspective. Our team included specialists in HIV prevention, sexology, computer science, computer systems administration, and ethics.
2. Start with an assessment of the risks involved in your study. Would it harm subjects if their participation was known? Would it harm subjects if their specific responses or activities were known? How controversial is your study and how likely is it that some people may attempt to attack or undermine it?
3. Develop a list of priorities and identify the key trade-offs up front. We found it easier to make decisions among alternatives when the trade-offs were explicit and acknowledged.
4. Develop your own set of pragmatic principles. We attempted to provide a set of principles in this chapter. However, to be useful in other study settings, consider adapting the principles for your environment and study needs.
5. Identify which data need particular protection; it is counterproductive to secure data that will then be released to the public.
6. Separate sensitive data from nonsensitive data.
7. Automate as much of the subject data monitoring as possible. This step improves consistency in identifying potential problems, such as repeat entries by the same subject. More important, it reduces the likelihood of exposure of sensitive data. The automated scripts can flag and bring to the attention of researchers particular cases without revealing unnecessary information.

12.6   REFLECTIONS: WHAT WORKED? WHAT SURPRISED US? WHAT LESSONS APPLY OFFLINE?

Our experience with MINTS, and a follow-up survey we recently conducted (called MINTS-II), both validate our approach and reveal areas where it was incumbent on the research team to adapt quickly to protect the integrity of the study.

To our knowledge, we completed the study without suffering attacks to our data and without the identity of any subjects being revealed. At the same time, we encountered a much greater level of repeat survey submission than we expected and found that automated monitoring is only part of the solution to this problem. In MINTS, we identified 119 repeat survey responses from 18 different subjects; one subject submitted 66 separate complete surveys (Konstan et al., 2005). We discovered that a $20 per survey reward motivates substantial creativity among prospective multiple respondents. Some of them created multiple e-mail addresses through which to receive PayPal payments. Others moved from computer to computer to avoid being detected by automated IP address matching.

One lesson from our experience is to carefully consider the trade-offs involved between protecting subject identity and response quality. On the one hand, we refused to automatically reject multiple submissions from an IP address (or with the same payment information) because it was important to allow cohabiting individuals to both participate. In addition, many dial-up users share the IP addresses of the parent ISP. However, on the other hand, we estimate that 10–20% of records will appear "suspicious" and will need to be validated by hand. Deduplication and subject validation efforts, therefore, often require extensive automated and manual safeguards that are costly in terms of personnel and time. Clearly, the level of concern depends heavily on both the population being studied and the compensation being offered.

While the experiences reported in this chapter are limited to online surveys, we recognize that some of the same principles apply to other forms of survey research where subject data privacy is of great concern. Both Internet-based and more conventional survey methods benefit from a highly trained interdisciplinary team approach. Whether subjects are called for telephone interviews, appear in person, or are reached online, a diverse team is better able to anticipate pitfalls and challenge the survey orthodoxy that may inadvertently compromise subject privacy and data quality.

Having an early and thorough assessment of risks, and identifying the priorities and key trade-offs would be wise in any study. We view our experience not as a set of results that should be copied, but rather as an example of the type of results that a thoughtful process may yield; the risks are different in other studies and the principles and trade-offs will be different.

Finally, the key issues of data integrity—from backup plans to separation and encryption of identifying data—apply to all research settings. Challenges faced with online survey techniques (see Pequegnat, et al., 2007) are in many ways reformulations of those which researchers have always struggled with. Proper planning and active monitoring are the most effective defenses in all cases.

REFERENCES

Bull, S. S., McFarlane, M., and Rietmeijer, C. (2001). HIV and sexually transmitted infection risk behaviors among men seeking sex with men on-line. *American Journal of Public Health, 91*, 988–999.

Cho, H., and LaRose, R. (1999). Privacy issues in Internet surveys. *Social Science Computer Review 17*(4), 421–434.

Horvath, K. J., Bowen, A. M., and Williams, M. L. (2006). Virtual and physical venues as context for HIV risk among rural men who have sex with men. *Health Psychology, 25,* 237–242.

Konstan, J. A., Rosser, B. R. S., Ross, M. W., Stanton, J., and Edwards, W. M. (2005). The story of subject naught: A cautionary but optimistic tale of Internet survey research. *Journal of Computer-Mediated Communication, 10*(2), article 11. http://jcmc.indiana.edu/vol10/issue2/konstan.html

Lenert, L., and Skoczen, S. (2002) . The Internet as a research tool: Worth the price of admission? *Annals of Behavioral Medicine, 24*(4), 251–256.

O'Neil, D. (2001). Analysis of Internet users level of online privacy concerns. *Social Science Computer Review, 19*(1), 17–31.

Pequegnat, W., Rosser, B. R. S., Bowen, A. M., Bull, S. S., Diclemente, R. J., Bockting, W. O., Elford, J., Fishbein, M., Gurak, L., Horvath, K. J., Konstan, J., Noar, S. M., Ross, M. W., Sherr, L., Spiegel, D., and Zimmerman, R. (2007). Internet-based HIV/STD prevention survey lessons: practical challenges and opportunities. *AIDS & Behavior.*

Reeves, P. M. (2001). How individuals coping with HIV/AIDS use the Internet. *Health Education Research,* 16(6), 709–719.

Ross, M. W., et al. (2000). Differences between Internet samples and conventional samples of men who have sex with men: implications for research and HIV interventions. *Social Science & Medicine 51*, 749–758.

Sills, S. J., and Song, C. (2002). Innovations in survey research: an application of Web-based surveys. *Social Science Computer Review, 20*(1), 22–30.

Stanton, J. M. (1998). An empirical assessment of data collection using the Internet. *Personnel Psychology 51*, 709–725.

CHAPTER 13

Surveys and Surveillance

Gary T. Marx
Massachusetts Institute of Technology, Cambridge, Massachusetts

13.1 INTRODUCTION

A survey is a form of surveillance. Survey and surveil are synonyms and share Latin roots. The former, however, does not usually conjure up Orwellian images. Rather the survey in its best light has been seen as a component of democratic society in which citizens can voluntarily inform leaders of their attitudes and needs and help clarify public policy questions and all under presumably neutral, scientific conditions that can be trusted. Survey respondents are encouraged to participate in order to "make your voice heard." A pollster observes, "polling is an integral part of the democratic process, it gives everybody a chance to have an equal voice" (Krehbel, 2006).

That lofty potential is present, but so is its opposite. There is risk of a kind of scientific colonialism in which various publics are expected to offer unlimited access to their personal data, but rather than serving the public interest, this can facilitate manipulation by elites pursuing selfish and/or undemocratic ends. We might ask the optimistic pollster of the paragraph above, "Does everyone also have an equal chance to determine what the questions are, who gets questioned and how the data will be used?" Consider the explicit merging of social control and social research in Henry Ford's 1910 Sociology Department. The department supervised a large number of informers on the factory floor and also visited the homes of workers, carrying out interviews on finances, sexual behavior, and related personal matters to be sure workers behaved in ways considered appropriate by the company (Hooker, 1997). The varied relation between these potentials in different contexts, the kinds of personal information sought, and the conditions under which it is gathered raise significant questions. In this chapter I draw from a broader approach for analyzing

Envisioning the Survey Interview of the Future, Edited by Frederick G. Conrad and Michael F. Schober
Copyright © 2008 John Wiley & Sons, Inc.

surveillance to consider some social and ethical implications of survey research, and how they may evolve in the survey interview of the future.

This approach analyzes the *structure* of surveillance settings and characteristics of the *means* used, *data collected* and *goals* sought, and a concern with ethics. This perspective is developed in Marx (forthcoming). Initial work on basic concepts and goals, means, type of data, and ethics is given in Marx (2005, 2004, 2006a, 1998, respectively) available at garymarx.net.

Distinct ethical questions may be raised about the data collection means and process (the topic most relevant to this book). This involves the inherent and/or applied characteristics of the means and the competence of its practitioners. In addition, ethical assessments need to consider the conditions of data analysis (including merging with other data apart from that directly collected) and storage and uses of the data. I note conditions under which surveillance is most likely to be questioned and the subject of controversy. I then ask how these relate to traditional and recent developments in the large scale survey.

Having benefited as a young scholar from the work of Paul Lazerfield, Sam Stouffer, and Hadley Cantril through direct contact with Charles Glock, Hannan Selvin, and S. M. Lipset, I am now pleased for the chance to revisit some of the social and ethical questions of concern to survey researchers in the 1960s and to consider new questions.[1]

A concern of some social researchers, particularly in the later 1960s and early 1970s, was with the impact of survey research as part of a broader critique of social science (Colfax and Roach, 1971; Gouldner,1970; Mills, 2000). Who did the surveys serve? Who sets the research agenda? Are potential users aware of the survey's sponsors and their reasons for seeking the information? What did the survey offer to those disadvantaged and beyond the mainstream who were disproportionately the subjects of research? What did surveys really tell us about social reality (Blumer, 1948)? Those issues have not gone away, but as this volume makes clear new issues are appearing.

Academic social researchers like other brethren in the discovery business (market researchers, investigative reporters, police, private detectives, national security agents) are surely more spies than spied upon, even if for academics this is usually in benign contexts. We all seek to find things out about others which they may, or may not, wish to reveal and which may hurt, harm, help, or be irrelevant to them—whether as individuals or more indirectly as group members (Marx, 1984).

---

[1] My master's thesis using American Institute of Public Opinion data on the 1930s radio priest Father Coughlin was part of a broader project seeking to understand correlates (and more optimistically causes) of presumed threats to democracy (Lipset, 1964; Marx, 1962, 2007b). *Protest and Prejudice* (Marx, 1967) sought to understand sources of black support for the civil rights movement. Both inquiries were in a positivist social engineering vein in which it was rather naively believed that survey analysis would yield instrumental results for specific goals (combating extremism and furthering civil rights militancy). In Marx (1972, 1984) a more tempered view stressing the possibility of surveys and related research as more likely to affect climates of opinion than to offer clear directives for intervention is suggested.

In their other roles, academics are of course also watched.[2] The researcher as agent of surveillance contrasts with the researcher as subject of surveillance as a citizen, consumer, communicator, and employee (e.g., database searches involved in promotion and tenure for faculty members at universities).

With respect to organization, problems seem more likely in the surveillance of an individual by an organization (and one that is minimally accountable) and when data collection is nonreciprocal and initiated by the surveillance agent. This is related to power and resource imbalances within stratified societies and organizations.

With respect to any technology, social and ethical concerns regarding surveillance are most likely to be raised by critics when the means are coercive, secret, involuntary, and passive and involve harm, risk, or other costs to the subject—whether during the data collection process or in subsequent uses, particularly when there is no formal juridical or policy review and when the playing field is inequitably tilted. Such review may apply to a general surveillance practice, as well as a given application.

Controversy is also more likely when the organization's goals are in conflict with the legitimate interests and goals of the subject. Consider, for example, selling lists of pharmaceutical drugs used to potential health insurers or employers.

Concerns about crossing informational borders are also more common when the subject's specific identity is known and he/she is locatable and when the data collected are personal, private, intimate, sensitive, stigmatizing, strategically valuable, extensive, biological, naturalistic, predictive, attached to the person, reveal deception, and involve an enduring and unalterable documentary record and when the data are treated as the property of the collector and are unavailable to the subject, whether to review and question or to use.

When compared to many forms of government, marketing and work surveillance, let alone that of freelance voyeurs, academic social studies including survey research, in particular, are in general less problematic. At least that is the case with respect to the structure of the data collection context, the means used, and the goals. However, it is potentially not the case when considering the kind of sensitive data the survey may collect.

With respect to structure, academic survey research can be categorized as *role relationship* surveillance carried out by an *organization* with an *individual subject.* The individual is part of an *external constituency.* The action is *surveillance agent initiated.* There is a *nonreciprocal* information *flow* from the subject to the agent (the respondent does not usually ask the interviewer about his/her income or birth control practices).[3] In the survey research context, see the work of Schober, Conrad, Ehlen, and Fricker (2003).

---

[2] Consider the FBI's interest in sociologists and anthropologists during the Cold War period (Keen, 1999; Price, 2004) and other encounters sometimes triggered by the researcher's proximity to dirty data topics involving crime and deviance. Psychologists were also no doubt of interest. The ironies can be striking. In an epilogue to his study of government welfare surveillance, Gilliom (2001) reports on his serendipitous encounter with some involuntary participant observation data, as he became the subject of unwarranted state surveillance.

[3] These distinctions are developed in Marx (2007a). For example, role relationship surveillance involves normatively patterned interactions—whether in a family or formal organizational setting. This contrasts

These structural conditions of the survey can be associated with misuse. However, this is lessened in general because the *goals* of the agent are likely to be either *neutral* or *supportive* of the subject's goals, rather than in conflict with them. Of course, questions can be raised about the goals and who determines and carries out the research agenda and has access to the results. Consider, for example, the aborted Project Camelot, a social science counterinsurgency effort (Horowitz, 1967) The data are also likely to become *publicly available* (both with respect to their content and for secondary use), rather than kept secret, or considered the exclusive property of the surveyor. Ironically, this can undercut the privacy of the data, particularly when the bin containing selected subgroups is very small (e.g., in a known location, such as a small town, when sorting by variables that successively narrow the size of the pool—age, gender, education, and religion). We also need to ask public to whom? Making the individual data available to the respondent differs from making it available to anyone who wants to do secondary analysis. There are irreconcilable conflicts and often competing goods between publicity and privacy. They can, however, be more sensitively managed when these concerns are acknowledged.

The major goal in academic research is the advancement of knowledge, whether for reasons of scholarship or public policy and planning. While there are ostensibly broad public interest goals (although that concept can be elusive)—serving health, education, and welfare, the economy, and ameliorating social problems (apart from whether or not these are met)—these are unlikely to be in direct conflict with the legitimate personal goals or interests of the subject (in contrast to some bank, insurance, or employment settings) that are similarly structured.

While the surveying organization has greater resources than the subject of the interview and data flow in only one direction, this need not lead to abuse. The survey is rooted in an institutional context, which provides standards and reviews. Universities and government agencies have expectations and procedures regarding responsible research. IRBs can serve as a brake on unrestrained research (that they can also break desirable research is a different issue). Public and private funding sources exert control. Peer networks, the professions, and their associations through socialization and codes of ethics also may serve to limit research excesses.

The survey subject is obviously aware of the data collection and the interviewer as surveillance agent (although not necessarily of the survey's full purposes or of what will count as data). The subject also consents. An ironic cost of this can be intruding into a person's solitude and taking his/her time away from other activities.

The respondent (Milgram effects to the contrary) is presumably free to refuse to be interviewed. According to a 2003 Pew Research Center survey, about seven out of ten telephone calls answered do not result in completed interviews (a significant increase over the seven years before). Contrast that with the difficulty of avoiding video, credit card, and internet surveillance. Respondents' decision to participate in surveys seems related to their perceptions of disclosure risk and risk of harm (see Fuchs, Chapter 4

---

with the nonrole relationship surveillance found with the voyeur whose watching is unnconnected to a legitmate social role.

in this volume, and the discussion of Bob Grooves's work in Schober and Conrad, Chapter 1 in this volume).

Beyond opting out when contacted, the individual may refuse to answer a question or terminate the interview at any point. The interview setting also offers (or at least traditionally has offered) room for the subject to cover and protect information he/she would prefer to keep private. This may permit dignity for the subject not found under coercive, secret, and involuntary conditions. The subject is free to decide just how much to reveal and how honest to be and even to walk away or hang up.

Survey data often involve sensitive material that could harm the individual. This harm may be appropriate or inappropriate in some larger sense. For example, if survey information on crime is reported to law enforcement, the harm is not to justice, but with the violation of the promise of confidentiality and data protection offered when the information was collected. The harm in data collection from being asked to recall and report sensitive information may differ from that with the involuntary and coercive crossing of a physical border as with taking blood, a urine sample, or body cavity search. However, the chance of harm is lessened because the goal is aggregate data on populations, rather than data on the subject per se. As a result, the individual is unlikely to suffer direct adverse consequences. Certainly very personal data offer temptations for misuse (whether commercial or prurient) to those in the survey research process. Yet the deindividualization of the data works against this (National Research Council, 1993).

Confidentiality and anonymity (with respect to both name and location) are likely to be promised. Identity is deleted or hidden via masking. The divorce of subject ID from the data can overcome the problems created in other similar structural and data settings, where the subject is identified and can be located. Preliminary evidence suggests there is relatively little chance of intruders identifying respondents by linking individual characteristics reported in the survey to other data sources.[4]

In summary, many of the correlates of surveillance abuse noted earlier are irrelevant to the typical survey context or if they apply, do so in a nonproblematic way. Considering the broad array of contemporary surveillance forms, the traditional survey seems a rather mild invasion and intrusion, which is not unduly or unreasonably costly to the subject. This does not call for a lessening of vigilance.

13.2   THE INTERVIEW OF THE FUTURE: NEW WINE NEW BOTTLES?

Yet what of new interviewing technologies involving paradata as discussed by Person, D'Mello and Olney in Chapter 10 in this volume and by Couper in Chapter 3 in this volume? These techniques are examples of the new surveillance. Along with video cameras, facial recognition, RFID chips, and drug testing or DNA analysis based on

---

[4] Discomfort in the *collection* of sensitive information may remain such as embarrassment and painful recall. New data collection techniques involving interaction with a computer may eliminate some of this (see Couper, Chapter 3 in this volume).

a strand of hair, they are automated and can be involuntary, passive, invisible, and unavailable to the subject, while extending the senses of the surveiller.[5]

It is easy to envision a sci-fi (or is it?) scenario involving duped or simply unaware respondents enveloped in a shield (or a vacuum?) of ambient intelligence. Consider a large interdisciplinary National Séance Foundation project done in the near future. This involves a collaborative relationship between police investigators and social scientists. The group had previously cooperated on improving methods of data collection in contexts of minimal or no cooperation, where there was reason to suspect at least some dishonesty, regardless of whether subjects are survey respondents or interrogated suspects.[6] The project involves five universities and is concerned with developing better tools to identify, in the hope of preventing, problems related to (1) drug use, (2) political extremism, (3) crime, and (4) sexual behavior.

Given the sensitive nature of the topics and the difficulty of obtaining adequate and valid data, researchers have sought unobtrusive methods (Hancock, 2004). Such methods can enhance the face-to-face interview by probing beneath the deceptive veneer of the seemingly "authentic intentionality" found with reliance exclusively on the subject's words. The methodology draws from recent communications research on the ease of conversational deception and the limits of any survey that gives the respondent space for impression management.[7] A mostly noninvasive multimodal approach (NIMMA) exploiting all channels of communication is used. This includes the methods Person et al. (Chapter 10) describe such as PEP, fMRI, EEG, EKG, BPMS, and Wmatrix and some additional means still under beta test such as ZOWIE, WAMIE, and BMOC[©].

Validity and completeness of response are greatly enhanced by access to unobtrusive measures and to comparisons to data found beyond the interview situation. This approach is also environmentally sound and efficient, as it doesn't waste data. It seeks to find new meaning through creating a mosaic of previously meaningless, unseen, unconnected, and unused data. The scientist who relies entirely on words is a rather unscientific, profligate, one trick *ancien* pony who needs to get with the program.

The interview and related detection occur in one of ten, ambient intelligence, pastel living rooms matched to the social and psychological characteristics of the subject (e.g., age, education, gender, lifestyle). This is inspired by the clustering of respondents into types pioneered by marketing research. One size hardly fits all. The rooms can be internally rearranged to accommodate 68 distinct types of respondent. Respondent characteristics are determined by a preliminary research encounter, which includes an electronically measured handwriting sample, a Google search, and search of public and (under carefully controlled conditions) restricted databases regarding

---

[5] Marx (2006b) considers a number of soft forms of surveillance.

[6] See Leo (forthcoming) for a highly informative account of other mergers of social science and law enforcement in a context of seeking withheld information.

[7] In their zeal to find the real deal, the researchers of course fail to note that deceptive data are none the less data worthy of analysis, apart from their empirical veracity.

sensitive behavior.[8] Music the respondent is likely to find pleasing is softly played and a subliminal voice repeats "be truthful." A barely noticeable scent of pine (believed to be calming) is delivered through the air duct system.

The subject is told that in order to accurately reflect his/her experiences and opinions a variety of "state of the art" technologies are being used. However, this message is nonspecific—the subject is not told that the chair seat measures wiggling, body temperature, and electrodermal response; that facial expression, eye patterns, voice microtremors, and brain wave emissions are recorded. Nor is there notice that all verbal utterances are recorded, as is the timing of responses. Some inferences are made about the respondent based on answers to questions unrelated to the topics of direct interest. The internal consistency of questions is analyzed and answers are compared to data found through cross checking a variety of public and (because this is such an important government-sponsored study) private databases.

The room is slightly warmer than most and subjects are provided with a full complement of free beverages. They are encouraged to use the restroom during or at the conclusion of the two hour interview. They are not told that a variety of automated biochemical urine and olfactory assays are performed, since such offerings by subjects are voluntary.

All of the data from the interview is available in real time via a password-protected Web page to professional observers at the cooperating research agencies. To reduce generative performance anxiety, respondents are not told about the remote observation. The interviewer wears a tiny earpiece device that permits feedback and suggestions from the remote observers. A case agent in another room monitors all data flows and quietly informs the interviewer if the respondent is deceptive, frustrated, stressed, fatigued, or doesn't understand the question. The agent is a kind of coach quietly offering information to the interviewer.

Respondents are of course promised confidentially and that only those who need to know their identity and their data will have them. However, they are given the chance to waive this protection (and a large percentage do, particularly those of lower status backgrounds), should they wish to benefit from a generous frequent shopper offer and the chance to win fabulous prizes. Funds for this are provided by leading marketing researchers and helping agencies who are eager to identify customers and clients in need of their goods and services and the sponsors don't mind the educational purpose tax benefit they receive.

13.3 QUESTIONS

How should this scenario be viewed? Is it merely satire? How far are we from such a situation? In their introduction to this volume, editors Frederick Conrad and Michael Schober note the importance of anticipating the issues raised by new methods. What questions should be asked about any surveillance? What questions seem specific to these new forms of data collection?

---

[8] For example, lists of magazines subscribed to, charities contributed to, Web usage, and criminal justice records, but not legally protected information such as medical records.

The implications of the new methods for sociology of knowledge and truth issues are fascinating. We know that variation may appear with differences in question wording, order, and the match between the respondent and the interviewer with respect to factors such as gender and ethnicity, as well as between a face-to-face versus a computer survey. But some of the new methods raise additional issues (Hancock, Curry, Goorha, & Woodworth, 2004; Holbrook et al., 2003; Schaeffer, 2000; Schober et al., 2003; Tourangeau and Smith, 1998).

In the traditional interview, the respondent has volitional space to appear knowledgeable and as a good citizen, whether warranted or not. It is not only that respondents may indeed have things to hide, but that they may be motivated to help or please the interviewer and hence not speak candidly. "Help" by giving untruthful answers that shorten the interview by leading to shorter pathways through the questionnaire (Hughes et al., 2002) and "please" by providing desirable answers that conform to what is socially desirable rather than what is necessarily truthful.

Traditionally, we accepted this as an inherent limitation of the method and as the cost of gaining cooperation, even as we sought to compensate for it by internal and external validity checks. But what happens when we have new techniques that may go further in identifying lying and in suggesting likely answers (or at least associated emotional responses) when no, or minimal information is directly and knowingly offered? One form involves powerful statistical models as with election and risk forecasting in which only modest amounts of information are required for very accurate aggregate predictions.[9]

Some of the techniques may yield better data in going beyond the deceptive, invalid, or incomplete information that may be found in the traditional interview. What if ways of characterizing and understanding respondents were possible that went far beyond the substantive content of an answer to a question in settings where respondents thought they were merely offering information on their attitudes and experiences? One wonders what Herbert Blumer (1969) and Erving Goffman (1959) might say about such research tools, given their emphasis on getting genuinely empirical and going beyond socially prescribed self-presentations that may be gathered/induced by the formal interview.

What is gained and what is lost when respondents are not as able to manage their impressions, as they could before these technical developments in communication and interrogation?[10]

---

[9] This relates to a broader conflict between economic rationality and efficiency associated with aggregate data on the one hand and expectations of individualized fair and accurate treatment on the other. The occasional errors found in the case of social research from judging a given case by rating it relative to a model based on a large N about which much more is known would be seen to balance each other out.

[10] In the best of all possible worlds, communication (Habermas, 1986) has a reciprocal, rather than a one-way quality. The scenario described (and indeed the framing of this book's concerns) fits within the perspective of communications. But it is a communication from the subject. To be sure, via targeted messages and propaganda, the results may then lead to communication back to the subject based on the interview results, but this is certainly asymmetrical communication compared to what is found in ordinary communication in the give and take of exchanges.

How does the interview situation—both empirically regarding the possible effect on results, and morally regarding appropriateness—change when there are hidden, or at least unseen, observers, when meaningless data are converted to information, when unwitting subject offerings are taken as cues to internal states and truthfulness, and when interviewer data is matched to other sources of data on the subject available beyond the interview setting?

How should the trade-offs between enhanced data quality be weighed against manipulating "subjects" in order to maximize what can be learned about them, when they may be unwilling, or unable, to tell us what we desire to know? How should the gains from overcoming the limitations to data quality be balanced with the harm to a respondent's self-image in being shown to be a liar, ignorant, or a bad citizen? That, of course, assumes respondents would be informed of the power of the technique to reach judgments about consistency, accuracy, and honesty and would have access to their own data and any subsequent additions, coding, and simulations. What would meaningful informed consent look like in such settings? Is it even possible to have this without boring people and significantly increasing refusal and incompletion rates? How can all possible uses/users of the data ever be known? Would offering informed consent significantly increase refusal rates from subgroups who may be central to the research (Singer, 2003)? For the broad public, would knowledge of the increased power of surveys spill over into even higher rates of refusal? Will future research be hurt if potential respondents come not to trust interviewer situations, after the inevitable appearance of newsworthy exposés? Should there be policy limits on the extent to which people may be seduced by rich rewards into selling their most intimate and sensitive information?

Clearly, in taking away some of the respondent's ability to offer a given presentation of self, something is lost. Yet much might be gained as well. The moral haze and the trade-offs here, as with other new surveillance techniques, are what makes the general topic so interesting and challenging.

13.4   MORE QUESTIONS

Broad ethical theories involving teleology, deontology, and casuistry are available, as are the ethical codes of the social science professions and principles such as those in the Belmont Report (National Commission, 1979). However, these are all rather abstract and in practice with the appearance of new means, it is often not at all clear when they apply. An alternative approach, while encouraging the researcher to be mindful of the above, asks for self-reflection (or better group reflection among the researchers and sponsors) with respect to specific questions about the data collection, analysis, and use contexts.

Table 13.1 lists questions that can be asked of surveillance generally. I have put an asterisk by those most relevant to the new survey means. I will argue that the more these can be answered in a way that affirms the underlying ethical principle, the more appropriate the means may be.

TABLE 13.1   Questions for Judging Surveillance

*1. Goals: Have the goals been clearly stated, justified, and prioritized?

*2. Accountable, public, and participatory policy development: Has the decision to apply the technique been developed through an open process, and if appropriate, with participation of those to be surveilled? This involves a transparency principle.

3. Law and ethics: Are the means and ends not only legal but also ethical?

4. Opening doors: Has adequate thought been given to precedent-creation and long term consequences?[a]

*5. Golden rule: Would the watcher be comfortable in being the subject rather than the agent of surveillance if the situation was reversed? Is reciprocity/equivalence possible and appropriate?

*6. Informed consent: Are participants fully apprised of the system's presence and the conditions under which it operates? Is consent genuine (i.e., beyond deception or unreasonable seduction) and can "participation" be refused without dire consequences for the person?

*7. Truth in use: Where personal and private information is involved, does a principle of "unitary usage" apply in which information collected for one purpose is not used for another? Are the announced goals the real goals?

*8. Means–ends relationships: Are the means clearly related to the ends sought and proportional in costs and benefits to the goals?

9. Can science save us? Can a strong empirical and logical case be made that a means will in fact have the broad positive consequences its advocates claim? (This is the "Does it really work? question.)

*10. Competent application: Even if in theory it works, does the system (or operative) using it apply it as intended?

11. Human review: Are automated results, with significant implications for life changes, subject to human review before action is taken?

*12. Minimization: If risks and harm are associated with the tactic, is it applied to minimize intrusiveness and invasiveness?

*13. Alternatives: Are alternative solutions available that would meet the same ends with lesser costs and greater benefits (using a variety of measures not just financial)?

14. Inaction as action: Has consideration been given to the "sometimes it is better to do nothing" principle?

*15. Periodic review: Are there regular efforts to test the system's vulnerability, effectiveness, and fairness and to review policies?

*16. Discovery and rectification of mistakes, errors, and abuses: Are there clear means for identifying and fixing these (and in the case of abuse, applying sanctions)?

*17. Right of inspection: Can individuals see and challenge their own records?

18. Reversibility: If evidence suggests that the costs outweigh the benefits, how easily can the surveillance be stopped (e.g., extent of capital expenditures and available alternatives)?

19. Unintended consequences: Has adequate consideration been given to undesirable consequences, including possible harm to watchers, the watched, and third parties? Can harm be easily discovered and compensated for?

TABLE 13.2 (*Continued*)

---

*20. Data protection and security: Can surveillants protect the information they collect? Do they follow standard data protection and information rights as expressed in the Code of Fair Information Protection Practices and the expanded European Data Protection Directive?[b] (See Bennett and Raab, 2006; European Union, 1995; United States HEW, 1973).

---

[a] The expansion of a technology introduced in a limited fashion can often be seen. Extensions beyond the initial use, whether reflecting *surveillance creep*, or in many cases *surveillance gallup*, are common. Awareness of this tendency requires asking of any new tactic, regardless of how benignly it is presented as both means and ends, "Where might it lead?" Consider examples of surveillance expansion such as the Social Security number that Congress intended only for tax purposes but that now has become a *de facto* national ID number and drug testing, once restricted to those working in nuclear-power facilities and the military, is now required of bank tellers and in some places even junior high schools. Once a surveillance system is established it may be extended to new subjects, uses, and users. Many factors contribute to this. Economies of scale are created that reduce the per-unit cost of such extensions. Experience is gained with the technique and precedent is established. What was initially seen as a shocking intrusion may come to be seen as business as usual and extensions may be taken for granted.
[b]These offer standards for the data gatherers with respect to the collection, retention, and use of personal information.

REFERENCES

Bennett, C., and Raab, C. (2006). *The Governance of Privacy.* Cambridge, MA: MIT Press.

Blumer, H. (1948). Public opinion and public opinion polling. *American Sociological Review, 13,* 542–549.

Blumer, H. (1969). *Symbolic Interaction: Perspective and Method.* Englewood Cliffs, NJ: Prentice Hall.

Colfax, J. D., and Roach, J. D. (1971). *Radical Sociology.* New York: Basic Books.

European Union (1995). *Directive 95/46/EC of the European Parliament and of the Council on the Protection of Individuals with Regard to the Processing of Personal Data and on the free Movement of Such Data* (OJ No. L281). Brussels.

Gilliom, J. (2005). *Overseers of the Poor.* Chicago: University of Chicago Press.

Goffman, E. (1959). *The Presentation of Self in Everyday Life.* New York: Doubleday.

Gouldner, A. (1970). *The Coming Crisis in Western Sociology.* New York: Basic Books.

Habermas, J. (1986). *The Theory of Communicative Action.* Boston: Beacon Press.

Hancock, J. T., Curry, L. E., Goorha, S. and Woodworth, M. T. (2004). Lies in conversation: An examination of deception using automated linguistic analysis. *Proceedings of the 26th Annual Conference of the Cognitive Science Society*, pp. 534–539.

Holbrook, A. L., Green, M. C., and Krosnick, J. A. (2003). Telephone versus face-to-face interviewing of national probability samples with long questionnaires: comparisons of respondent satisficing and social desirability response bias. *Public Opinion Quarterly, 67,* 79–125.

Hooker, C. (1997). Ford's sociology department and the Americanization campaign and the manufacture of popular culture among assembly line workers c.1910–1917. *The Journal of American Culture, 20*(1), 47–53.

Horowitz, I. L. (1967). *The Rise and Fall of Project Camelot.* Cambridge, MA: MIT Press.

Hughes, A., Chromy, J., Giacoletti, K., and Odom, D. (2002). Impact of Interviewer Experience on Respondent Reports of Substance Use. In J. Gfroerer, J. Eyerman and J. Chromy (eds.), *Redesigning an Ongoing National Household Survey: Methodological Issues.* DHHS Publication No. SMA 03-3768. Rockville, MD: Substance Abuse and Mental Health Services Administration, Office of Applied Studies.

Keen, M. (1999). *Stalking the Sociological Imagination.* Westport, CT: Greenwood Press.

Krehbel, R. (2006, July 9). Poll takers count on public for answers. *Tulsa World.*

Lipset, S. M. (1964). Coughlinites, McCarthyites and Birchers: radical rightists of three Generations. In D. Bell (Ed.), *The Radical Right.* New York: Doubleday.

Leo, R. (forthcoming) *Police Interrogation and American justice.* Cambridge, MA: Harvard University Press.

Marx, G. T. (1962). *The Social Basis of the Support of a Depression Era Extremist: Father Coughlin,* Monograph No. 7. Berkeley: University of California Survey Research Center.

Marx, G. T. (1967). *Protest and Prejudice,* New York: Harper and Row.

Marx, G. T. (1972). *Muckraking Sociology.* New Brunswick: Transaction.

Marx, G. T. (1984). Notes on the discovery, collection, and assessment of hidden and dirty data. In J. Schneider and J. Kitsuse (Eds.), *Studies in the Sociology of Social Problems.* Norwood, NJ: Ablex Publishing Corporation.

Marx, G. T. (1998). An ethics for the new surveillance.*Information Society, 14,* 171–186.

Marx, G. T. (2004). What's new about the new surveillance? Classifying for change and continuity. *Knowledge, Technology and Policy, 17*(1), 18–37. (Earlier version in *Surveillance and Society*, 1, X–X.)

Marx, G. T. (2005). Surveillance and society. In G. Ritzer (Ed.), *Encyclopedia of Social Theory.* Thousand Oaks, CA: Sage Publications.

Marx, G. T. (2006a). Varieties of personal information as influences on attitudes toward surveillance. In R. Ericson and K. Haggerty (Eds.), *The New Politics of Surveillance and Visibility.* Toronto: University of Toronto Press.

Marx, G. T. (2006b). Surveillance: the growth of mandatory volunteerism in collecting personal information—Hey buddy can you spare a DNA? In T. Monahan (Ed.),*Surveillance and Security.* New York: Routledge.

Marx, G. T. (2007a). Desperately seeking surveillance studies: players in search of a field. *Contemporary Sociology, 36,* 125–130.

Marx, G. T. (2007b). Travels with Marty: Seymour Martin Lipset as mentor. *The American Sociologist, 37,* 76–83.

Marx, G. T. (forthcoming). *Windows into the Soul: Surveillance and Society in an Age of High Technology.* Chicago: University of Chicago Press.

Mills, C. W. (2000). *The Sociological Imagination.* New York: Oxford.

National Commission. (1979). *The Belmont Report: Ethical Principles and Guidelines for the Protection of Human Subjects and Research.* Washington DC: HEW.

National Research Council. (1993). *Private Lives and Public Policies.* Washington DC: National Academy Press.

Price D. (2004). *Threatening Anthropology: McCarthyism and the FBI's Surveillance of Activist Anthropologists.* Durham, NC: Duke University Press.

Schaeffer, N. C. (2000). Asking questions about threatening topics: a selective overview. In A. A. Stone et al. (Eds.), *The Science of Self-report: Implications for Research and Practice* (pp. 105–122). Mahwah, NJ: Lawrence Erlbaum.

Schober, M. F., Conrad, F. G., Ehlen, P. and Fricker, S. S. (2003). *How Web Surveys Differ From Other Kinds of User Interfaces. Proceedings of the American Statistical Association, Section on Survey Research Methods*, Alexandria, VA: American Statistical Association.

Singer, E. (2003). Exploring the meaning of consent: Participation in research and beliefs about risks and benefits. *Journal of Official Statistics*, 19, 273–285.

Tourangeau, R., and Smith, T. (1998). Collecting sensitive information with different modes of data collection. In M. P. Couper, R. P. Baker, J. Bethlehem, C. Z. F. Clark, J. Martin, W. L. Nicholls II, and J. M. O'Reilly (Eds.), *Computer Assisted Survey Information Collection*. Hoboken, NJ: John Wiley and Sons.

United States HEW. (1973). Secretary's advisory committee on automated personal data systems, records, computers, and the rights of citizens.

CHAPTER 14

Survey Interviews with New Communication Technologies: Synthesis and Future Opportunities

Arthur C. Graesser, Moongee Jeon, and Bethany McDaniel
University of Memphis, Memphis, Tennessee

14.1 INTRODUCTION

There undeniably has been a revolution in new communication technologies during the last decade. Some of these advances in technology will end up having a radical impact on the process of collecting survey interview data, as the chapters in this edited volume make abundantly clear. A change in interview technology may end up being a necessity, not merely a convenience or a matter of cutting costs. For example, the survey world will need to make some adjustments very soon as the public experiences a shift from landline telephones to mobile telephones or to the Web. Surveys will need to be very different in ten years if the mainstream human–computer interface ends up having animated conversational agents, graphical user interfaces, speech recognition, and multimodal interaction. We can only speculate, of course, what the world will be like in ten years.

This chapter has two primary goals. Our first goal is to summarize the existing state of research on interview technologies. We will not attempt to give an exhaustive and comprehensive summary, given the broad coverage of the field in the introductory chapter by Schober and Conrad (Chapter 1) and by Couper (Chapter 3). Instead, we provide a succinct snapshot of where we are now in the field in order to address important themes, obvious research gaps, and unresolved trade-offs on the impact of technologies on psychological and practical dimensions. Our second goal is to identify some important research areas for the future.

We entirely recognize the need to be cautious when exploring the impact of technology on survey interview practice. Consider, for example, the role of technology in training and education. Cuban (1986, 2001) has documented that technology has historically had a negligible impact on improvements in education. As a case in point, visionaries predicted that radio and television would have a revolutionary impact on education, but this never materialized, as we all know. Clark (1983) argued that it is the pedagogy underlying a learning environment, not the technology per se, that typically explains learning gains. That conclusion, of course, motivates the need to investigate how particular technologies are naturally aligned with particular psychological principles, theories, models, hypotheses, or intuitions. These lessons learned from the learning sciences have relevance to the present book on interview technologies.

The *universal interface* of any communication system is the human face, with accompanying speech, gesture, facial expressions, posture, body movements, and human intelligence. This face-to-face (FTF) "inter-face" has sometimes been considered the gold standard for survey interviews. All other technologies are mediated, if not degenerate, forms of FTF communication, whether they be surveys with paper and pencil, telephone, Web, or animated conversational agents. One fundamental question is what gets lost or is gained from these alternatives to FTF interviews. For example, compared to Web surveys, FTF has an advantage of being better able to detect whether the respondent understands the survey questions by virtue of paralinguistic cues (e.g., pauses, intonation, speech disfluencies, gaze, posture, and facial expressions). These cues provide a *social presence* that is minimal or nonexistent on Web surveys. However, on the flip side, the social presence makes respondents more reluctant to share information on sensitive topics such as sexual activities and consumption of drugs and alcohol (Currivan et al., 2004; Tourangeau and Smith, 1998). There is a fundamental trade-off between technologies that afford social presence and the revelation of sensitive information (Sudman and Bradburn, 1974). Such a comparison in technologies suggests that it is illusory to believe that there is a perfect universal interview technology. Instead, there are trade-offs between evaluation measures when considering the landscape of technologies. We need to develop a science that takes these trade-offs into account when selecting the optimal technology for a survey that is targeted for a particular topic, objective, population, budget, and margin of accuracy.

14.2   CURRENT STATE OF RESEARCH ON INTERVIEW TECHNOLOGIES

Table 14.1 lists the interview technologies that have been used in the survey field or investigated by researchers of survey methodology. Chapters in this volume focus on particular technologies, as in the case of video conferencing (Anderson, Chapter 5), interactive voice response (Bloom, Chapter 6; Couper, Chapter 3), mobile and landline telephone surveys (Fuchs, Chapter 4; Hancock, Chapter 9), instant messaging (Hancock, Chapter 9), Web surveys (Couper, Chapter 3; Fuchs, Chapter 4; Konstan et al., Chapter 12), computer-mediated conversation (Fussell et al., Chapter 11),

TABLE 14.1   Interview Technologies

1. Face to face (FTF)
2. Video conferencing
3. Animated conversational agents
4. Video telephony
5. Telephone, including landlines and mobile phones
6. Interactive voice response (IVR)
7. Self-administered questionnaire via mail or directly administered
8. Email
9. Instant messaging
10. Web
11. Computer assisted survey interview (CASI), including audio CASI and video CASI
12. Computer assisted personal interview (CAPI)
13. Computer assisted telephone interview (CATI)
14. Advanced sensing capabilities, including speech, gesture, and handwriting recognition
15. Multimedia and multimodal interfaces

multimodal interfaces (Johnston, Chapter 7), computer-assisted interviews (Bloom, Chapter 6; Couper, Chapter 3), advanced sensing technologies (Bloom, Chapter 6; Cassell and Miller, Chapter 8; Johnston, Chapter 7; Person et al., Chapter 10), and animated conversational agents (Cassell and Miller, Chapter 8; Person et al., Chapter 10). Most of these chapters compare two or more interview technologies in an attempt to assess the impact of such contrasts on performance measures and also to test general psychological principles.

Table 14.2 lists the characteristics on which the interview technologies have been specified, measured, and evaluated. The table identifies the technology's affordances (Clark and Brennan, 1991; Fussell et al., Chapter 11 in this volume), modalities and media (Anderson, Chapter 5; Couper, Chapter 3; Couper et al., 2004; Johnston, Chapter 7; Johnston and Bangalore, 2005; Mayer, 2005), conversational capabilities (Conrad and Schober, 2000; Schober and Conrad, Chapter 1; Schaeffer and Maynard,

TABLE 14.2   Technology affordances, modalities, media, conversation capabilities, errors, costs, and ethics

Affordances (Clark and Brennan, 1991): Co-presence, visibility, audibility, cotemporality, simultaneity, sequentiality, reviewability, revisability.

Modalities/media (Mayer, 2005): Visual, graphical, spatial, linguistic, print, auditory, spoken.

Conversation capabilities (Conrad and Schober, 1997; Schaeffer and Maynard, 2002; Chapters 1 and 2): Social presence, scripted, interactive, grounded, interviewer-entered, mixed-initiative.

Errors (Groves, 1989): validity, reliability, measurement error, processing error, coverage error, sampling error, non-response error.

Costs: Development, implementation.

Ethics: Privacy, informed consent, ethical principle.

2002, and Chapter 2 in this volume; Schober and Conrad, 1997, 2002; Suchman and Jordan, 1990), errors (Groves, 1989), costs, and ethical ramifications (Konstan et al., Chapter 12; Marx, Chapter 13). Some of these characteristics have been investigated in considerable detail, such as validity and reliability, whereas data are conspicuously absent for costs of development and implementation.

At this point in the science, it would be worthwhile to fill out all of the cells in a matrix that we call the *technology-characteristic landscape* (TC landscape). In essence, the affordances, modalities, media, conversational capabilities, and other characteristics in Table 14.2 are specified in tables that map these characteristics onto particular interview technologies in Table 14.1. Researchers have filled out parts of the landscape. For example, Fussell et al. (Chapter 11) and Hancock (Chapter 9) identify the affordances for FTF, video conferencing, telephone, and e-mail versus instant messaging. The conventional technologies (FTF, self-administered questionnaires, telephone) have been compared on reliability and validity (Groves, 1989). However, cells are sparse on the more advanced interview technologies such as animated conversational agents and multimodal interfaces.

Regarding costs, it is informative that quantitative data are either missing or difficult to locate for the advanced interview technologies. The comparative costs can be estimated at a crude level. For example, development costs are high for developing IVR and highly interactive Web surveys compared with FTF and telephone surveys with humans, but the implementation costs would yield the opposite trend. Without precise costs on development and implementation, no solid conclusions can be made on returns on investments (ROIs) for alternative interview technologies. An econometric analysis would be worthwhile that allows an assessment of the gains in validity or reliability as a function of increments in costs for particular technologies.

Psychological theories and principles offer predictions on the comparative impact of alternative technologies on measures of interest. *Social presence* (Reeves and Nass, 1996) is a theoretical principle that has received considerable attention. An interview technology has progressively more social presence to the extent that it resembles human characteristics of the gold standard FTF interview. This would predict the following gradient on social presence: FTF > telephone > IVR > self-administered questionnaire. As mentioned earlier, an interview technology with high social presence has the liability of lowering the likelihood that respondents will accurately report information on sexual habits, consumption of alcohol, and other sensitive topics. Precisely where animated conversational agents fit into this continuum is unclear compared to telephones because agents have a face, unlike the telephone interviews, but the speech and conversational intelligence is limited. The data collected by agents may be indistinguishable from those collected in highly scripted, rigid FTF interviews that have few degrees of freedom, but the agents would be akin to IVR when respondents desire a more flexible conversation that permits clarification questions. Conversational interviews require human intelligence that successfully achieves deep comprehension, language generation, and the processing of pragmatically appropriate paralinguistic cues, all of which are beyond the horizon of current agent technologies. Of course, this does not stop researchers from attempting to build agents that emulate humans. Some reasonably convincing agents have been developed in the context of tutoring and advanced learning environments, as in the case of AutoTutor (Graesser et al., 2005a; Graesser et al., 2001; VanLehn et al., 2007) and Tactical Iraqi (Johnson and Beal, 2005).

Conversational grounding is another theoretical principle that has salient relevance to interviews (Conrad and Schober, 2000; Schaeffer and Maynard, 2002, and Chapter 2 in this volume; Schober and Conrad, Chapter 1; Suchman and Jordan, 1990). Grounding is permitted when the respondent can ask questions or otherwise receive help that clarifies the intended meaning of terms and ideas expressed in survey questions. This is impossible on self-administered questionnaires but possible in FTF and telephone interviews that allow interviewers to answer respondents' clarification questions. It is also possible on Web surveys with facilities that answer respondent questions (Schober et al., 2003). Conversational interviewing increases the likelihood of there being an alignment between the plans, intentions, and common ground of interviewer and respondent. One danger in conversational interviewing, however, is that the interviewer runs the risk of implicitly unveiling attitudes, leading the respondent, and thereby increasing error. An agent would of course provide a more controlled conversational interaction, but this does not necessary reduce the error. A quirky agent would conceivably create pragmatic irregularities that also result in sources of error.

Discourse theories emphasize the importance of identifying the structure, ground rules, and constraints of specific conversational registers. In the workshop that led to this volume, Norman Bradburn suggested that there are four core characteristics of the traditional survey interview (see Fowler and Mangione, 1990): (1) it is a special kind of conversation with its own rules, (2) there is a sequence of questions posed by the interviewer and answers provided by respondents, (3) the interviewer's questions control the conversation, and (4) there is a stereotypical question format. Conversational interviewing loosens the constraints by allowing the interviewer to answer respondents' clarification questions (a limited form of mixed initiative dialogue), and for the question formats to vary in an attempt to achieve conversational grounding and alignment. It is important for the researcher to specify the characteristics of the discourse register of any interview technology.

Contributors to this volume have meticulously identified many of the benefits and liabilities of particular interview technologies. It is quite apparent that there are trade-offs when examining many of the characteristics and measures. There is a trade-off between social presence and the reporting of sensitive information, between the use of computer technologies and implementation costs, between complex interfaces with many options and the need for training, between speech and interpretation accuracy, between advanced technologies and errors of coverage and nonresponse—the list goes on. It is important to document these trade-offs as researchers fill in the cells of the TC landscape.

Survey researchers have developed coding systems to track both the verbal content of the answers and the unconscious reactions, paralinguistic communication cues, and other forms of paradata (Fowler and Cannell, 1996; Lessler and Forsyth, 1996; Presser et al., 2004). The verbal codes include categories of speech acts, such as interviewer reads survey item verbatim, interviewer repeats survey item, respondent requests clarification, respondent repeats answer, interviewer signals return to script, interviewer or respondent laughs, and so on (Schaeffer and Maynard, 2002, and Chapter 2 in this volume). The unconscious codes are pauses, restarts, speech disfluencies, intonation contours, eye gaze, facial expressions, gestures, posture, and so on (Cassell and Miller, Chapter 8 in this volume). These unconscious codes, measured by trained judges or instruments, are sometimes diagnostic of respondents' lying (Hancock, Chapter 9 in this volume), but so are the verbal codes. One question for the Institutional Review Board (IRB) committees is whether informed consent is needed when researchers analyze the unconscious communication channels (Konstan et al., Chapter 12; Marx, Chapter 13; Person et al., Chapter 10 in this volume). Informed consent is normally required for the verbal answers and actions that respondents intentionally provide, but not currently for the unconscious streams of activity.

Matters of ethics, of course, need to be scrutinized for all of the interview technologies. There are *seven critical issues* relevant to ethical violations in clinical and medical research according to Emmanuel, Wendler, and Grady (2000): (a) social or scientific value, (b) scientific validity, (c) fair subject selection, (d) favorable risk–benefit ratio, (e) independent review, (f) informed consent, and (g) respect for enrolled subjects. Fair subject selection is called into question when subsets of the target population tend to be excluded and thereby create coverage error or nonresponse error. If elderly and low SES populations tend not to use the Web, then a Web-based survey would be prone to this ethical violation (Couper, 2000). Informed consent is not collected in Web surveys that are routinely part of the practice of some marketing research firms and for the videotaped surveillance of citizens. The ubiquitous and noninvasive methods of data collection are not always submitted to IRBs so they are prone to ethical violations.

The current state of survey interview research is underdeveloped when it comes to examining differences in culture, languages, community norms, and demographic characteristics other than race, gender, and SES. Cross-cultural research is conspicuously nonexistent in the more advanced interview technologies. Chapter 11 by Fussell et al. is an initial attempt to extend findings from cross-cultural research on computer-mediated communication to survey interviews through new media. The TC landscape would be tremendously expanded as we add these sociocultural dimensions and conduct studies that directly involve survey interviews.

14.3   RESEARCH NEEDS IN THE NEAR HORIZON

This section identifies five important directions for survey researchers to pursue in the future. Some of these would help fill gargantuan gaps in the TC landscape. Others would increase the quantitative and analytical sophistication of the research.

14.3.1   Fine-Grained Engineering and Quantitative Modeling

There comes a point in a social science when researchers shift from inferential statistics that compare conditions on dependent measures to quantitative models that would

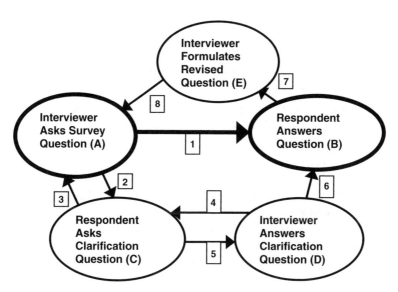

FIGURE 14.1   A Simple State-Transition Network for Survey Interactions.

appeal to engineers. It is time to wear some engineering hats as we move forward to the next stage of analyzing interview technologies. There are a suite of engineering models that have been used in the fields of human–computer interaction, cognitive engineering, and human factors.

One class of models is state-transition networks. There are a finite set of states and a set of arcs that designate transitions between states. Figure 14.1 shows a simple state transition network for survey interactions. There are five states that designate the speech acts of the interviewer and respondent (nodes A through E) and eight arcs that designate transitions between the speech acts (arcs 1 through 8). All surveys require node A (interviewer asks survey question), transition 1, and node B (respondent answers question); this can be signified as either A → B, A1B, or simply AB.

Interview technologies vary on what speech acts are admissible or available in the interview. A self-administered, paper questionnaire or strictly standardized interview would afford only the AB path because there is no possibility of respondents asking clarification questions (node C), the interviewer answering these questions (node D), and the interviewer formulating revised questions (node E). The only option available to the survey researcher is to optimize the context and wording of the questions to maximize the reliability and validity of the responses. This is accomplished by pretesting the questions (Presser et al., 2004) or by using computer tools, such as Question Understanding Aid (QUAID) (Graesser et al., 2006), to minimize the respondents' misunderstanding the questions.

A conversational interview, on the other hand, would allow speech acts in nodes C, D, and E in an effort to enhance conversational grounding and alignment (Conrad and Schober, 2000; Schaeffer and Maynard, 2002, and Chapter 2 in this volume; Schober and Conrad, 1997, and Chapter 1 in this volume; Suchman and Jordan, 1990). This is illustrated in the following hypothetical exchange.

Interviewer Question (A): *How many vehicles are in your household?*
Respondent Answer (B): *20 or 30.*
Interviewer Revised Question (E,A): *How many vehicles during the last year?*
Respondent Question (C): *Do you include motorcycles and bicycles?*
Interviewer Answer (D): *Vehicles includes motorcycles, but not bicycles.*
Respondent Answer (B): *Okay, 4.*

This example illustrates that the initial answer of *20 or 30* would not be valid, that the self-administered questionnaire would fall prey to the invalid response, and that the conversational interview would unveil a more valid response. The potential downside of the conversational interview is that the interviewer's speech acts run the risk of telegraphing expectations of the interviewer and indirect feedback, another source of error. So there is a trade-off between validity and these types of biases.

Ideally, the survey methodologist would be able to collect quantitative parameters, metrics, and measures for the nodes and arcs in Fig. 14.1. For each of the interview technologies, researchers would have an estimate that *any* answer is produced by the respondent (node B), that a *reliable* answer is produced, and that a *valid* answer is produced. Survey methodologists collect such data in their research, although it is widely acknowledged that estimates of validity are extremely difficult to collect in practice. What is being advocated is an even more refined analysis. We can quantify the likelihood of interviewers expressing speech acts $\{C \mid D \mid E\}$, the extent of different types of errors, correlations between different types of error and $\{C \mid D \mid E\}$, and transaction times for completing C, D, and E. With such quantitative indices at hand, we can simulate survey completion times and the extent that errors will be committed with a new technology before it is designed. For example, one could imagine a Web survey that answered respondent's clarification questions through a word definition help facility (allowing C and D), but would not be able to formulate revised questions (node E). Performance measures could be simulated for such a system. The designer could also change the interface to encourage the respondent to ask clarification questions (node C) by instructions or highlighting the help facility. Designers could then explore how that interface enhancement would influence survey completion times, the validity of answers, and other performance measures. The state transition network and affiliated parameters would generate estimates in its simulations. The model would be validated by comparing the predicted quantities with empirical data that is eventually collected.

Another example of a quantitative model is the *GOMS* model of human–computer interaction (Card et al., 1983; Gray et al., 1993). GOMS stands for Goals, Operators, Methods, and Selection rules. This model was originally used to simulate performance of users of word processing software but has since been applied to dozens of tasks with system–person interfaces, including telephone operator systems. Performance is simulated at multiple levels of grain size: the keystroke level, completion of subgoals, and completion of major tasks. Another virtue of GOMS is that it uses a theoretical model of human information processing to predict execution times for basic operations of perception, action, and memory. It would predict, for example, how long it takes to: (1) process one eye fixation in a perception–cognition–motor cycle (approximately 240 milliseconds), (2) move the eyes from one location on a computer screen to another location, (3) move a finger from one location to a target location of a particular size, and (4) make a decision among N alternatives.

One could imagine the benefits of such a model in the arena of animated conversational agents. The GOMS model would simulate how long it takes for an agent to produce a spoken answer to a respondent's question versus how long the same answer would take for the respondent to read. Given that a spoken message is difficult to ignore, and that the respondent would sometimes skip reading a printed message, the researcher can estimate the comparative advantage of the two media when measuring the likelihood that the respondent attends to the message. Many other questions could be answered by the modeling tool. How would the likelihood of attending to the message be influenced by the placement of the printed message on a screen? How would listening times and the respondents' requests for repeated messages be influenced by agents speaking at different speeds, with different dialects, and different paralinguistic channels? Answers to such questions, and literally thousands of other questions, can be estimated with a GOMS model. It is important to acknowledge that there is never enough time for survey methodologists to test empirically a staggering number of alternative interfaces. We would argue, therefore, that such a quantitative model is a necessity—not merely a luxury—for survey methodologists during this age of rapid changes in technologies.

14.3.2   Analysis of Dialogue Modules

There is a need to perform a deep analysis of the discourse that underlies an interview technology because all surveys and human–computer interfaces are fundamentally forms of discourse. It is convenient to decompose the discourse into dialogue modules. A dialogue module is defined, for the present purposes, as a dialogue exchange between the interviewer and respondent that accomplishes a particular discourse function. The dialogue moves in Fig. 14.1, for example, would include the following dialogue modules: (a) respondent answers survey question asked by interviewer, (b) interviewer clarifies meaning of expression for respondent, and (c) interviewer aligns question with knowledge of respondent. Each module has its own participant roles, knowledge, and beliefs of participants, goals or intentions of participants, and pragmatic ground rules. We believe that it is important for survey researchers to understand the cognitive, pragmatic, and social components of the important dialogue modules in interviews and to systematically analyze how these components are realized (if at all) in each interview technology under consideration.

Consider the main dialogue module below that would apply to all surveys and interview technologies.

Dialogue Module 1: *Respondent answers survey question asked by interviewer.*

1. I = interviewer; R = respondent; Q = question
2. I does not know answer to Q at time t_1
3. I believes that R knows the answer to Q at time t_1
4. I believes that R understands the meaning of the Q at time t_1
5. I asks R the Q at time t_1
6. R understands the meaning of the Q at time t_2
7. R knows the answer to the Q at time t_2
8. R answers the Q at t_2
9. R believes I understands R's answer to the Q at time t_3
10. I believes that R understands the meaning of the Q at time t_3
11. I understands the meaning of the answer at time t_3
12. I believes that R believes that the answer to the Q is true at time t_3
13. I acknowledges R's answer to the Q at time t_4
14. I accepts R's answer to the Q at time t_4
15. R believes that I accepts R's answer to the Q at time t_5

At first blush this may seem to be a tedious specification of obvious epistemological and pragmatic assumptions of a simple survey question and its answer. However, each of these (as well as others if we were to include intentions) has important ramifications on interview technologies and the success of a conversational exchange. Fortunately, computational linguists have developed computer tools to assist researchers in keeping track of the beliefs, knowledge plans, intentions, and other states of participants in dialogue modules. Two of these are COLLAGEN (Rich et al., 2001) and the TRINDI toolkit (Larsson and Traum, 2000).

The 15 components in the example dialogue module would apply to a successful exchange with satisfactory alignment between the interviewer and respondent. However, the verbal answers and paradata in a FTF conversational interview often suggest that one or more of the assumptions are false. If the respondent pauses, frowns, or looks confused at time t_2, then assumptions 6 and 10 are suspect; this would motivate the interviewer to revise the question in a conversational interview and launch a subordinate dialogue module: interviewer aligns question with knowledge of respondent. If the interviewer pauses, has a hesitation prosody in the acknowledgment, looks skeptical, or asks a revised question at time t_4, then assumptions 6, 9, 10, 14, and/or 15 are suspect; this would motivate the respondent to ask a clarification question and the following subordinate dialogue module would be launched: interviewer clarifies meaning of expression for respondent. The subtleties of the paradata are often diagnostic of which assumptions are suspect.

Communication breaks down when there is misalignment at different levels of communication (Clark, 1996; Pickering and Garrod, 2004; Walker et al., 2003). Misalignments would of course compromise the validity of the responses in surveys, as has been shown in the research by Conrad and Schober (Conrad and Schober,

2000; Schober and Conrad, 1997). In the example dialogue module, there can be misalignments in components 4-6-10, 9-11, 7-14, and so on. Potential misalignments become visible in some interview technologies more than others. FTF and video conferencing allow visual and auditory paradata from the respondent, telephone allows only auditory, whereas self-administered questionnaires and the Web allow neither. Limitations in paradata of course have an impact on errors and other measures. Designers of computer technologies might provide special facilities to compensate for the lack of paradata in those technologies without contact with a human interviewer. For example, a Web survey might have a help facility that answers the respondents' clarification questions (nodes C and D in Fig. 14.1) or that generates revised questions (node E). However, it is well documented that individuals rarely ask questions and seek electronic help in most computer systems (Carroll, 1987) and learning environments (Graesser et al., 2005b). The likelihood of help seeking is no doubt lower in a survey environment because the respondent is voluntarily *providing* information rather than having the goal of *obtaining* information (Schober et al., 2003). It is not sufficient for a computer technology to simply offer electronic assistance for conversational grounding and alignment because there is a low likelihood that the user will use the facilities when they are needed. Conversational agents hold some promise in encouraging the respondent to seek help or in barging in to offer help. However, there is a risk of the agents committing false alarms when the agent cannot accurately understand the respondent and diagnose misalignments.

14.3.3 Measuring the Accuracy of Interpretation and Sensing Components

How accurately does the interviewer, computer, or instrument measure the respondents' intended messages and unintended paradata? This is a pervasive research question in survey methodology, and indeed the field of communications in general. Speech recognition has received considerable attention because this is a technology that would have enormous practical value in the world of surveys (Bloom, Chapter 6; and Johnston, Chapter 7 in this volume). Although the quality of speech recognition is improving (Cole et al., 2003), the accuracy of speech-to-text translation is substantially higher for humans than computers.

The accuracy of spoken messages being recognized correctly apparently improves quite a bit when the speech can be accompanied by pointing to interface elements via touch panel or electronic pens (Johnston, Chapter 7 in this volume). One example of a successful multimodal interface is the MATCH (Multiple Access to City Help) system (Johnston, Chapter 7 in this volume). More research is needed on how well respondent messages can be automatically interpreted in a multimodal system with various combinations of speech recognition, handwriting recognition, gesture recognition, recognition of facial expressions, detection of eye gaze directions, detection of posture, and so on. In addition to the accuracy of the verbal messages, there needs to be research on the accuracy of sensing the paradata, such as pauses, intonation, direction of gazes, gesture, posture, and facial expressions. In our own lab, we have attempted to classify learners' emotions while college students interact with AutoTutor (D'Mello et al., 2006; Person et al., Chapter 10 in this volume). Emotions such as confusion, frustration, boredom, engagement, delight, and surprise are being classified on the basis of the verbal dialogue, facial expressions, speech intonation, and posture. Our hypothesis is that emotion classification will require a combination of communication modalities, not any one alone.

The messages and paradata of the respondent are of course influenced by those of the interviewer. Such interviewer–respondent interactions need into be incorporated into any assessment of the accuracy of interpreting respondent messages and paradata. Respondents' conversational style and paradata can model, mirror, or otherwise respond to what the interviewer does. One of the virtues of conversational agents is that the messages and paralinguistic cues are controlled, whereas those of human interviewers are variable or intractable. Future research with agents in interviews need to document the impact of the agent's conversational, physical, stylistic, and personality characteristics on the behavior of the respondents (see Cassell and Miller, Chapter 8; and Person et al., Chapter 10 in this volume).

The accuracy of automated comprehension of messages is limited at this point in the fields of computational linguistics and artificial intelligence (Allen, 1995; Jurafsky and Martin, 2000; Rus, 2004). There have been impressive advances in computer analyses of words (such as *Linguistic Inquiry and Word Count*, Pennebaker and Francis, 1999), shallow semantics (such as *Coh-Metrix*, Graesser et al., 2004), automated grading of essays (such as *e-Rater*, Burstein, 2003; *Intelligent Essay Assessor*, Landauer et al., 2000), and question answering (Dumais, 2003; Harabagiu et al., 2002; Voorhees, 2001). However, the performance of automated systems is far from adequate whenever there is a need for the construction of context-specific mental models, knowledge-based inferences, deictic references, resolution of anaphoric references, and other components of deep comprehension. Surveys require a high precision in comprehending messages, unlike other conversational registers that can get by with less rigorous grounding and alignment of meanings, such as tutoring (Graesser et al., 2001, 2005a,b) and small talk (Bickmore and Cassell, 1999; Cassell et al., 2000). An interviewer agent would require a fairly accurate mental model of the respondent (Walker et al., 2003), which is at least a decade away from being technically achieved.

On a technical note, the various fields that have investigated communication systems have adopted somewhat different quantitative foundations for measuring communication accuracy. Social psychologists are prone to collecting ratings from multiple judges as to whether a message is understood by a respondent or system. Computational linguists collect recall scores, precision scores, and F-measures in comparisons between a computer's output and judgments of human experts (Jurafsky and Martin, 2000). Cognitive psychologists collect hit rates, false alarm rates, d' scores, and other measures from signal detection theory in comparisons between computer output and judgments of humans. It would be worthwhile for colleagues in the survey methodology world to converge on some agreement on measurements of the accuracy when analyzing interpretation and sensing components.

14.3.4 Animated Conversational Agents in the Survey World

Our forecast is that animated conversational agents will eventually become ubiquitous in human–computer interfaces, in spite of the cautions that problems will arise when humans attribute too much intelligence to the agents (Norman, 1994; Shneiderman and Plaisant, 2005). Adults have the ability to gauge the intelligence, personality, and limitations of other humans and classes of humans. Adults can discriminate what is true in the real world from what they see in movies. In essence, they can make fine discriminations about the capabilities of humans and their presence in technologies. This being the case, we can imagine that there will be a new genre of agents that are *Auto-Interviewers* and that these agents will be eventually understood and accepted by the public. However, it is uncertain how the public will view the Auto-Interviewer's intelligence, believability, trust, pragmatic ground rules, personality, and other dimensions of social presence (Cassell and Miller, Chapter 8 in this volume; Cassell et al., 2000; Reeves and Nass, 1996). An agent can announce the limits of its protocol ("Sorry I am not allowed to answer your questions") or capabilities ("Sorry but I am not understanding you", "Could you speak more clearly?"), just as interviewers and other people do. We could imagine an Auto-Interviewer with an engaging and endearing personality that respondents enjoy interacting with for a sustained period of time, despite its limitations on various human dimensions. We could imagine a new class of agents with its own identity: the Auto-Interviewers.

It will be important to conduct parametric studies that investigate the impact of agent features on respondents. Is it best to have agents matched to the respondent in gender, age, ethnicity, and personality, or is it better to have contrasts (e.g., male agents with female respondents) or high status prototypes (earnest older white males)? How do these features and the agent's attractiveness influence the engagement, completion rates, and response validity of respondents? How and when should the agent give back-channel feedback (uh huh, okay, head nod) that the agent has heard what the respondent has said? Should the agent deviate from realism, as with cartoon agents or caricatures, or be very close to a realistic depiction of humans?

The roboticist Masahiro Mori (1970) claimed that people have an increasing level of comfort and even affection for robots as they increasingly resemble a human. However, there reaches a point on the robot–human similarity continuum that the human feels uneasy or even disgusted when the robot mimics a human very closely, but not quite perfectly. Humans become more comfortable again when there is a perfect robot–human match, a state that is theoretical because no robots are that good. Mori calls the uneasy trough the *zone of the uncanny valley*. The technology is now available to measure humans' reactions to agents as a function of the different values on the agent–human similarity continuum (Schober and Conrad, Chapter 1 in this volume). There is software developed by Neven Vision (recently acquired by Google) that has agents mirror what a human agent does by mapping the facial features of the agent onto the features of the human. This will allow researchers to empirically test the uncanny valley and other mappings between agent characteristics and human emotions.

A successful animated agent is not limited to its appearance. It will need to co-ordinate the speech, facial expressions, gesture, eye gaze, body posture, and other paralinguistic channels. The conversational dialogue will need to generate the appro-priate speech acts at each turn. A totally natural agent would need to generate speech disfluencies that reflect the difficulty of its message planning mechanism and its con-versational floor management (i.e., maintaining its turn or taking the conversational floor). There is a rich terrain of research avenues for those interested in animated conversational agents.

Once again, skeptics may ask "Why bother?" Our view is that a good Auto-Interviewer will represent the human–computer interface of the future, will be cheaper to implement (although expensive to initially develop), and will provide more control over the interview than will FTF and video conference interviews with humans.

14.3.5   Culture

It is perfectly obvious that our research on interview technologies needs to be replicated in different cultures in order to assess the scope and generality of scientific claims. This gives all of us free reign to tour the world in the interest of multicultural research on survey methodology. We can explore different cultures, languages, age groups, socioeconomic strata, and so on.

The role of technology in different cultures is particularly diverse. The younger generations tend to use advanced technologies, but not the elderly, so there is the worry of coverage and nonresponse errors in the elderly populations. People in most cultures do not understand the semiotics of visual icons that bloat the interfaces of most Web sites, so they tend to be excluded from Web interviews and many other interview technologies. The study of icon semiotics is indeed an important area for exploration because the computer interfaces of today are too arcane and unintuitive for the majority of the U.S. population, a culture that values technology. Designers often justify their poor human–computer interfaces by claiming that users can read instructions or learn from their peers. However, the data reveal that U.S. adults quickly abandon any Web site or inexpensive software product if they encounter a couple of obstacles within a few seconds of receiving it (Ives et al., 2000). The truth is that the open market of software users is unforgiving of shabby human–computer interfaces.

The face is the universal interface, as we proclaimed at the beginning of this chapter. As such, the face presents very few obstacles in human–computer interaction for members of different cultures. Those of us who conduct research with animated conversational agents know that the time it takes to instruct participants how to use a computer system with an agent is on the order of a few seconds, with few if any user errors. In contrast, it takes several minutes to instruct participants how to read and use a conventional interface, often with troublesome bottlenecks that require retraining. As expressed earlier, however, there are potential liabilities with these agents (e.g., technical imperfections, social presence lowering rates of sensitive disclosure) and these probably get magnified when considering differences among cultures. However, this latter claim about differences among cultures is speculative and awaits empirical research.

The younger generation in the United States is currently a generation that thrives on computer games (Gee, 2003) and instant messaging (Kinzie, Whitaker, and Hofer, 2005). The older generation is virtually alienated from these two technologies. What are the implications of this fact? Should we embed our surveys in games and instant messaging if we want to reach the next generation? The younger generation is also in a culture of media that cultivates short attention spans. Should we distribute the survey questions over time, locations, and contexts in order to improve response rates? Should the surveys fit within these constraints?

Citizens in the workforce live in different cultures than the youth and the elderly. Those in the workforce have very little time and that may prevent some from being willing to complete a survey. Workforce respondents may be more open to completing a survey if they are viewed as consultants or experts on a topic. The solution to reducing nonresponse rates may lie in creating more engaging interviews that are perceived as being important to members of a culture.

14.4   CLOSING COMMENTS

This chapter has provided a snapshot of the current state of research in learning technologies and has proposed some research directions that are needed in the near horizon. This is an unusual point in history because we are in a revolution of communications technologies and the communication media of choice are extremely diverse among different sectors of the population. Mail is no longer reliable, nor is the telephone, nor the Web. The media of choice are entirely different for the young and old, for the rich and poor, and for those in different languages and cultures. The future of survey research is destined to be entirely different than it was thirty years ago. We have no choice but to pursue novel solutions and to shed some of the sacred guidelines from the past.

ACKNOWLEDGMENTS

This research was partially supported by the National Science Foundation (ITR 0325428). Any opinions, findings, and conclusions or recommendations expressed in this chapter are those of the authors and do not necessarily reflect the views of NSF.

REFERENCES

Allen, J. (1995). *Natural Language Understanding.* Redwood City, CA: Benjamin/Cummings.

Bickmore, T., and Cassell, J. (1999). Small talk and conversational storytelling in embodied conversational characters. In *Proceedings of the American Association for Artificial Intelligence Fall Symposium on Narrative Intelligence* (pp. 87–92). Cape Cod, MA: AAAI Press.

Burstein, J. (2003). The E-rater scoring engine: automated essay scoring with natural language processing. In M. D. Shermis and J. C. Burstein (Eds.), *Automated Essay Scoring: A Cross-Disciplinary Perspective* (pp. 122–133). Mahwah, NJ: Erlbaum.

Card, S., Moran, T., and Newell, A. (1983). *The Psychology of Human–Computer Interaction.* Hillsdale, NJ: Erlbaum.

Carroll, J. M. (Ed.) (1987). *Interfacing Thought: Cognitive Aspects of Human–Computer Interaction.* Cambridge, MA: MIT Press/Bradford Books.

Cassell, J., Sullivan, J., Prevost, S., and Churchill, E. (2000). *Embodied Conversational Agents.* Cambridge, MA: MIT Press.

Clark, H. H. (1996). *Using Language.* Cambridge: Cambridge University Press.

Clark, R. E. (1983). Reconsidering research on learning from media. *Review of Educational Research, 53,* 445–460.

Clark, H. H., and Brennan, S. E. (1991). Grounding in communication. In L. Resnick, J. Levine, and S. Teasely (Eds.), *Perspectives on Socially Shared Cognition* (pp. 127–149). Washington DC: American Psychological Association.

Cole, R. van Vuuren, S., Pellom, B., Hacioglu, K., Ma, J., Movellan, J., Schwartz, S., Wade-Stein, D. Ward, W., and Yan, J. (2003). Perceptive animated interfaces: first steps toward a new paradigm for human computer interaction. *Proceedings of the IEEE, 91,* 1391–1405.

Conrad, F. G., and Schober, M. F. (2000). Clarifying question meaning in a household telephone survey. *Public Opinion Quarterly, 64,* 1–28.

Couper, M. P. (2000). Web surveys: a review of issues and approaches. *Public Opinion Quarterly, 64,* 464–494.

Couper, M. P., Singer, E., and Tourangeau, R. (2004). Does voice matter? An interactive voice response (IVR) experiment. *Journal of Official Statistics, 20*(3), 1–20.

Cuban, L. (1986). *Teachers and Machines: The Classroom Use of Technology Since 1920.* New York: Teachers College.

Cuban, L. (2001). *Oversold and Underused: Computers in the Classroom.* Cambridge, MA: Harvard University Press.

Currivan, D., Nyman, A. L., Turner, C. F., and Biener, L. (2004). Does telephone audio computer-assisted survey interviewing improve the accuracy of prevalence estimates of youth smoking? Evidence from the UMass Tobacco Study. *Public Opinion Quarterly, 68,* 542–564.

D'Mello, S. K., Craig, S. D., Sullins, J., and Graesser, A. C. (2006). Predicting affective states through an emote-aloud procedure from AutoTutor's mixed-initiative dialogue. *International Journal of Artificial Intelligence in Education, 16,* 3–28.

Dumais, S. (2003). Data-driven approaches to information access. *Cognitive Science, 27*(3), 491–524.

Emmanuel, E. J., Wendler, D., and Grady, C. (2000). What makes clinical research ethical? *Journal of the American Medical Association, 283,* 2701–2711.

Fowler, F. J., and Cannell, D. F. (1996). Using behavioral coding to identify problems with survey questions. In N. Schwarz and S. Sudman (Eds.), *Answering Questions: Methodology for Determining Cognitive and Communicative Processes in Survey Research* (pp. 15–36). San Francisco: Jossey-Bass.

Fowler, F. J., and Mangione, T. W. (1990). *Standardized Survey Interviewing: Minimizing Interview-Related Error.* Newbury Park, CA: Sage.

Gee, J. (2003). *What Video Games Have to Teach Us About Learning and Literacy.* New York: Palgrave/Macmillan.

Graesser, A. C., Cai, Z., Louwerse, M., and Daniel, F. (2006). Question Understanding Aid (QUAID): a Web facility that helps survey methodologists improve the comprehensibility of questions. *Public Opinion Quarterly, 70,* 3–22.

Graesser, A. C., Chipman, P., Haynes, B. C., and Olney, A. (2005a). AutoTutor: an intelligent tutoring system with mixed-initiative dialogue. *IEEE Transactions in Education, 48,* 612–618.

Graesser, A. C., McNamara, D. S., Louwerse, M. M., and Cai, Z. (2004). Coh-Metrix: analysis of text on cohesion and language. *Behavioral Research Methods, Instruments, and Computers, 36*, 193–202.

Graesser, A. C., McNamara, D. S., and VanLehn, K. (2005b). Scaffolding deep comprehension strategies through Point&Query, AutoTutor, and iSTART. *Educational Psychologist, 40*, 225–234.

Graesser, A. C., Person, N., Harter, D., and TRG (2001). Teaching tactics and dialog in AutoTutor. *International Journal of Artificial Intelligence in Education, 12*, 257–279.

Groves, R. M. (1989). *Survey Errors and Survey Costs*. Hoboken, NJ: Wiley & Sons.

Gray, W. D., John, B. E., and Atwood, M. E. (1993). Project Ernestine: validating a GOMS analysis for predicting and explaining real-world performance. *Human–Computer Interaction, 8*(3), 237–309.

Harabagiu, S. M., Maiorano, S. J., and Pasca, M. A. (2002). Open-domain question answering techniques. *Natural Language Engineering, 1*, 1–38.

Ives, Z. G., Levy, A. Y., and Weld, D. S. (2000). Efficient evaluation of regular path expressions on streaming XML data. Technical report (UW-CSE-2000-05-02). University of Washington.

Johnson, W. L., and Beal, C. (2005). Iterative evaluation of a large-scale intelligent game for language learning. In C. Looi, G. McCalla, B. Bredeweg, and J. Breuker (Eds.), *Artificial Intelligence in Education: Supporting Learning Through Intelligent and Socially Informed Technology* (pp. 290–297). Amsterdam: IOS Press.

Johnston, M., and Bangalore, S. (2005). Finite-state multimodal integration and understanding. *Journal of Natural Language Engineering, 11*, 159–187.

Jurafsky, D., and Martin, J. H. (2000). *Speech and Language Processing: An Introduction to Natural Language Processing, Computational Linguistics, and Speech Recognition*. Upper Saddle River, NJ: Prentice-Hall.

Kinzie, M. B., Whitaker, S. D., and Hofer, M. J. (2005). Instructional uses of instant messaging (IM) during classroom lectures. *Educational Technology and Society, 8*(2), 150–160.

Landauer, T. K., Laham, D., and Foltz, P. W. (2000). The Intelligent Essay Assessor. *IEEE Intelligent Systems 15*, 27–31.

Larsson, S., and Traum, D. (2000). Information state and dialogue management in the TRINDI dialogue move engine toolkit. *Natural Language Engineering, 6*(3–4), 323–340.

Lessler, J. T., and Forsyth, B. H. (1996). A coding system for appraising questionnaires. In N. Schwarz and S. Sudman (Eds.), *Answering Questions: Methodology for Determining Cognitive and Communicative Processes in Survey Research* (pp. 259–292). San Francisco: Jossey-Bass.

Mayer, R. E. (2005). *Multimedia Learning*. Cambridge, UK: Cambridge University Press.

Mori, M. (1970). Bukimi no tani (the uncanny valley). *Energy, 7*, 33–35 (in Japanese).

Norman, D. A. (1994). How might people interact with agents. *Communications of the ACM, 37*, 68–71.

Pennebaker, J. W., and Francis, M. E. (1999). *Linguistic Inquiry and Word Count (LIWC)*. Mahwah, NJ: Erlbaum.

Pickering, M. J., and Garrod, S. (2004). Toward a mechanistic psychology of dialogue. *Brain and Behavioral Sciences, 27*, 169–190.

Presser, S., Couper, M. P., Lessler, J. T., Martin, E., Martin, J., Rothgeb, J. M., and Singer, E. (2004). Methods for testing and evaluating survey testing. *Public Opinion Quarterly, 68,* 109–130.

Reeves, B., and Nass, C. (1996). *The Media Equation.* New York: Cambridge University Press.

Rich, C., Sidner, C. L., and Lesh, N. (2001). COLLAGEN: applying collaborative discourse theory to human–computer interaction. *AI Magazine, 22*(4), 15–25.

Rus, V. (2004). A first exercise for evaluating logic from identification systems. In *Proceedings Third International Workshop on the Evaluation of Systems for the Semantic Analysis of Text (SENSEVAL-3)*, at the Association of Computational Linguistics Annual Meeting, July 2004. Barcelona, Spain: ACL.

Schaeffer, N. C., and Maynard, D. W. (2002). Occasions for intervention: interactional resources for comprehension in standardized survey interviews. In D. W. Maynard, H. Houtkoop-Steenstra, N. C. Schaeffer, and J. van der Zouwen (Eds.), *Standardization and Tacit Knowledge: Interaction and Practice in the Survey Interview* (pp. 261–280). Hoboken, NJ: John Wiley & Sons.

Schober, M. F., and Conrad, F. G. (1997). Does conversational interviewing reduce survey measurement error? *Public Opinion Quarterly, 60,* 576–602.

Schober, M. F., and Conrad, F. G. (2002). A collaborative view of standardized survey interviews. In D. Maynard, H. Houtkoop-Steenstra, N. C. Schaeffer, and J. Van der Zouwen (Eds.), *Standardization and Tacit Knowledge: Interaction and Practice in the Survey Interview* (pp. 67–94). Hoboken, NJ: John Wiley & Sons.

Schober, M. F., Conrad, F. G., Ehlen, P., and Fricker, S. S. (2003). How Web surveys differ from other kinds of user interfaces. In *Proceedings of the American Statistical Association, Section on Survey Research Methods.* Alexandria, VA: American Statistical Association.

Shneiderman, B., and Plaisant, C. (2005). *Designing the User Interface: Strategies for Effective Human–Computer Interaction*, 4th ed. Reading, MA: Addison-Wesley.

Suchman, L., and Jordan, B. (1990). Interactional troubles in face-to-face survey interviews. *Journal of the American Statistical Association 85*(409), 232–241.

Sudman, S., and Bradburn, N. M. (1974). *Response Effects in Surveys.* Chicago: Aldine.

Tourangeau, R., and Smith, T. W. (1998). Collecting sensitive information with different modes of data collection. In M. P. Couper, R. P. Baker, J. Bethlehem, J. Martin, W. L. Nicholls II, and J. M. O'Reilly (Eds.), *Computer Assisted Survey Information Collection* (pp. 431–453). Hoboken, NJ: John Wiley & Sons.

VanLehn, K., Graesser, A. C., Jackson, G. T., Jordan, P., Olney, A., and Rose, C. P. (2007). When are tutorial dialogues more effective than reading? *Cognitive Science, 31,* 3–62.

Voorhees, E. (2001). The TREC question answering track. *Natural Language Engineering, 7,* 361–378.

Walker, M., Whittaker, S., Stent, A., Maloor, P., Moore, J., Johnson, M., and Vasireddy, G. (2003). Generation and evaluation of user tailored responses in multimodal dialogue. *Cognitive Science, 28,* 811–840.

Author Index

Abramson, P. 105
Adair, W.L. 227, 229, 231
Adams, R.E. 89
Aha, D. 206
Ahrenberg, L. 144
Allen, D.R. 61
Allen, J. 278
Allgayer, J. 144
Al-Tayyib, A.A. 127, 169
Alwin, D. 9, 22
Ambady, N. 227
Anderson, A.H. 8, 99, 100, 102, 104, 111, 113, 224, 231
Anderson, B. 105
Anderson, J.R. 11
Anderson, W.N. 219, 229
André, E. 145, 153, 198, 210
Aquilino, W. 183, 218, 230
Argyle, M. 171, 172, 226
Armenakis, A.A. 201
Armsby, P.P. 173
Asch, B. 62
Asting, T. 229
Athey, E.R. 163
Atkinson, R.K. 198
Atwood, M.E. 274
Ayling, K. 101, 227

Bacon-Shone, J. 83
Baker, R.P. 63, 65, 68
Bandilla, W. 84
Bangalore, S. 142, 144, 146, 151, 269
Bard, E. 100
Bargh, J. 186

Barlow, T. 198
Bartlett, M.S. 207
Batageli, Z. 81
Bates, I. 84, 85
Batliner, A. 204
Baylor, A.L. 174, 197
Beal, C. 271
Beattie, G. 104
Beatty, P. 26
Bekker, M.M. 224
Bell, L. 106, 107
Belli, R.F. 36
Benney, M. 163
Bennett, C. 264
Berand, F. 204
Berck, J. 129
Bergweiler, S. 145
Bhogal, R. 198
Bianchi-Berthouze, N. 199
Bickmore, T. 170, 173, 278
Biemer, P.P. 13, 61
Biener, L. 127, 169, 183, 268
Bienenstock, H. 70, 127
Bingham, W. 40
Binson, D. 87
Bischoping, K. 164
Bishop, G. 182
Blair, J. 59
Blixt, S. 49
Bloom, J.E. 9, 26, 51, 125, 130, 168
Blumer, H. 255, 261
Blyth, W.G. 60
Bockting, W.O. 252
Boehner, K. 189

Envisioning the Survey Interview of the Future, Edited by Frederick G. Conrad and Michael F. Schober
Copyright © 2008 John Wiley & Sons, Inc.

Subject Index

administration mode effects (see mode effects)
affect (see also rapport, relational communication)
  detecting users', 196, 199, 200, 203–207, 210
animated agents 50, 71–72, 84, 196–199, 210,
    218
  (see also embodied conversational agents,
    interviewing agents)
anonymity, 100, 103, 186, 189, 243, 249–250
  and detecting repeat responders, 252
  perceived, 186, 189
audio conferencing, 222
Auto-Interviewer, 279
automated systems, 61, 127–129, 132,
    195–197, 204
  dialogue module, 275–276
  vs. human interviewers, 139
automatic speech recognition, 142

body posture (see posture)

cell phone, 82 (see also mobile interaction)
cognitive interviewing, 32
communication environment, features of (see
    affordances)
communication mode (see mode of
    communication)
communication technology
  cell phone (see mobile interaction, cell phone)
  changes in use of, 3
  considerations for adopting for interviews,
    23–34
  conversational agent (see embodied
    conversational agent),
  internet (see web)

obstacles to adopting for interviews, 16
speech, 119 (see also speech recognition,
    spoken dialogue systems)
video (see video conferencing, video telephony)
communication
  face-to-face (FTF), 7, 8, 10, 13, 23–26, 26,
    198–202, 268
  non-verbal interaction, 8, 95–96, 110–111, 172,
    174, 233, 236
  relational (see relational communication)
  taxonomy of situations, 6–7
  video mediated (see video conferencing)
comparison of survey technologies
completion rate (see also incentives to participate)
  break-offs, 13
  relationship to payment type, 245
composite multimodality, 144
computer assisted interviewing (CAI), 58, 61,
    141
  audio-CASI, 60, 61, 63, 222, 227
  CAPI, 60, 61, 92, 269
  CASI, 5, 10, 23, 50, 55–56, 60, 61, 63, 100,
    173–174, 178, 192, 194, 227–228, 238–239
  CATI, 4–5, 60, 61, 93, 126–127, 129–130,
    132–133
  video-CASI, 142
conceptual misalignment, 123–124, 141, 142
confidentiality, 180–182, 240–253, 257–258
consent (see informed consent)
consumer price index, 140
control, 68–69
conversation analysis, 38, 50
conversational agents (see embodied
    conversational agents)

Envisioning the Survey Interview of the Future, Edited by Frederick G. Conrad and Michael F. Schober
Copyright © 2008 John Wiley & Sons, Inc.

295

WILEY SERIES IN SURVEY METHODOLOGY
Established in Part by WALTER A. SHEWHART AND SAMUEL S. WILKS

Editors: *Robert M. Groves, Graham Kalton, J. N. K. Rao, Norbert Schwarz, Christopher Skinner*

A complete list of the titles in this series appears at the end of this volume.

WILEY SERIES IN SURVEY METHODOLOGY
Established in Part by WALTER A. SHEWHART AND SAMUEL S. WILKS

Editors: *Robert M. Groves, Graham Kalton, J. N. K. Rao, Norbert Schwarz, Christopher Skinner*

The **Wiley Series in Survey Methodology** covers topics of current research and practical interests in survey methodology and sampling. While the emphasis is on application, theoretical discussion is encouraged when it supports a broader understanding of the subject matter.

The authors are leading academics and researchers in survey methodology and sampling. The readership includes professionals in, and students of, the fields of applied statistics, biostatistics, public policy, and government and corporate enterprises.

*Now available in a lower priced paperback edition in the Wiley Classics Library.

MAYNARD, HOUTKOOP-STEENSTRA, SCHAEFFER, VAN DER ZOUWEN ·
    Standardization and Tacit Knowledge: Interaction and Practice in the Survey Interview
PORTER (editor) · Overcoming Survey Research Problems: New Directions for
    Institutional Research, No. 121
PRESSER, ROTHGEB, COUPER, LESSLER, MARTIN, MARTIN, and SINGER
    (editors) · Methods for Testing and Evaluating Survey Questionnaires
RAO · Small Area Estimation
REA and PARKER · Designing and Conducting Survey Research: A Comprehensive
    Guide, *Third Edition*
SARIS and GALLHOFER · Design, Evaluation, and Analysis of Questionnaires for
    Survey Research
SÄRNDAL and LUNDSTRÖM · Estimation in Surveys with Nonresponse
SCHWARZ and SUDMAN (editors) · Answering Questions: Methodology for
    Determining Cognitive and Communicative Processes in Survey Research
SIRKEN, HERRMANN, SCHECHTER, SCHWARZ, TANUR, and TOURANGEAU
    (editors) · Cognition and Survey Research
SUDMAN, BRADBURN, and SCHWARZ · Thinking about Answers: The Application
    of Cognitive Processes to Survey Methodology
UMBACH (editor) · Survey Research Emerging Issues: New Directions for Institutional
    Research No. 127
VALLIANT, DORFMAN, and ROYALL · Finite Population Sampling and Inference: A
    Prediction Approach

DATE DUE
